Advances in Remote Sensing of the Inland and Coastal Water Zones

Advances in Remote Sensing of the Inland and Coastal Water Zones

Editors

Jacek Lubczonek
Paweł Terefenko
Katarzyna Bradtke
Marta Wlodarczyk-Sielicka

MDPI • Basel • Beijing • Wuhan • Barcelona • Belgrade • Manchester • Tokyo • Cluj • Tianjin

Editors
Jacek Lubczonek
Maritime University of Szczecin
Szczecin, Poland

Paweł Terefenko
University of Szczecin
Szczecin, Poland

Katarzyna Bradtke
University of Gdynia
Gdynia, Poland

Marta Wlodarczyk-Sielicka
Maritime University of Szczecin
Szczecin, Poland

Editorial Office
MDPI
St. Alban-Anlage 66
4052 Basel, Switzerland

This is a reprint of articles from the Special Issue published online in the open access journal *Remote Sensing* (ISSN 2072-4292) (available at: https://www.mdpi.com/journal/remotesensing/special_issues/Remote_Sensing_Inland_Coastal_Water).

For citation purposes, cite each article independently as indicated on the article page online and as indicated below:

LastName, A.A.; LastName, B.B.; LastName, C.C. Article Title. *Journal Name* **Year**, *Volume Number*, Page Range.

ISBN 978-3-0365-6674-0 (Hbk)
ISBN 978-3-0365-6675-7 (PDF)

Contents

About the Editors

Jacek Lubczonek

Jacek Lubczonek is Associate Professor at the Maritime University of Szczecin. He obtained his PhD in 2003 in technical sciences in the field of Geodesy and Cartography at the Maritime University. The subject of his thesis was the use of artificial neural networks to create digital bathymetric models. He is currently Head of the Chair of Geoinformatics and Hydrography at the Maritime University of Szczecin. In terms of research, he has carried out projects involving machine learning, geospatial modelling and analysis, and the development of geoinformatics systems. Project topics have included the use of machine learning methods in geodata processing, spatial planning of sensors, creation of databases for navigational charts, and modelling of bathymetric surfaces using remote sensing data. His current work focuses on the use of geoinformatics and machine learning methods in remote sensing, hydrography, and GIS.

Paweł Terefenko

Paweł Terefenko is Assistant Professor at the Institute of Marine and Environmental Sciences, University of Szczecin, where he leads the Coastal Ecosystems Monitoring Group. He holds a master's degree in Marine Geography and a Ph.D. and habilitation in Earth Sciences (University of Szczecin). His main research interest is on the analysis of remote sensing datasets, including aerial and drone images, for marine and coastal applications. His expertise includes the automatic detection of coastal morphology indicators, image processing, and remote sensing modeling of the coastal environment. His current research focuses on coastal ecosystems monitoring through innovative approaches and novel applications, including the use of Unmanned Aerial Vehicles in 2D and 3D mapping of spatiotemporal phenomena in the coastal cliffs' environment.

Katarzyna Bradtke

Katarzyna Bradtke is Assistant Professor at the Institute of Oceanography at the University of Gdańsk. She graduated from Physical Oceanography at the University of Gdańsk and completed postgraduate studies in the field of databases and data science at the Poznań University of Technology and Gdańsk University of Technology. In 2004, she obtained a PhD in Earth Science. Since then, working in the Laboratory of Remote Sensing and Spatial Analysis, she has been dealing with the development of satellite data processing algorithms for the purposes of monitoring the water quality and temperature of the Baltic Sea. She participated in the implementation of, among others the SatBałtyk project, which developed an operational monitoring system for the Baltic Sea based on satellite observations. Her current research interests include the object-oriented classification of surface signatures of dynamic processes in the coastal zone.

Marta Wlodarczyk-Sielicka

Marta Wlodarczyk-Sielicka received a B.S. degree in Hydrographic Survey and Navigation Aids and an M.Sc. degree in Navigation from Maritime University of Szczecin, Poland. Her Ph.D. thesis was on a geodata reduction method for bathymetric map creation. She is currently working as Assistant Profesor and researcher in the Chair of Geoinformatics and Hydrography, Maritime University of Szczecin. Her main research interests are related to the field of geoinformatics and hydrography. The main topics of her research are spatial data processing and their comprehensive analysis. She works on bathymetric geodata modelling, which can be included in the spatial big data group. She is also involved in research on adapting artificial intelligence methods for processing of bathymetric data. She has research experience in designing mobile systems, systems of harmonized river information services, systems for the automatic identification of vessels, as well as the development of electronic navigation charts. She is currently managing a project related to multidimensional and multitemporal coastal zone monitoring systems using an autonomous unmanned vessel.

remote sensing MDPI

Editorial

Advances in Remote Sensing of the Inland and Coastal Water Zones

Jacek Lubczonek [1,*], Paweł Terefenko [2], Katarzyna Bradtke [3] and Marta Wlodarczyk-Sielicka [1]

1 Chair of Geoinformatics and Hydrography, Maritime University of Szczecin, 70-500 Szczecin, Poland
2 Institute of Marine and Environmental Sciences, University of Szczecin, 70-453 Szczecin, Poland
3 Institute of Oceanography, University of Gdańsk, 80-309 Gdańsk, Poland
* Correspondence: j.lubczonek@pm.szczecin.pl

Citation: Lubczonek, J.; Terefenko, P.; Bradtke, K.; Wlodarczyk-Sielicka, M. Advances in Remote Sensing of the Inland and Coastal Water Zones. *Remote Sens.* **2022**, *14*, 5346. https://doi.org/10.3390/rs14215346

Received: 10 October 2022
Accepted: 19 October 2022
Published: 25 October 2022

Coastal zone areas are important parts of the environment, very often subject to constant change. This applies to both marine and inland coastal zones. Factors for these changes can include the natural impact of marine and inland waters on land, anthropogenic activities, environmental changes affecting the ecosystem, or climate change. Given that coastal zone areas cover diverse types of regions, including urbanised as well as non-urbanised regions, they often have an impact on the functioning of ports, natural or recreational areas, or on human activities. Therefore, these two environments require constant monitoring. Increasingly detailed and diverse remote sensing data of the aquatic environment and adjacent areas enable comprehensive studies, often related to the increasing resolution, completeness, availability, or the development of the survey instruments themselves. This Special Issue presents scientific papers related to studies of the coastal zone environment and inland waters, addressing complex and current trends in solving scientific problems. The research presented in the publications deals with the application of geospatial techniques such as the classification of submerged vegetation and wetlands, the geomodelling of remote sensing data to create combined bathymetric and topographic models, atmospheric corrections of airborne hyperspectral images, the creation of bathymetric surfaces from unmanned aerial vehicle (UAV) and unmanned surface vehicle (USV) data, image processing for monitoring underwater objects, the determination of sound velocities in shallow-water areas, analyses of multi-temporal shoreline changes in the Baltic Sea using aerial imagery, supporting navigation processes using modern geospatial environments in the form of augmented and virtual reality, assessments of adjacency correction using Sentinel-2 MSI images, applications of optical and synthetic aperture radar images and data analysis techniques to quantify vertical ground displacements, and applications of the Google Earth Engine for monitoring sand split variability. Twelve publications providing new insights into the subject of remote sensing of inland and coastal water zones are presented, detailed below.

The research in the first publication [1] addresses the issue of estuary degradation, which is rapidly increasing due to anthropogenic pressure. The twin Sentinel-2 A/B satellites of the Copernicus program were used to study the Guadiaro River estuary (Spain) over a 4-year period (2017–2020). Mapping of marine sites was performed using the Normalised Differential Water Index (NDWI) in the Google Earth Engine (GEE) platform, selecting Sentinel-2 Level 2A imagery. In this study, an optimised data processing method was developed. Through research using the optimised data processing method, the considerable potential of Sentinel-2 as well as GEE images for estuarine monitoring was found. In addition, the practical aspect of the developed image processing method was shown, related to possible continuous and detailed monitoring of the estuary.

Observations from the C-band Envisat/ASAR and Sentinel-1A sensors (2004–2010 and 2014–2016) enabled the study of vertical ground displacements and assessment of their contribution to coastline erosion along the Karachi Coastline (Pakistan) [2]. The results suggest that some coastline parts are subsiding at comparable rates to, or even much higher than, the relative sea-level rise. Landsat images from 1989 to 2018 (10-year temporal resolution) were further used to examine the state of coastline erosion using different statistical approaches. The results indicate that rapid urbanization, construction on reclaimed land, coastline erosion favouring seawater intrusion, failed drainage/sewerage networks, and soil liquefaction are contributing to the site-specific variations in land displacement in Karachi.

Quantifying water quality using medium-resolution optical satellite sensors for small inland lakes is challenging due to distortions of the water signal by radiation reflected from the surrounding land surface and scattered in the atmosphere towards the radiometer. In [3], the authors used Sentinel-2/MSI images and in situ optical measurements to assess the uncertainty of reflectance derived from satellite data due to adjacency effects for lakes surrounded by both natural and urbanized environments and characterized by different optical characteristics of water. They assessed the importance of factors such as aerosol load, water body shape and size, differences in the reflectance of water and land cover on the strength of this effect, and the accuracy of its correction. Using a physical approach based on the Atmospheric Point Spread Function (APSF) to correct the influence of the surrounding land on the reflectance determined from the satellite data, the authors achieved good agreement with in situ observations, especially in the visible part of the spectrum (MAPE close to 50%), while pointing out that proper determination of the horizontal range of the adjacency effect is crucial for the correction.

A remarkably interesting solution is the use of augmented and virtual reality to support the safe navigation of sailors in inland shallow waters proposed in [4]. The authors identified a new solution used during bottom modelling. Nowadays, these models are mainly represented by visualizations on standard displays. The proposed solution, augmented reality (AR) technology, is a new and interesting approach for their presentation. The authors proposed the use of AR on inland and coastal waterways to enhance the safety of boating and other water-sports-related tourist activities. They showed the detailed architecture and prototype of an augmented reality mobile application (MAR). Notably, the authors also demonstrated that the real-time AR capabilities enabled the observation of other users on the water in conditions of limited visibility and at night. Their study shows that the performance of the prototype has been verified during tests on the waters of the lakes of Warmia and Mazury in Poland. The tests clearly indicated that the proposed solution can be used in practice.

One example of the analysis of coastal changes on a protected section of the southern Baltic coast, free from anthropogenic impacts, is presented in [5], based on historical and contemporary aerial photographs and orthophotomaps. The authors determined the size and intensity of morphodynamic changes in 1951–2016 (a period of 65 years) and indicated the intensification of erosion and accumulation processes as a result of the occurrence of severe storms. The obtained results confirm that despite the periodical changes in the regime from accumulative to erosive, and vice versa, the coastal processes on such naturally formed sections of the shoreline remain in balance.

Unmanned surface vessels are very often equipped with single-beam echo sounders (SBES) and multi-beam echo sounders (MBES), which measure depths in bodies of water. These devices need information about the speed of sound in the water. Makar, in [6], proposed a simplified method for determining the speed of sound in water based on temperature measurements and salinity prediction for shallow-water bathymetry. His main objective was to develop a method to determine the speed of sound based on temperature measurements made using an ordinary laboratory, low-cost thermometer with a probe fitted with a long cable. This is an interesting approach to the problem under study. Makar assumed that a change in the salinity of the water relative to its depth has no significant effect on the distribution of the speed of sound determined from changes in temperature. The study only assessed shallow water. The speed of sound in the water was determined from commonly accepted formulae, and the salinity was obtained via a web service from the nearest measuring station. The author compared the results with those obtained from measurements made with the Conductivity/Salinity Temperature Depth (CTD/STD) probe. The impact of the inaccuracy of the sound velocity determination in relation to the SBES immersion depth was calculated. Measurements were taken in the low-salinity Baltic Sea and then verified with measurements in the Mediterranean Sea, representing a high-salinity environment. The author indicated that the proposed method could be an alternative to the bar check calibration of SBES.

Performing frequent monitoring activities, especially when applying new nature-friendly coastal defence methods, is a major challenge. In their manuscript [7], Śledziowski et al. propose a pipeline for performing low-cost monitoring using RGB images, accessed by an UAV and a four-level analysis architecture of an underwater object detection methodology. First, several colour-based pre-processing activities were applied. Second, contrast-limited adaptive histogram equalization and the Hough transform methodology were used to automatically detect the underwater, circle-shaped elements of a hybrid coastal defence construction. An alternative pipeline was used to detect holes in the circle-shaped elements with an adaptive thresholding method; this pipeline was subsequently applied to the normalized images. Finally, concatenation of the results from both the methods and the validation processes was performed. The results indicate that automated monitoring tools work for RGB images captured by a low-cost consumer UAV.

Despite being recognized as a key component of shallow-water ecosystems, submerged aquatic vegetation (SAV) remains difficult to monitor over large spatial scales. In their paper [8], authors propose an example of an adaptable open-source object-based image analysis (OBIA) workflow to generate SAV cover maps. They applied radiometric calibration, atmospheric correction, a de-striping correction, and hierarchical iterative OBIA random forest classification to generate SAV cover maps. The workflow was applied to images taken over two spatially complex fluvial lakes in Quebec, Canada, using Quickbird-02 and Worldview-03 satellites. Classification performance based on training sets revealed conservative SAV cover estimates with less than 10% error across all classes except for lower SAV growth forms in the most turbid waters. The results indicated that it is possible to monitor SAV distribution using high-resolution remote sensing within an open-source environment with a flexible and functional workflow.

Today, more and more data are being acquired using unmanned platforms. One interesting case is the combination of bathymetric data acquired with hydroacoustic and photogrammetric sensors. Research in this area was carried out in [9]. This study presents a method for integrating data acquired by unmanned surface vehicles and unmanned aerial vehicles, based on data geoprocessing using a bathymetric reference surface. The aim was to create a uniform bathymetric surface extending to the shoreline in shallow and ultra-shallow areas. Numerical bathymetric models created by interpolation methods confirmed the usefulness of the proposed method.

Aerial hyperspectral sensors, such as CASI-1500, are used in bio-optical research in inland waters due to their high and adjustable spatial and spectral resolutions. The use of data for quantitative assessments requires, as in the case of satellite data, correction of the influence of the atmosphere. Yang et al., in their study [10], compared the results of three commonly used atmospheric correction approaches, i.e., Polymer, 6S, and FLAASH, for CASI-1500 data recorded over the shallow waters of the Uljin coast in Korea, where primary production is dominated by seaweed, and adverse environmental changes affect coral bleaching. The results showed that only the Polymer algorithm (POLYnomial algorithm based on MERIS), originally developed for the MERIS satellite sensor, yielded reliable reflectance (R_{rs}) spectra, which was further confirmed by comparison with satellite data recorded with the Moderate Resolution Imaging Spectroradiometer sensor (MODIS). The other two algorithms calculated negative reflectance values in the blue spectral range. However, the authors noted the differences in R_{rs} derived from CASI-1500 and MODIS data. The recordings from both sensors were separated by 1 day; therefore, the authors identified the need for further validation based on in situ measurements and the possible modification of the Polymer algorithm in order to improve its results.

To develop the final product, remotely sensed elevation data often require processing to create a digital geographic surface. In [11], research related to the use of interpolation and data reduction techniques was carried out. A novel approach in the field of research is the interpolation method based on processing datasets with different degrees of spatial data reduction. Surface modelling was carried out using data acquired by modern remote sensing sensors, such as a single-beam echo sounder and an RGB camera mounted on an unmanned aerial platform. The developed method of data processing and surface modelling allows for minimizing the error of the surface approximation of combined bathymetric and topographic models.

Natural wetlands can be categorized as coastal zone areas. They have an impact on the environment and the ecosystems that exist there due to the strong variations in hydrological and biogeochemical processes. They are heterogeneous and consist of a mosaic of ecosystem patches with different vegetation types. Ju and Bohrer [12] proposed a method for the classification of wetland vegetation based on NDVI time series from the HLS Dataset. The authors used very-high-resolution multispectral imagery to classify wetland vegetation. They used NDVI time series, generated from NASA's HLS dataset, to classify vegetation patches. The authors conducted their study on the coastal estuarine, mineral soil marsh of Lake Erie. It should be mentioned that a sudden rise in water levels was observed in the study area. The article presents the classification process, which consists of several steps. The authors obtained results with a classification accuracy rate of more than 73% for all patch types, except water lily. It was shown that the accuracy of the proposed classification was higher in pixels with a more uniform composition.

Author Contributions: Conceptualization, J.L., P.T., K.B. and M.W.-S.; writing—original draft preparation, J.L., P.T., K.B. and M.W.-S.; writing—review and editing, J.L., P.T., K.B. and M.W.-S. All authors have read and agreed to the published version of the manuscript.

Funding: This research received no external funding.

Conflicts of Interest: The authors declare no conflict of interest.

References

1. Roca, M.; Navarro, G.; García-Sanabria, J.; Caballero, I. Monitoring Sand Spit Variability Using Sentinel-2 and Google Earth Engine in a Mediterranean Estuary. *Remote Sens.* **2022**, *14*, 2345. [CrossRef]
2. Kanwal, S.; Ding, X.; Wu, S.; Sajjad, M. Vertical Ground Displacements and Its Impact on Erosion along the Karachi Coastline, Pakistan. *Remote Sens.* **2022**, *14*, 2054. [CrossRef]
3. Paulino, R.S.; Martins, V.S.; Novo, E.M.L.M.; Barbosa, C.C.F.; de Carvalho, L.A.S.; Begliomini, F.N. Assessment of Adjacency Correction over Inland Waters Using Sentinel-2 MSI Images. *Remote Sens.* **2022**, *14*, 1829. [CrossRef]
4. Templin, T.; Popielarczyk, D.; Gryszko, M. Using Augmented and Virtual Reality (AR/VR) to Support Safe Navigation on Inland and Coastal Water Zones. *Remote Sens.* **2022**, *14*, 1520. [CrossRef]
5. Michałowska, K.; Głowienka, E. Multi-Temporal Analysis of Changes of the Southern Part of the Baltic Sea Coast Using Aerial Remote Sensing Data. *Remote Sens.* **2022**, *14*, 1212. [CrossRef]
6. Makar, A. Simplified Method of Determination of the Sound Speed in Water on the Basis of Temperature Measurements and Salinity Prediction for Shallow Water Bathymetry. *Remote Sens.* **2022**, *14*, 636. [CrossRef]
7. Śledziowski, J.; Terefenko, P.; Giza, A.; Forczmański, P.; Łysko, A.; Maćków, W.; Stępień, G.; Tomczak, A.; Kurylczyk, A. Application of Unmanned Aerial Vehicles and Image Processing Techniques in Monitoring Underwater Coastal Protection Measures. *Remote Sens.* **2022**, *14*, 458. [CrossRef]
8. De Grandpré, A.; Kinnard, C.; Bertolo, A. Open-Source Analysis of Submerged Aquatic Vegetation Cover in Complex Waters Using High-Resolution Satellite Remote Sensing: An Adaptable Framework. *Remote Sens.* **2022**, *14*, 267. [CrossRef]
9. Lubczonek, J.; Kazimierski, W.; Zaniewicz, G.; Lacka, M. Methodology for Combining Data Acquired by Unmanned Surface and Aerial Vehicles to Create Digital Bathymetric Models in Shallow and Ultra-Shallow Waters. *Remote Sens.* **2022**, *14*, 105. [CrossRef]
10. Yang, M.; Hu, Y.; Tian, H.; Khan, F.A.; Liu, Q.; Goes, J.I.; Gomes, H.d.R.; Kim, W. Atmospheric Correction of Airborne Hyperspectral CASI Data Using Polymer, 6S and FLAASH. *Remote Sens.* **2021**, *13*, 5062. [CrossRef]

11. Lubczonek, J.; Wlodarczyk-Sielicka, M.; Lacka, M.; Zaniewicz, G. Methodology for Developing a Combined Bathymetric and Topographic Surface Model Using Interpolation and Geodata Reduction Techniques. *Remote Sens.* **2021**, *13*, 4427. [CrossRef]
12. Ju, Y.; Bohrer, G. Classification of Wetland Vegetation Based on NDVI Time Series from the HLS Dataset. *Remote Sens.* **2022**, *14*, 2107. [CrossRef]

remote sensing

MDPI

Article

Methodology for Developing a Combined Bathymetric and Topographic Surface Model Using Interpolation and Geodata Reduction Techniques

Jacek Lubczonek, Marta Wlodarczyk-Sielicka *, Malgorzata Lacka and Grzegorz Zaniewicz

Faculty of Navigation, Maritime University of Szczecin, ul. Waly Chrobrego 1, 70-500 Szczecin, Poland; j.lubczonek@am.szczecin.pl (J.L.); m.lacka@am.szczecin.pl (M.L.); g.zaniewicz@am.szczecin.pl (G.Z.)
* Correspondence: m.wlodarczyk@am.szczecin.pl

Abstract: The research in this paper is concerned with the development of a continuous elevation model in the coastal zones of inland waters. The source data for the creation of numerical terrain models were data acquired by contemporary sensors, such as such as a single-beam echosounder and an unmanned aircraft system. Different interpolation methods were tested in the study. A new approach in the research field is an interpolation method based on the processing of datasets with different degrees of spatial data reduction. The authors call it the Spatial Interpolation Method based on Data Reduction (SIMDR). The choice of method is based on quantitative and qualitative analysis, taking into account the type of interpolation and the method of geodata reduction. A proposal for the practical implementation of the method involves script processing, which automates the processes of modeling and error calculation.

Keywords: DTM; seabed; hydrography; UAV; topobathymetry; data reduction; spatial interpolation; terrain modeling

Citation: Lubczonek, J.; Wlodarczyk-Sielicka, M.; Lacka, M.; Zaniewicz, G. Methodology for Developing a Combined Bathymetric and Topographic Surface Model using Interpolation and Geodata Reduction Techniques. *Remote Sens.* **2021**, *13*, 4427. https://doi.org/10.3390/rs13214427

Academic Editor: Danilo Orlando

Received: 20 September 2021
Accepted: 1 November 2021
Published: 3 November 2021

1. Introduction

Increasing attention is being paid to the creation of elevation models of coastal zones, in which the bottom is naturally combined with the terrain. In the literature, one can find various cases of combined models, which are usually determined by the type of sensor used. Ideally, the product is a continuous model, which requires the use of a topobathymetric sensor. The product in this case is a point cloud from which a height model can be created. Another case is the integration of models using different sensors. An example would be the integration of UAV (Unmanned Aerial Vehicle) and USV (Unmanned Surface Vessel) sensor data [1] or UAV and mobile laser scanner data [2]. The last option is the integration of models from GIS (Geographic Information System) databases [3]. Taking into account the variety of spatial data acquisition technologies, the creation of a uniform elevation model often requires a customized approach due to the spatial distribution, extent, density, or presence of areas with missing data [2].

Elevation and bathymetric data can be acquired from various sensors. Some of them allow the acquisition of bathymetric and topographic data (LIDAR (Light Detection and Ranging) and UAS (Unmanned Aircraft System)), and some of them only allow the acquisition of bathymetric data (single-beam and multibeam echosounder). Data acquisition techniques have their advantages and disadvantages. Usually, a sensor dedicated to acquiring bathymetric and topographic data separately provides better quality data and has fewer limitations. LIDAR sensors and UASs with multispectral sensors are examples. The topographic and bathymetric LIDAR sensor has different pulse rate, flight altitude, vertical and horizontal accuracy, vertical datum, resolution, footprint, swath, and data processing [2]. On the other hand, the limitations of a UAS with an RGB (Red, Green, Blue) camera usually include its dependence on water transparency, cloudiness, water ripples, or the obscuration of part of the water surface by trees [4]. Today, acoustic sensors, such as single- and multibeam echosounders, are used to acquire bathymetric data. The data acquired with this technology are used in the production of nautical charts, dredging, or offshore installation works. Their application can also be seen in research and development work related directly to the effects of wave and water levels on coastlines [5,6]. Currently, acoustic sensors are the most accurate systems for acquiring depth of water data, and models [7,8] generated from these measurement are used as references for other measurement techniques besides acoustics [9,10]. In coastal zones, characterized by shallow depths, a single-beam echosounder is most often used in combination with a floating platform, such as a boat or an Unmanned Surface Vehicle [11]. The use of USVs will not always be possible, due to very shallow depths or natural limitations in the form of submerged and partially submerged aquatic plants, submerged plants with floating leaves, or various marsh plants present in the shallow water zone. Other obstacles associated with shallow-water measurements include the presence of local shallows or underwater obstructions, such as submerged trees.

During the processing of spatial data, attention must be paid to data reduction. Modern surveying systems collect very large datasets belonging to Big Data. Their reduction is necessary in the elaboration process [12–14], similar to the process of surface reconstruction by triangulation. This method creates an irregular network of triangles, the vertices of which are the measurement data. Given the dependence of the number of triangles on the number of measurement points, their reduction is particularly important. A model generated on a smaller set of data can more easily display and perform various spatial analyses. According to Osowski [15], data reduction is a method of optimally representing large quantities of data in a much smaller representation. The purpose of reduction is to contain the core of information in a reduced volume of data. He also defines selection, which is closely related to reduction. Osowski describes selection as the operation of finding the optimal data representations for a specific task. Point sets can be reduced at the data preprocessing stage (by reducing the number of observation sets through point removal) or during numerical model generation (by interpolating point depths on grid nodes) [16].

The use of interpolation techniques that create a continuous surface is also an important consideration when generating numerical elevation models. Today, various interpolation methods can be used to calculate heights in a GRID or TIN (Triangulated Irregular Network) structure. These methods have been the subject of many studies, involving datasets characterized by different densities or spatial distributions [8,17–21]. In the literature, we can find numerous examples of authors studying the influences of different aspects, such as the interpolation method, resolution, data density, and slope, on the construction of a numerical terrain model or a numerical bottom model. In his work, Desment [22] evaluated the impact of different interpolation methods on, among other things, the preservation of the shape of a topographic surface. Curtarelli [23] studied four spatial interpolation methods for mapping the bathymetry of an Amazonian water body, and, as the authors concluded, each method was able to map important bathymetric features. Steranczyk [24] investigated the effects of algorithms available in four different pieces of software on the accuracy of DEM (Digital Elevation Model) models created from LIDAR data. The authors of [25] compared four interpolation methods, concluding on the greatest efficiency of the IDW (Inverse Distance Weighting) method. Habib [26] investigated three deterministic interpolation algorithms used in the ArcGIS software—TIN, IDW, and ANUDEM (Australian National University's Digital Elevation Model)—to generate reliable DEMs for large-scale mapping. Additionally, the authors also addressed the qualitative evaluation of DTM (Digital Terrain Model) data. As Podobnikar [27] writes, the application of the visual method depends on the knowledge and experience of the operator and the visual inspection of the basic derivatives of the DTM focusses on the visualization of slope, aspect, curvature, and terrain roughness. Łubczonek [28] examined the methods used to build a numerical bottom model and determined the accuracy of the modeled surface depending on the interpolation algorithm used. In this paper, surfaces were modeled on data of different densities using surfaces generated from mathematical functions (simulated surfaces and datasets). Authors of [29] investigated the utility of UAV data reduction for an urbanized area via random sampling using CloudCompare software. The authors used kriging for spatial interpolation. On the other hand, in [30], a data reduction algorithm was developed for large LIDAR datasets, wherein the study areas were flat land and land with moderate inclination and steep slopes. Modeling was performed by the IDW method. In contrast, the authors in [25] studied reduction via random sampling, data resolution, and interpolation methods for a desert area of 13 by 13 m (micro topographic level), and they pointed to IDW as an effective method. A shared characteristic of the above studies is that, with a higher degree of data reduction, the surface approximation errors increased. In our approach, we use different data reduction algorithms for modeling topobathymetric surfaces and show that the approximation errors of elevation models can be reduced with increased data reduction.

In this paper, the authors investigated the influence of three elements in the process of creating a continuous bathymetric and topographic surface: data reduction, interpolation technique, and the presence of areas with missing data (which is an innovative approach for such studies). The final result is the proposal of a Spatial Interpolation Method based on Data Reduction (SIMDR). The analyzed cases concern bathymetric data acquired with a single-beam echosounder sensor (SBES), topographic data acquired with low-altitude photogrammetry methods, and the occurrence of data gaps in both. The inability to acquire data in this study was due to the very shallow depth and transparency of the water, making it impossible to gather data with the bathymetric sensor or UAS. In addition, filtering out land cover points and the processes of their development both resulted in missing data, which should be considered during the process of developing height models using interpolation techniques. The proposed methodology will be used during work on the MORGAV project "Development of technology for acquisition and exploration of gravimetric data of foreshore and seashore of Polish maritime areas" and can be applied in practice during the development of coastal area models for inland waters. The testing options include seven interpolation methods: triangulation [31], natural neighbor [32], nearest neighbor, inverse distance to a power [33], kriging [34], the RBF (Radial Basis Function) method [35], and minimum curvature [36]. Different degrees and types of data reduction were also taken into account. The motivation for choosing these methods is their implementation in many GIS software types and, particularly, those dedicated to creating numerical terrain models. Despite the recommendation of a particular method, the operator always encounters a problem in the choice of a particular one, especially in the current era of changes in the methods of geodata acquisition, which now generate large datasets with high density. The paper also pays attention to the correctness of terrain plasticity, taking into account terrain microforms related to the correctness of surface reconstruction. The practically developed models can be used in many applications. However, due to the current data acquisition capabilities of smaller hydrographic units, UAVs will be employed. The first of these is the development of better quality photogrammetric products that allow more accurate data to be acquired. The more accurate the numerical elevation model, the smaller the situational errors in objects imaged on the orthophotomap. In the analyzed area, such objects may be dikes, coastlines, or vegetation. These materials can be used to create and update topographic and navigational databases for inland navigation. In the case of dikes, 3D spatial models and image data can be used for monitoring, especially in areas where lower resolution images have lower photointerpretation potential. From the point of view of inland navigation and dredging, it is important to determine the exact depth. In the first case, this is related to navigation safety and involves depth mapping, while in the second case, it involves the accurate calculation of earth masses. Accurate elevation models can also be used in studies related to determining the impact of resolution on gravimetric terrain correction [37]. Combined models allow for comprehensive analyses that address both bathymetric and topographic features.

2. Study Area and Materials

The study area included a section of Lake Dabie, located in Poland's West Pomeranian Province, along with an adjacent inland area. Lake Dabie is a shallow-water area with an average depth of 2.61 m [38]. The maximum depth in the study area was 3.82 m. In the direction of the shore, the depths decrease significantly. The bottom is characterized by moderate regularity, and there is one depression and quite considerable shallowing in the strip adjacent to the shoreline. The surrounding land area is characterized by a flat surface. Approximately 5 m from the shoreline, there is a floodbank with a hardened surface covered with mineral aggregate. On the water side, part of the site is overgrown with vegetation characteristic of aquatic areas, dominated by reeds. On the land side, part of the site is overgrown with shrubs (medium vegetation) and grass (meadow area). Single trees are also present. The geographical location of the study area is illustrated in Figure 1.

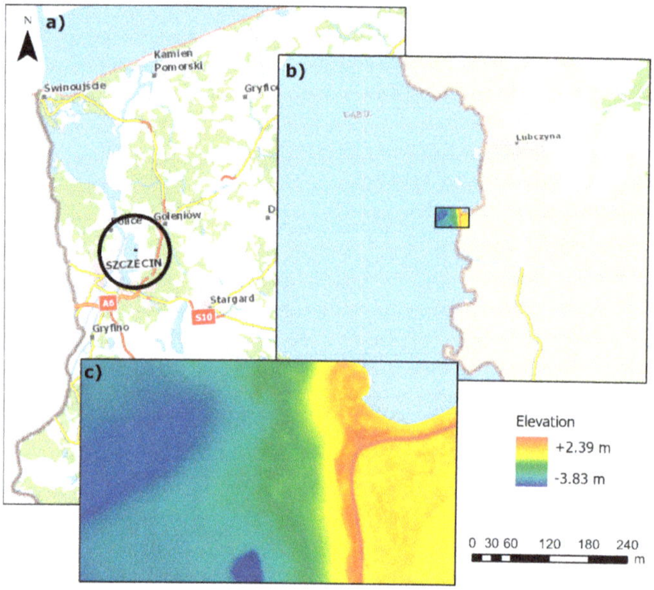

Figure 1. Study area: (**a**,**b**) location of the study area, (**c**) detailed survey area with measured elevations.

Topographic data were collected using the UAS platform. The flight took place on 24 September 2020. The weather conditions were moderately good, partly sunny with wind gusts. The overcast was variable, from partial to total. Before the flight, the photogrammetric reference network was stabilized, and consisted of 6 points distributed evenly over the data processing area. Three points of the photogrammetric matrix were located on the hard surface (on the embankment), two in the area covered with low vegetation, and one on the beach. Control points were also measured to verify the height of the obtained terrain model, and these were used to calculate the errors as an independent set of measurements. The flight was performed at an altitude of 120 m with a DJI Phantom 4 Pro drone using a polarizing filter according to a predesigned flight plan. During the flight, 111 images were acquired, and the size of the compiled area was 0.109 km^2. The last two rows of photos were acquired over a water area. The images from these series were not calibrated (merged with the other images) due to the lack of unambiguous binding points, and therefore, the percentage of calibrated images is 64%. The basic parameters of the equipment used and the data acquired are also summarized in Table 1.

Table 1. UAS parameters and basic data related to image processing.

UAS	DJI Phantom 4 Pro
Satellite Positioning Systems	GPS/GLONASS
Sensor	1″ CMOS, effective pixels: 20 M
Photo Format	JPEG
Flight Altitude	120
Front/Side Overlap	80%/70%
Percentage of Calibrated Images	64%
Georeferencing (mean RMS error of 6 GCP)	1.1 cm
Ground Control Points RMS error (m)	X = 0.7 cm, Y = 0.5 cm, Z = 2.3 cm
Coverage Area	0.109 km^2
Number of Photos	111

The low-level data processing of the acquired images was performed in the PIX4D Mapper, a piece of software dedicated to this type of data. The preprocessing of the data was performed to obtain products such as an automatically classified dense point cloud, which was further processed. Georeferencing was performed based on the points measured with the GNSS–RTK (Global Navigation Satellite Systems–Real Time Kinematic) receiver, and the obtained errors are shown in Table 1. A further processing step was performed in the Global Mapper software, in which a qualitative assessment of the automatic photogrammetric classification of the point cloud was carried out and a manual classification was performed.

In order to obtain bathymetric data, a Hydrograf XXI survey boat was used. This vessel, with a shallow draft, is dedicated to shallow water measurements, especially in inland waters. In the first stage, 79 survey lines were set at a distance of 5 m from one another within the designated survey area. The average depth in the area was 2.74 m.

The survey setup consisted of a Kongsberg EA400 single-beam echo sounder operating at two frequencies, 33 kHz and 200 kHz, with the second frequency used to generate a numerical bottom model from the acquired depths. A Trimble R6 GNSS receiver operating with RTK corrections derived from the VRSNET base station was used as the positioning system. Additionally, in order to eliminate the effects of weather conditions and wave movement, the IMU (Inertial Measurement Unit) SMC-108 motion sensor was used for roll, pitch and heave motion correction. The parameters of the measuring equipment (Table 2) allowed for a total measurement accuracy within the special category defined by the IHO (International Hydrographic Organization) in specification S-44 [39]. All data integration and acquisition was performed in the Hypack hydrographic software. Data processing was performed in the SBMAX64 module. First, after the data were imported, navigational data representing the boat's movements through the survey profiles were analyzed. In the next step, the motion sensor data were subjected to verification. The main focus of the analysis was the depth data. Incorrect readings were eliminated using manual data processing. In most cases where the correct depth reading was uncertain, raw data from digital echograms were used. Before exporting to the ASCII file, the XYZ dataset was corrected via the vertical correction of water level, employing the water level gauge in Szczecin harbor.

The spatial distribution of the UAS and SBES data after processing and their statistical information are illustrated in Figure 2. On the left, points obtained with the SBES are shown in blue. On the right, points acquired with the UAS are shown in orange. The final sonar dataset contains 1,019,276 measurement points, and the UAS dataset contains 101,035 points. Negative values of UAS data in the study area have resulted from local depressions (there are water ditches near the dike from the land side).

Table 2. SBES parameters and basic data associated with bathymetric data processing.

SBE Model	Kongsberg EA 400
Frequency	33/200 kHz
Satellite Positioning System	Trimble R6 GPS/GLONASS
GNSS Positioning Accuracy	H: 8 mm + 0.5 ppm RMS ǀ V: 15 mm + 0.5 ppm RMS
Motion Sensor	IMU SMC—108
Roll and Pitch/Heave Accuracy	0.03° RMS/5 cm or 5%
Coverage Area	160,166 m^2

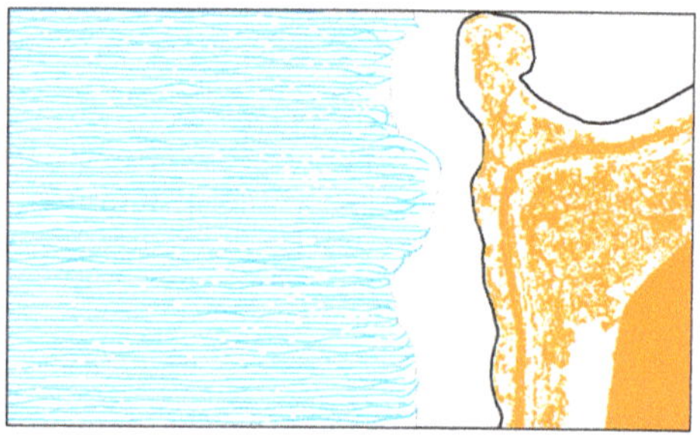

Statistics of measurement points before reduction for SBES and UAS

UAS (orange)	
Number	**1,019,276**
Min. Z	−0.59 m
Max. Z	2.39 m
Density	57.43 points/m^2
SBES (blue)	
Number	**101,035**
Min. Z	−3.83 m
Max. Z	−0.87 m
Density	0.82 points/m^2

Figure 2. Spatial distribution and statistics of measurement data.

3. Methods

The input data for this study were two sets of measurement data acquired with a single-beam echosounder and using an UAS, as described in Section 2 (study area and materials). After processing the measurement data, their reduction was carried out, and the result was 39 sets of points with different scales of reduction for data acquired using SBES and UAS separately.

The data in this form were further processed by adding an additional set representing the shoreline to each set of points. This set was created by generating shoreline points at 1 m intervals in the vector data. The purpose of modifying the datasets was to take into account the natural constraint of the modeled surface, which was the shoreline. Modeling was not carried out on combined bathymetric and topographic datasets because, in the shoreline area, point search algorithms can create mixed datasets from which the value at a grid node is calculated. In this case, the height of the topographic area or the depth of the bathymetric area can be calculated from topographic and bathymetric points. This distorts the plasticity of the model, which is different for the bottom and the terrain. The creation of separate models ensures that the shape of the topographic surface is created from the topographic points, while the shape of the bottom surface is created only from the bathymetric points. The shoreline in this case is a natural link between the models.

In the next step, surfaces were created in the GRID structure with 1 m resolution. The authors used the following interpolation methods: triangulation (referred to as TRI), natural neighbor (referred to as NAN), nearest neighbor (referred to as NEN), inverse distance weighted to a power (referred to as IDW), kriging (referred to as KRI), the radial basis function method (referred to as RBF), and minimum curvature (referred to as MIC). In addition, in regard to searching in the interpolation process, two basic methods were investigated: sectorless and four-sector. The sectorless method consists of searching for the measurement points used for interpolation from the nearest neighborhood of a grid node. The selection of points is based on the calculated distance between a measurement point and a grid node. In the case of, for example, interpolation from the 16 nearest measurement points, those with the shortest distances between them are selected. The four-sector method works in a similar way, except that the selection of points takes place separately in each sector. The method used in this study divides the space around a grid node into four sectors. Geometrically, this is achieved by drawing two perpendicular lines intersecting at the grid node. In this way, it is possible to force a search for points in each sector. If, for example, we use a sector search, indicating four points per sector, the search algorithm will search for four points in each sector. In total, we will derive an interpolation from 16 measurement points, but these will be evenly distributed around the interpolated point. In the case of a sectorless search, the distribution of measurement points can vary, which can affect the quality of the modeled surface. In both cases, in order to ensure that the desired number of points is searched, the search radius was set at 150 m. In the first stage of the research, the selection of interpolation variants was made on the basis of the quality analysis. Then, the obtained results were subjected to a visual evaluation of the shape of the modeled surface. This approach made it possible to exclude those options that prevented the correct reconstruction of the surface and to determine the final variants of the study related to the interpolation methods.

In the second stage, the final study options were determined. Qualitative and quantitative analyses were conducted for each modeled surface. For the quantitative analysis, errors were calculated using test points, which constituted an independent set of measurements. Test points were acquired using the GNSS–RTK technology. In the case of the whole water area, a test point was selected from the main bathymetric measurement set, while in the water area with missing data and the land part, measurements were made using a GNSS–RTK geodetic receiver. These points provided the reference measurements for calculating the deviations from the modeled surfaces, from which the absolute values of the mean and maximum errors were calculated. For the bottom models, errors were analyzed for the whole area using SBES data with 19 test points (mean value: −2.70 m; minimum: −3.70 m; maximum: −1.16 m) and for the area with no data using 14 test points (mean value: −0.70 m; minimum: −0.96 m; maximum: −0.40 m). In the UAS data, there were nine test points (mean value: 1.05 m, minimum: −0.27 m, maximum: 2.09 m).

In the third stage of the research, different methods of combining models based on raster data mosaicking techniques—bathymetric and topographic—were analyzed. These analyses provided our final conclusions. The research methodology is presented schematically in Figure 3.

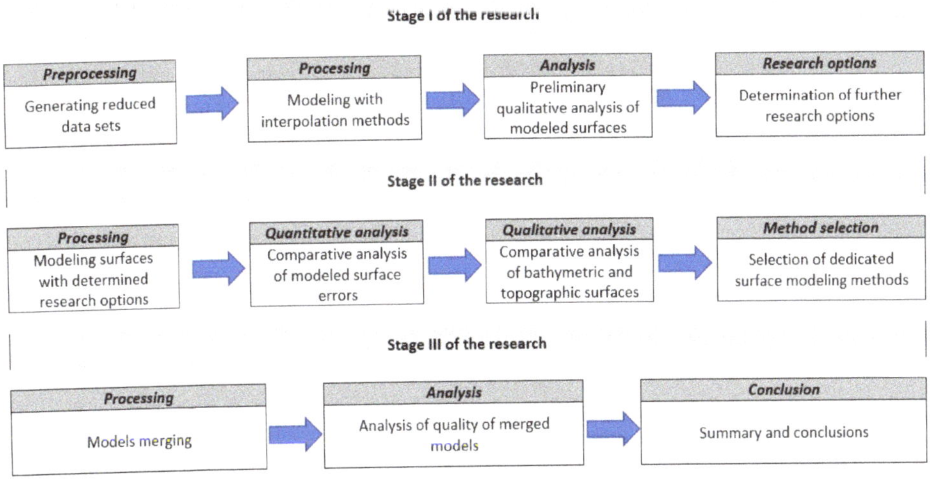

Figure 3. The research methodology.

Data with different spatial distributions and point densities were used for this study. Bathymetric data were characterized by a regular spatial distribution, which resulted from the methods we used for their acquisition from the planned survey profiles. The UAS datasets were generated from aerial photographs as a point cloud. A characteristic of the data used is their dispersed spatial distribution, which makes the surface interpolation process more difficult [17]. In addition, smaller areas with missing points belonging to land cover (high and medium levels vegetation) were created through data processing after filtering out points belonging to land cover. The lack of ground data in areas of dense vegetation is related to the specifics of obtaining these data; a dense point cloud is created on the basis of photographs, so there is no possibility of obtaining height information about parts of the ground that are not visible on the images (under dense vegetation, buildings). Data were also lacking in a narrow strip near the shoreline. An important area in the interpolation process was the shallow-water area. These shallow depths made it impossible to take measurements with a hydrographic vessel; as such, here, the data had to be interpolated based on extreme measurement points, which comprised shoreline points and bathymetric data. In addition, the surface characteristic elements were analyzed. For the bathymetric model, these were the slope of the bottom surface and the depression, and for the topographic model, these were the floodwall and the areas of medium and low vegetation. All the mentioned datasets and surface elements are presented in Figure 4.

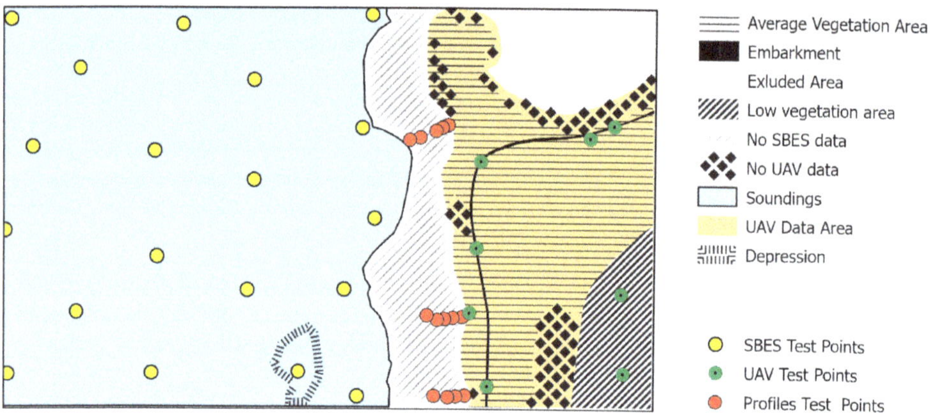

Figure 4. Analyzed test areas.

The datasets used during the study were reduced using the Reduce Point Density method, which is implemented in ArcGIS software. This reduces the number of points in the sets depending on the Nominal Start Thinning Radius (R) parameter and the method of selecting individual points. The first parameter specifies the radius of the circle within which a particular point will be selected. The remaining points are reduced. The second parameter is the selection method. Two variants were used in this research: MAX and MIN. In the first, the reduction is based on retaining the data with maximum values (MAX variant), while in the second, the minimum values are retained (MIN variant). Geodata in the form of points can be transformed via the following algorithms [40]:

- Aggregation—combining more points into one;
- Simplification—removing objects in order to properly present those remaining;
- Typography—retaining the predominant source symbolization of point objects while removing points;
- Regionalization—drawing a boundary around a group of point objects and creating a new surface object from it;
- Selective omission—selecting objects that are more significant while omitting objects of lesser significance.

The used method can be classified as selective omission. In this case, points in a certain area (circle) with the smallest or largest Z values were selected. The data were reduced over a range of R parameter values, from 0.2 m to 4 m, with an interval of 0.1 m. This gave a total of 39 sets of points, with different scales of reduction for data acquired using SBES and UAS for the two variants MIN and MAX. In total, 156 reduced datasets were obtained for further analysis. Figure 5 shows an example of reduction for data acquired using SBES.

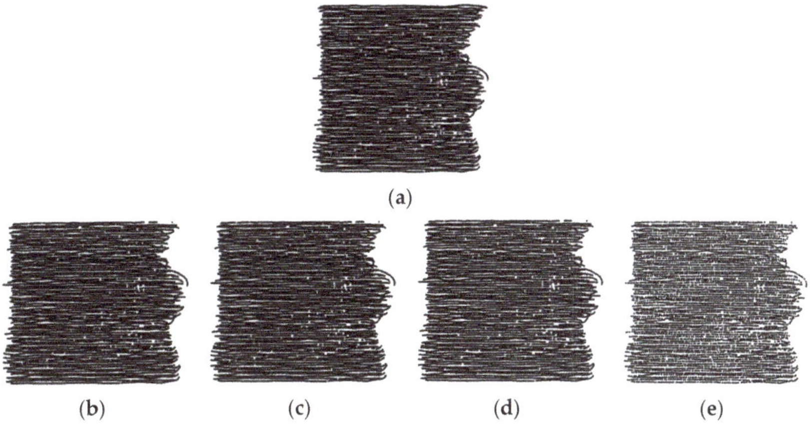

Figure 5. Example of reduced sets of SBES data: (**a**) raw SBES data, (**b**) reduced datasets with R = 0.2 m, (**c**) reduced datasets with R = 1 m, (**d**) reduced datasets with R = 2 m, (**e**) reduced datasets with R = 4 m.

The figure above shows the spatial distribution of the input data and the reduced sets with radius values (0.2 m, 1 m, 2 m, and 4 m) specifically selected for retaining the minimum values. The initial set contains 101,035 points, with a minimum depth of −3.83 m, a maximum depth of −0.87 m, and an average depth of −2.79 m. The reduction results for the SBES data with the main characteristics are shown in Table 3.

Table 3. The reduction results for SBES data.

	Raw Data	R = 0.2 m	R = 1 m	R = 2 m	R = 4 m
Number of points	101,035	94,151	27,544	14,339	7245
Minimum depth (m)	−3.83	−3.83	−3.83	−3.83	−3.83
Maximum depth (m)	−0.87	−0.87	−0.88	−0.88	−0.89
Average depth (m)	−2.79	−2.81	−2.84	−2.85	−2.86

Figure 6 shows an example of reduced data acquired with the UAS system. The example uses the same parameters as Figure 5.

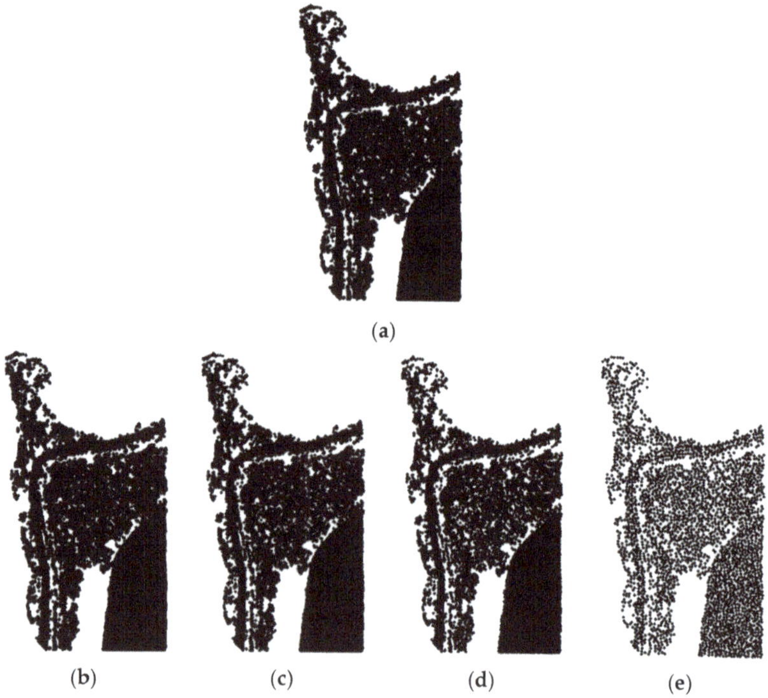

Figure 6. Example of reduced sets of UAS data: (**a**) raw UAS data, (**b**) reduced datasets with R = 0.2 m, (**c**) reduced datasets with R = 1 m, (**d**) reduced datasets with R = 2 m, (**e**) reduced datasets with R = 4 m.

The initial set contains 1,019,276 points, wherein the minimum height is −0.59 m, the maximum height is 2.39 m, and the average height is 0.32 m. The reduction results for the UAS data along with their characteristics are shown in Table 4.

Table 4. The reduction results for UAS data.

	Raw Data	**R = 0.2 m**	**R = 1 m**	**R = 2 m**	**R = 4 m**
Number of points	1,019,276	405,624	27,472	9107	3000
Minimum hight (m)	−0.59	−0.59	−0.59	−0.59	−0.59
Maximum hight (m)	2.39	2.39	2.38	2.21	2.12
Average hight (m)	0.32	0.29	0.32	0.32	0.28

A total of 156 reduced datasets were obtained for further analysis. Different surface interpolation methods were selected for the study. These included parameter-free, parameter setting, local, and global surface interpolation methods. For methods interpolating the surface locally, the effect of the search method was also investigated. Automated interpolation was performed using the Surfer 20 software with a scripting language. The methods, along with a brief description and the parameters chosen here, are summarized in Table 5.

Table 5. Interpolation methods used in the research.

Method	Interpolation Type		Point Search Type for Local Methods		Total Number of Points Used for Interpolation	Method Parameters Used	Number of Models Developed Including All Options
	Global	Local	Sector Less	4 Sectors			
TRI	*	-	-	-	All	-	156
NAN	*	-	-	-	All	-	156
NEN	*	-	-	-	All	-	156
IDW	-	*	*	*	8, 16, 24, 32, 40, 48, 56, 64, 72	Weighting Power: 2	2808
KRI	-	*	*	*	8, 16, 24, 32, 40, 48, 56, 64, 72	Semivariogram Model: Linear Kriging Type: Point, Polynomial Drift Order: 0 (Ordinary Kriging)	2808
RBF	-	*	*	*	8, 16, 24, 32, 40, 48, 56, 64, 72	Basis Kernel Type: Multiquadric, Shape Factor (R^2) calculated according to the formula: (length of diagonal of the data extent)2/(25 * number of data points)	2808
MIC	*	-	-	-	ALL	Maximum Residual: 0.004 (SBES), 0.003 (UAS). Maximum Iteration: 100,000. Internal Tension: 0. Boundary Tension: 0. Relaxation Factor: 1.	2808

* type of interpolation and interpolation options of of reaearched methods.

4. Research

In the first stage of the research, all modeled surfaces were analyzed. Preliminary qualitative analysis, consisting of the assessment of the correctness of the modeled surface, allowed us to reject some of the options. In this respect, the locally interpolating methods (IDW, kriging, RBF) without data search sectors were found to have low usefulness. The reason for this is the poor surface representation in areas with few data, which manifested in the form of artifacts. A comparison of models used in the area without bathymetric data is shown in Figure 7. A color-coded depth scale is used for each height within the model. As can be seen, in the center of the area without bathymetric data, the surface is characterized by a clear lack of continuity. The next step was to determine the number of points used in the local interpolation. In this respect, the plasticity of the terrain representation was used as a criterion of correctness. While the number of points was found to have less influence on the surface reconstruction in the area with data, the number of points was important in areas with no data. Based on a visual assessment, 12 points per sector was selected as the correct number of points, resulting in an interpolation of 48 survey points in total. Following visual comparative analysis, it is also possible to estimate the qualitative differences in surface modeling using sectorless and sector search points. In the case of sector searching, the surface in the no-data area is characterized by continuity, i.e., there are no artifacts representing sudden depth changes.

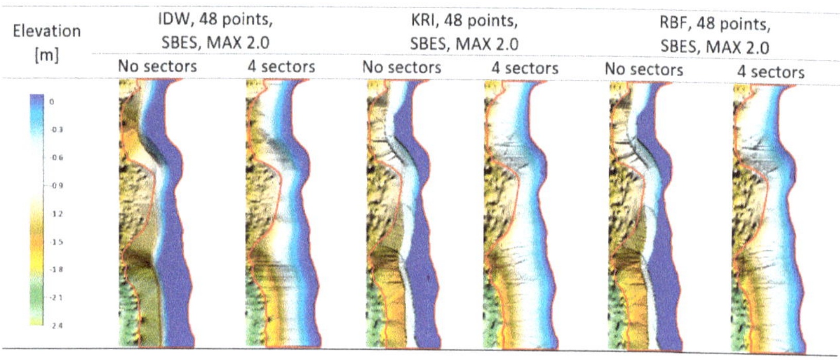

Figure 7. Comparison of surface modeling in the area with a lack of data using sectorless and sector search options.

In interesting cases, problems with complete surface modeling via the natural neighborhood method were noted. For the UAS data with a very small degree of data reduction (level 0.2–1.0), pixels were found in the surface with missing data (Figure 8). In this case, we also used a color-coded depth scale for each height in the model. The lack of data was particularly evident on the flood embankment as well as in the upper part of the area. Therefore, this method was not considered further in the study. Other methods that were not qualified for further study were the minimum curvature method and the nearest neighbor method. These methods generated artifacts in the area with no SBES data (Figure 8).

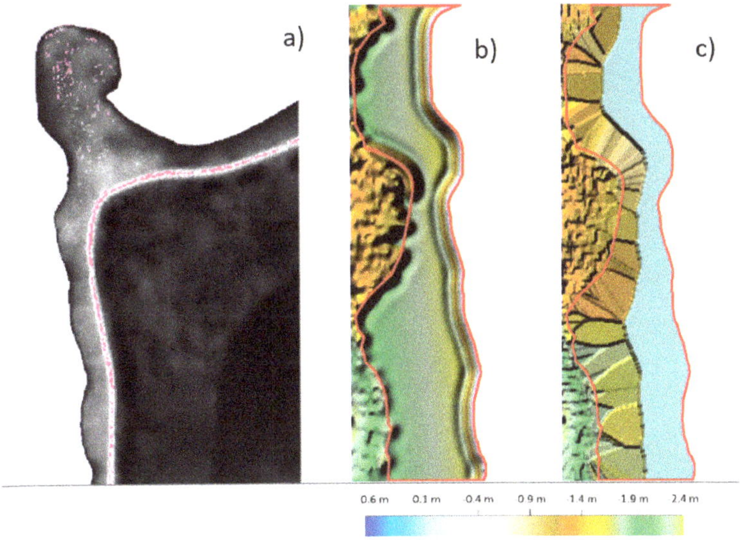

Figure 8. Cases of inappropriate surface modeling for: (**a**) missing data in the surface under the natural neighborhood method (marked in pink), (**b**) artifacts in the area missing bathymetric data under the minimum curvature method, (**c**) artifacts under the nearest neighbor method.

In summary, after the first stage of research, the analysis of interpolation options using sectorless point search and natural neighborhood, nearest neighbor and minimum curvature methods was abandoned. The triangulation, kriging, RBF, and IDW methods were qualified for the next stage of research. Of the rejected methods, natural neighborhood can also be considered, as no deficiencies were seen in this method's modeled surface, and it showed a higher degree of data reduction. However, this requires additional qualitative analyses of the modeled surface in terms of the data gaps present.

The second stage of the research involved quantitative and qualitative analysis. The quantitative analysis was based on a comparison of mean and maximum errors related to the four methods. Basing on the graphs for all the samples, in the final stage, the quantitative results of four levels of data reduction—0.2, 1, 2, 3, and 4—were compiled. The choice of these levels was preceded by the analysis of errors via graphs constructed for all the samples, which allowed us to select representative results reflecting the characteristic relations between method and level of data reduction. The errors are summarized in Figures 9–11.

Analyzing the SBES data, it can be concluded that the errors increase with data reduction for both MIN and MAX methods. The errors have smaller values under the MIN reduction method. The mean and maximum errors both differ depending on the interpolation method. The highest values are derived from the inverse distance method, followed by triangulation, kriging, and the RBF method. It should be mentioned here that the errors for each variant differ within the range of about 1 cm, which may be a negligible value when choosing a method. Greater differences become apparent in the case of the maximum errors, which take much smaller values under the MIN reduction method. The mean values for all trials are 47% smaller (2 cm). Under the MIN method, on the other hand, the mean errors are 17% (0.01 cm) smaller, which in principle could also be a negligible value. Another observation is that, under the MIN method, the maximum errors derived by the kriging and RBF methods change slightly from 1 cm to 3 cm as the level of data reduction increases. According to the quantitative criterion, the RBF and kriging methods perform best.

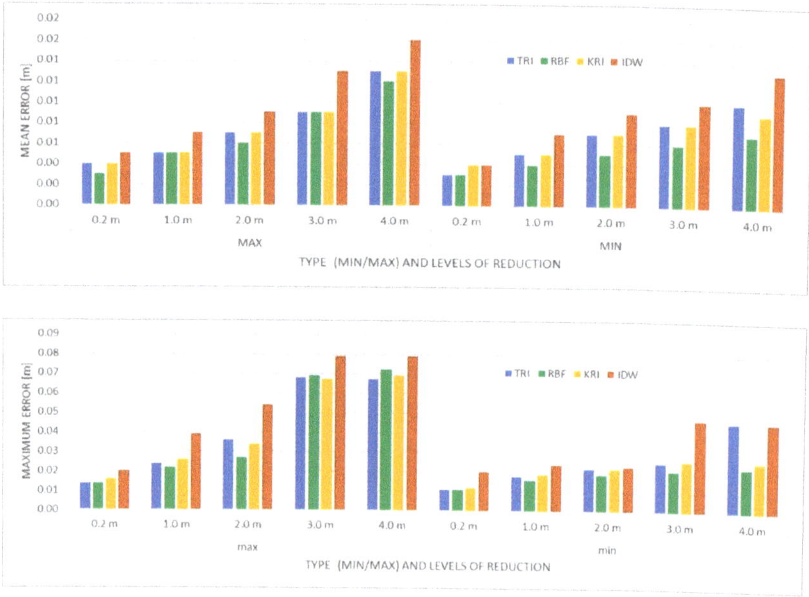

Figure 9. Summary of mean and maximum errors for SBES.

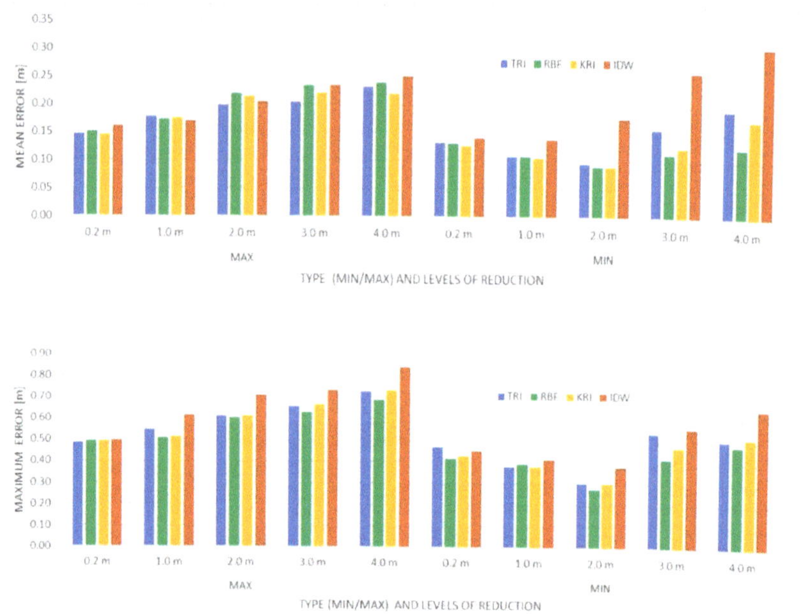

Figure 10. Summary of mean and maximum errors for UAS.

For UAS data, a linear increase in the mean and maximum error values can be observed under the MAX reduction method. The relationship is slightly different for data reduction via the MIN method. In this case, a decrease in error values up to a reduction rate of 2 can be observed, followed by an increase. This applies to the values of both mean and maximum errors. Comparing the errors for the MAX and MIN methods, a similar relationship can be observed as in the SBES data—the errors are smaller under the MIN reduction method. In regard to the MAX reduction method, the best results can be achieved using the triangulation and kriging methods, while for the MIN reduction method, the RBF and kriging methods are best. By far the worst results are obtained with the inverse distance method, especially for the MIN method. Smaller errors can be obtained with the MIN reduction method, with the average errors reduced by 27% (6 cm) and maximum errors reduced by 29% (22 cm).

For the area of the water body with no data, the results for all levels of reduction are similar, meaning its influence is practically negligible. It can also be seen that the average errors for all samples are much larger than for the area with data, at 42 cm. In contrast, for the SBES data in the area with data, this value is 1 cm, while for the UAS data, it is 17 cm. It can be concluded that the best results can be achieved using the triangulation method, followed by kriging and RBF (comparable results) and then the inverse distance method.

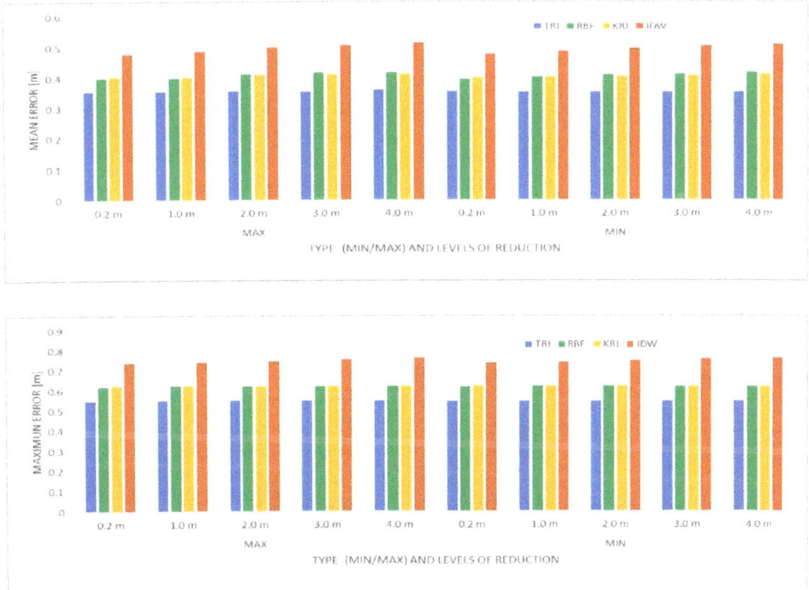

Figure 11. Summary of mean errors for the water area without data.

The next step during the research was a qualitative analysis of the previously selected models. The modeled surfaces are shown in Figures 12–19. DBM is the Digital Bottom Model, DTM is the Digital Terrain Model, MIN and MAX denote the data reduction methods, the numbers next to MIN and MAX denote parameter of reduction (R), and a colored depth scale is applied to the surface models. Figure 12 shows the surfaces created from the SBES data using the triangulation method. Each of the created models presents a smooth surface with correct terrain plasticity. However, some of the areas exhibit profile lines related to the source data, which make the perception and interpretation of the created models easier. The terrain slopes are modeled correctly. In the area lacking data, a surface was created based on elevation values derived from polygon vertices. Due to the lack of available data in this area, the correctness of the created surface cannot be assessed, but the visual appearance of the model is acceptable. The impact of reduction for this model and this type of data is negligible.

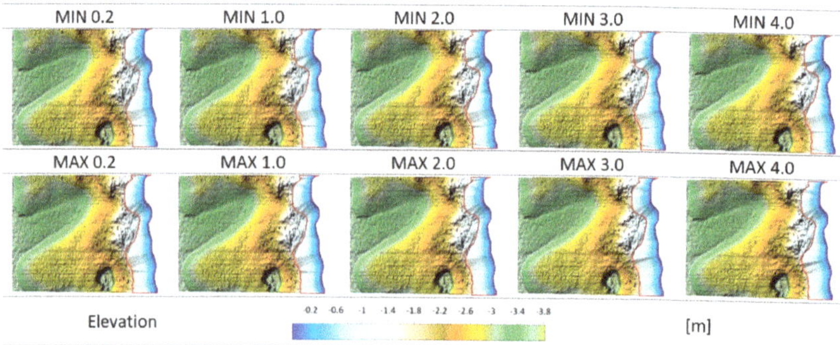

Figure 12. Comparison of DBMs for the triangulation method.

Figure 13 shows surfaces created from photogrammetric data derived from UAS using the triangulation method. First, it is worth noticing that both the type of reduction (MIN, MAX) and its level (0.2, 1.0, 2.0, 4.0) have a significant impact on the final surface obtained. The level of reduction has a significant impact on the visual perception and generalizability of the final model. The type of reduction has an influence on elevation modeling, especially in the area with vegetation cover (low and medium). Models obtained with the MIN method are rougher. It is also important to note the modeling of the embankment, which becomes increasingly fractured as the input data decreases. Areas of high vegetation, i.e., small number of ground points, were modeled slightly differently on each of the analyzed models, with the most correct modeling offered by the MAX 4.0 method. All models obtained with this method are plastic and interpretable; the models lack artifacts and other deformations.

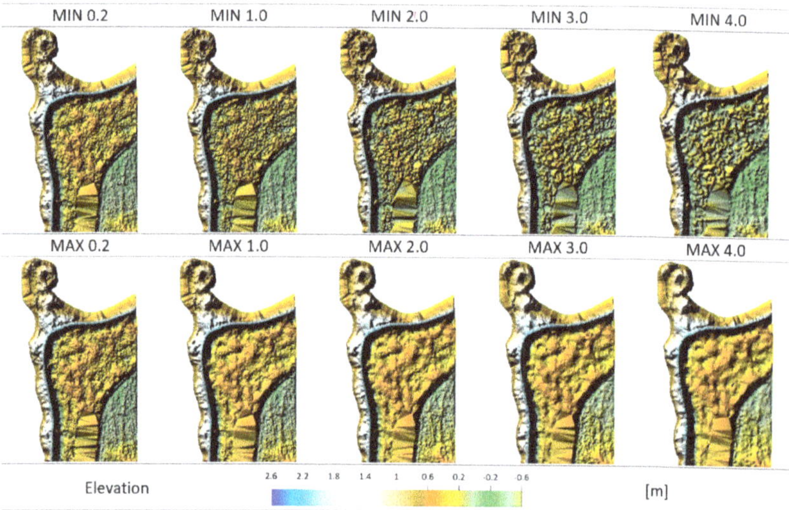

Figure 13. Comparison of DTMs used for the triangulation method.

Figure 14 shows the surfaces created from the SBES data using the inverse distance method. Qualitative analysis was performed on models with four-sector interpolation using 12 points per sector. The overall plasticity of the terrain is correct but becomes worse as the level of reduction increases; at the highest reduction, and especially at points of depression, small artifacts (short horizontal lines) are visible, making interpretation difficult. The representation of the area with missing data varies with the level of reduction.

The higher the degree, the smoother the area of missing data. Under the MIN method, the nearshore area (near the line of missing data) is rougher.

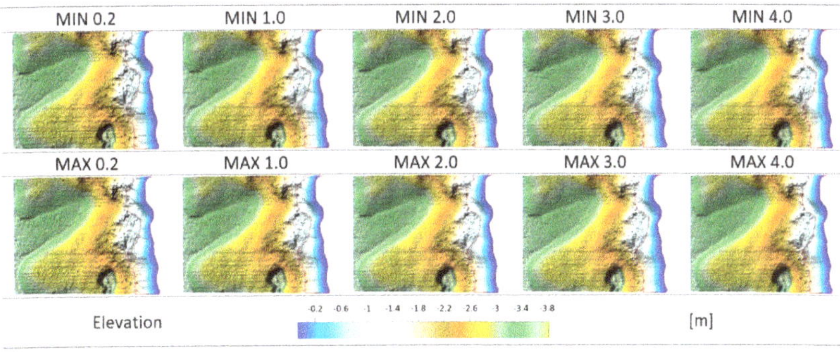

Figure 14. Comparison of DBMs for IDW.

Figure 15 shows the surfaces created from the UAS-derived data using the IDW method. A qualitative analysis was performed on models with four-sector interpolation using 12 points per sector. The effect of the level of data reduction on the results obtained is significant. High reduction eliminates characteristic terrain forms. The MIN method with a reduction level of 4.0 causes significant terrain deformations, which make correct interpretation of the surface impossible. The results obtained with the MAX method for the same degree of reduction are significantly better, although the representation of the dike is also incorrect in this case. For small levels of reduction, the terrain plasticity values are correct—smoother for the MAX method, rougher for the MIN method.

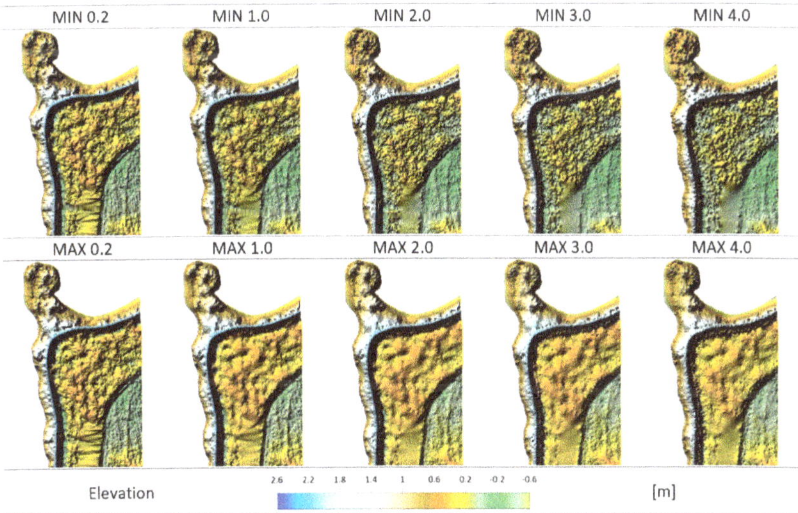

Figure 15. Comparison of DTMs for IDW method.

Figure 16 shows the surfaces created from the SBES data using kriging methods. The qualitative analysis was performed on models with four-sector interpolation, 12 points per sector. The MIN and MAX methods produce very similar results when modeling surfaces from the profile data. The modeling is also very similar when performed on data with different levels of reduction. The obtained models are plastic and allow for good interpretation.

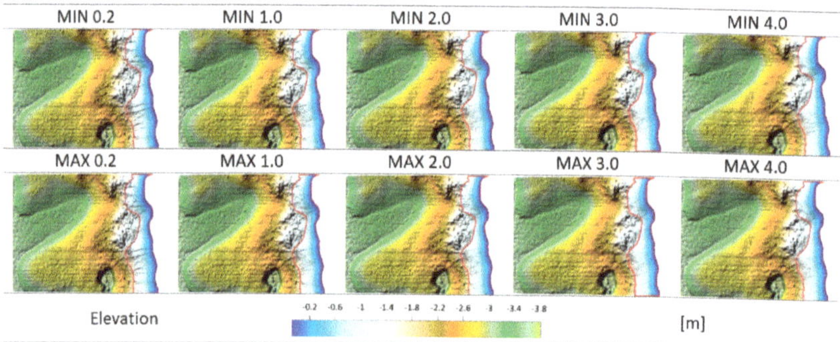

Figure 16. Comparison of DBMs under the kriging method.

Figure 17 shows surfaces created from the UAS data using the kriging method. Qualitative analysis was performed on models with four-sector interpolation, 12 points per sector. The main difference between the MIN and MAX methods is the occurrence of greater terrain roughness in coastal areas and average vegetation in the MIN method; the MAX method produces a smoother surface. There is a considerable loss of terrain detail at a high reduction level, which is most visible in the embankment area. The overall plasticity of the modeling is correct.

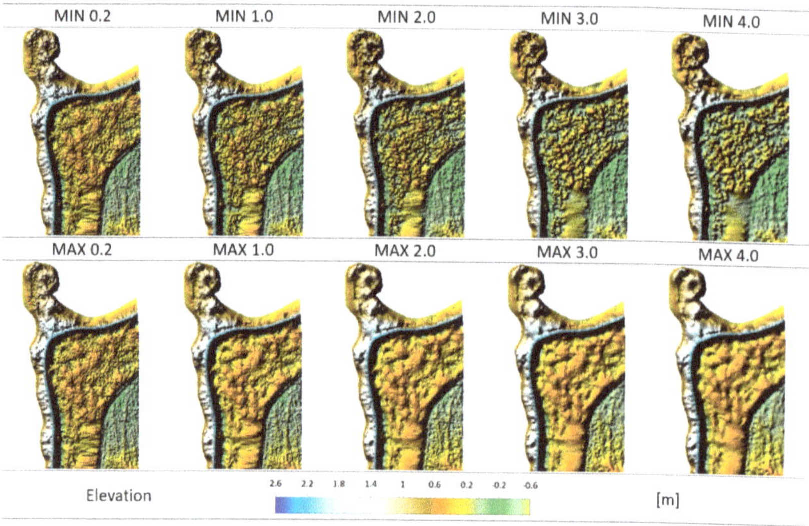

Figure 17. Comparison of DTMs under the kriging method.

Figure 18 shows the surfaces created from SBES-derived data using the RBF method. There is little difference between the MIN and MAX reduction methods; the most noticeable differences appear at the high reduction level (4) in the area bordering the missing data area; there is greater roughness under the MAX method. At a lower level of reduction, the changes between MIN and MAX are unnoticeable. The lack of reduction has the greatest effect on the modeling of the area with missing data and the boundary area. The plasticity of the area is correct.

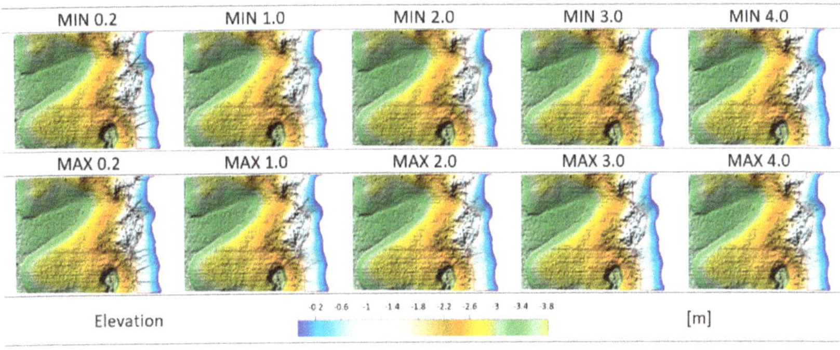

Figure 18. Comparison of DBMs under the RBF method.

Figure 19 shows the surfaces created from the UAS data using the RBF method. The qualitative analysis was performed on models with four-sector interpolation with 12 points per sector. This method provides very similar results to the kriging method for this dataset. The main difference between the MIN and MAX methods is the presence of greater terrain roughness in the coastal areas and medium levels of vegetation in the MIN method; the MAX method also provides a smoother surface. The plasticity of the modeling is correct; there is a loss of details with the increase in data reduction. The area lacking data is modeled correctly—a smoother surface is obtained with a higher level of data reduction.

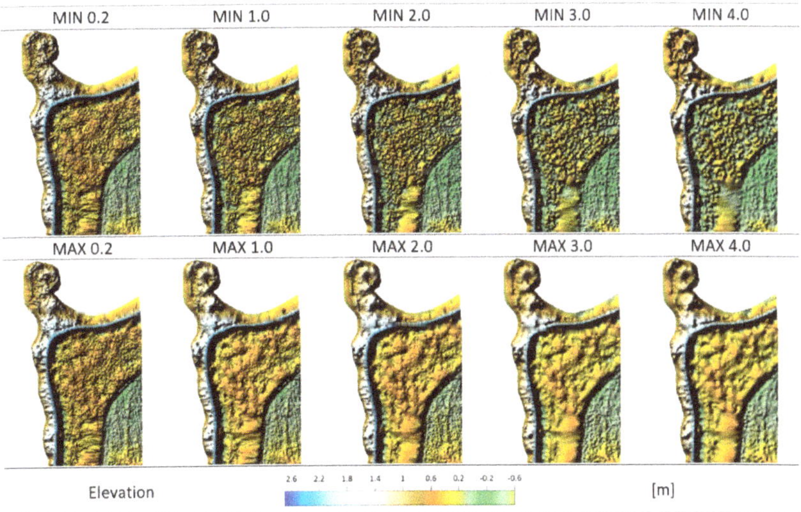

Figure 19. Comparison of DTMs under the RBF method.

The qualitative analysis considering the topographic surface elements is summarized in Appendix A (Qualitative analysis for surfaces created from SBES data) and B (Qualitative analysis for surfaces created from UAV data). Based on the qualitative analysis, it can be concluded that the modeling of SBES data is achieved comparably well between the triangulation, kriging, and RBF methods. Only the models obtained via the IDW method give slightly worse results, which is reflected in the formation of artifacts at the higher levels of data reduction on the higher parts of the bottom's slope. In the case of triangulation, a tendency toward the slight mapping of profiles was noted, perhaps related to the interpolation technique's preserving of the source survey points in the structure. Similar conclusions can be drawn from the UAS data. All the methods—triangulation, kriging, RBF, and IDW with sector searching—provided visually correct surfaces. However, it should be noted that the IDW method again performed the worst with the given level of reduction (above the second degree can result in a significant loss of detail in terrain microforms). The triangulation method, on the other hand, performed less well in terms of creating a uniform surface in the area lacking UAS data. The MIN reduction type produced a rougher surface. Based on the qualitative analysis, it can be concluded that the methods produce comparable qualitative surfaces, but artifacts can be identified, as is well demonstrated by the IDW method. This method tends to produce artifacts and lose significant detail at higher levels of data reduction. In conclusion, visual analysis should be supported by results obtained from quantitative analysis, and the choice of the appropriate method should be made accordingly.

Combining Models

Models were developed separately for bathymetric and topographic data, and, therefore, did not overlap. The option of combining models in overlapping areas was omitted in this case; therefore, the effectiveness of three resampling methods was evaluated: nearest neighbor, bilinear interpolation, and cubic convolution. Here, the merging line was not recognizable, which qualifies all the methods for resampling this type of raster. Interpolation within the boundaries of the area is important in this case, and this was achieved by inputting the corresponding shapefile. In this case, the area outside the polygon had the status of "No Data". A 2D illustration of the combination of models using the kriging method (four-sector search, 12 points per sector, minimum reduction method at the 2 m level) is given in Figure 20, and a 3D view is given in Figure 21.

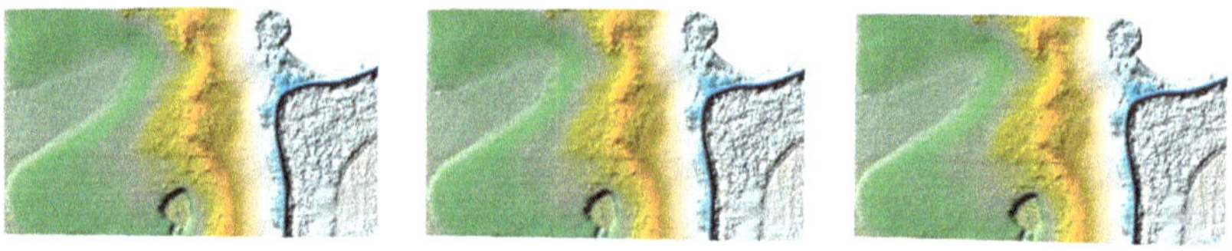

Figure 20. Combination of models constructed by nearest neighbor (**left**), bilinear interpolation (**middle**), and cubic convolution (**right**) method.

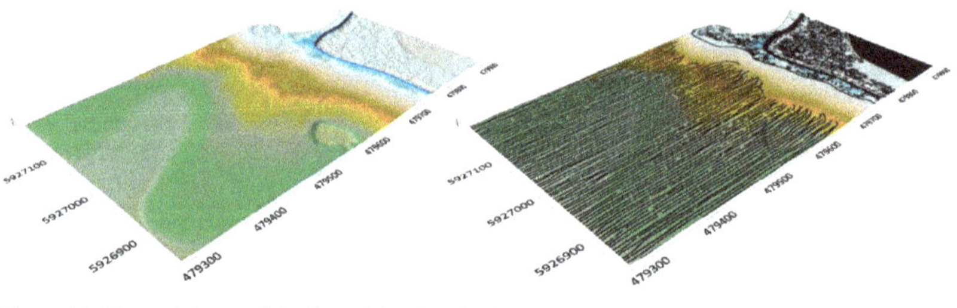

Figure 21. View of the combined model using the bilinear interpolation method. On the right, view with overlaid measurement data.

5. Discussion

The aim of the study was to develop a combined bathymetric and topographic model. As can be seen, the choice of interpolation method and the degree of data reduction are not straightforward. As far as interpolation methods are concerned, various studies can be found identifying the best. However, it is important to realize that each data type has a different density and may have a different spatial distribution, and so a different method may be required in each case. The research conducted in this paper exemplifies this. For the SBES data, the differences in the errors obtained are so insignificant that practically any method can be used. In the case of UAS data, the best results can be obtained using the RBF and kriging methods, while in the case of a body of water without bathymetric data, the triangulation method is favored. Quite clearly, the IDW method can be abandoned, as it gives larger errors and is more prone to artifacts and significant losses of detail with greater degrees of data reduction. Based on these observations, the question can be raised as to whether there is really a universal method. Certainly, the choice of such a method will entail some compromises. Therefore, a qualitative analysis was performed to evaluate the methods, seeking to determine the best method via the correctness of the plasticity of the reconstructed surface. An incorrectly created surface can lead to misleading conclusions or the qualification of artifacts created by the method as surface microforms. Given the final results, the authors decided not to choose a universal interpolation method, and they selected a method on the basis of the results of the modeling process. This is particularly relevant given the results derived using the MIN reduction method for UAS data. In this case, as the reduction increased, the errors started to decrease up to the 2 m reduction level and then started to increase. Would such a relationship occur in a similar but different dataset? It is difficult to answer this question, and so it is reasonable to employ automated data processing and the automated selection of an appropriate method and degree of data reduction. Such a process can be implemented using scripting programming languages. As can be seen in the results of the study, such a process should be carried out separately for bathymetric and topographic data, due to the different levels of reduction that can minimize the errors obtained by specific interpolation methods.

The proposed method was developed for data obtained from a small area, which typically resulted from the operational limitations of the measuring vehicles used. However, taking into account the similar structure of data obtained by modern remote sensors, such as topobathymetric LIDAR, i.e., high-density sets with a regular or scattered spatial distribution, it is possible to use the method for other types of areas, including those with a larger size. The proposed calculation algorithm includes a qualitative and quantitative assessment module; hence, when testing the method on other areas with different topographic features, it is possible to determine whether the method can be used to develop numerical height models. The only limitations are the test points, the measurement of which in the case of larger areas may be problematic and, therefore, may exclude the application of the proposed solution.

6. Conclusions

Based on the results obtained, the authors propose an automated surface modeling method, which they have called the Spatial Interpolation Method based on Data Reduction (SIMDR). Based on their analysis, the number of samples can be reduced using three interpolation methods—triangulation, kriging, and RBF. If it is necessary to preserve the survey data in the structure of the elevation model, the triangulation method should be chosen. In the process of surface modeling, the datasets should be prepared for a reduction level from 0.2 to 4 m using the MIN method, with an interval of 1, starting from the value of 0.2. The proposed method also allows for the precise selection of the data reduction level or intervals. The proposed process is presented below in the form of three steps:

Step 1—Preparation of the datasets

1. Reduce data by MIN method at 1 m interval from 0.2 to n, where the proposed starting value is 4;
2. Add the shoreline points with height values at the 1 m interval to the SBES and UAS datasets prepared in step 1.

Step 2—Modeling

1. Perform modeling process via three methods, triangulation, kriging, and RBF;
2. Create 3D surface models and relief drawings for each option (needed for qualitative analysis);
3. Calculate mean and maximum errors on the test dataset (needed for quantitative analysis).

Step 3—Analysis and development of the final model

1. Select the reduction level and the bathymetric surface model for which quantitative and qualitative errors are the smallest;
2. Select the reduction level and the topographic surface model for which quantitative and qualitative errors are the smallest;
3. Merge the selected models.

On the basis of the research carried out, it can also be concluded that almost no method can provide certainty about the correctness of the construction in areas without data. There will be no problem in the case of planar surfaces, but for the area of a body of water with no data, where there is a change in the shape of the surface due to shallowing, relatively large errors were obtained in comparison with the errors in an area assessed via bathymetric data. The elevation data in such an area should certainly be regarded as uncertain, and the only way to create a correct model is to measure the real values. It should also be realized that this will not always be possible. The reason for this may be the presence of aquatic vegetation (submerged or partially submerged), which precludes measurement via unmanned units such as ASVs (Autonomous Surface Vehicles) or UASs. The above method can be used practically for modeling combined surfaces in coastal zones, using data of different types, characterized by different spatial distributions or data densities. It is also possible to obtain smaller errors via the reduction of data, which may be important in cases related to the development of high-accuracy models. The research also points to three specific methods (triangulation, kriging, and RBF) that can be used in the process of geographic surface reconstruction, which significantly reduces the number of analyses to be performed and the choices regarding the method that gives the best results.

Author Contributions: Conceptualization, J.L.; methodology J.L. and M.W.-S.; validation, J.L., M.W.-S. and M.L.; formal analysis, J.L., M.W.-S. and M.L.; investigation, J.L., M.L. and M.W.-S.; resources, J.L., M.L., M.W.-S. and G.Z.; data curation, J.L., M.L., M.W.-S. and G.Z.; writing—original draft preparation, J.L., M.L., M.W.-S. and G.Z.; writing—review and editing, J.L., M.W.-S. and M.L.; visualization, J.L., M.L., M.W.-S. and G.Z.; supervision, J.L. and M.W.-S.; project administration, J.L. All authors have read and agreed to the published version of the manuscript.

Funding: The study presented was cofinanced by the European Union from the European Regional Development Fund under the 2014–2020 Operational Programme Smart Growth. The project titled "Development of technology for acquisition and exploration of gravimetric data of foreshore and seashore of Polish maritime areas" was implemented as part of the National Centre for Research and Development competition: 1/4.1.4/2018 "Application projects".

Institutional Review Board Statement: Not applicable.

Informed Consent Statement: Not applicable.

Data Availability Statement: Not applicable.

Conflicts of Interest: The authors declare no conflict of interest.

Appendix A

Table A1. Qualitative analysis for surfaces created from ABES data.

Interpolation	Surface Plasticity	Roughness/Smoothness	Artifacts, Deformations	Shape on Profiles	Slope	Depression	Interpolation in the Area of Lack of Data	The Change Rate as the Reduction Increases
TRI MIN	good	4	none	slightly visible profiles	correct	correct	correct by visual evaluation, sharp edges	unnoticeable
TRI MAX	good	4	none	slightly visible profiles	correct	correct	correct by visual evaluation, sharp edges	unnoticeable
IDW MIN	good	3	visible at high reduction rate on high-slope surfaces	nearly invisible profile	correct	correct, but with a high degree of reduction, there are small artifacts	correct, impact of sectoral search, number of points in search, and level of reduction	noticeable, most significantly in high-slope areas
IDW MAX	good	4	visible at high reduction rate on high-slope surfaces	nearly invisible profile	correct	correct, but with a high degree of reduction, there are small artifacts	correct, impact of sectoral search, number of points in search, and level of reduction	noticeable, most significantly in high-slope areas
KRI MIN	good	4	none	nearly invisible profile	correct	correct	correct, impact of sectoral search, number of points in search, and level of reduction	unnoticeable
KRI MAX	good	4	none	nearly invisible profile	correct	correct	correct, impact of sectoral search, number of points in search, and level of reduction	unnoticeable
RBF MIN	good	4	none	nearly invisible profile	correct	correct	correct, impact of sectoral search, number of points in search, and level of reduction	unnoticeable
RBF MAX	good	4	none	nearly invisible profile	correct	correct	correct, impact of sectoral search, number of points in search, and level of reduction	slightly noticeable, in the boundary area water–no data

Appendix B

Table A2. Qualitative analysis for surfaces created from UAV data.

Interpolation	Surface Plasticity	Roughness/Smoothness	Artifacts, Deformations	Slope (from the Shoreline to the Points)	Dike Plasticity	Area of Low and Medium Vegetation	Interpolation in the Area of Lack of Data	The Change Rate as the Reduction Increases
TRI MIN	good	3 (roughness in the area of medium vegetation)	none	correct, sharp edges present	generally correct, worsens slightly as data reduction increases	rougher in areas of medium vegetation	no surface smoothness	significant; increase in surface roughness
TRI MAX	good	4	none	correct, sharp edges present	generally correct, worsens slightly as data reduction increases	in areas of medium vegetation, smoother than with the MIN option	no surface smoothness	significant; increased surface smoothness
IDW MIN	good, only up to a specified level of data reduction (2)	3 (roughness in the area of medium vegetation)	none	correct	good, only up to a specified level of data reduction (2)	rougher in areas of medium vegetation	correct, mainly dependent on sector search selection and number of points per sector	significant, the highest levels of reduction making it impossible to interpret correctly
IDW MAX	good, only up to a specified level of data reduction (2)	4	none	correct	good, only up to a specified level of data reduction (2)	in areas of medium vegetation, smoother than with the MIN option	correct, mainly dependent on sector search selection and number of points per sector	significant, the highest levels of reduction making it impossible to interpret correctly
KRI MIN	good	3 (roughness in the area of medium vegetation)	none	correct	at high reduction level, too much loss of detail	rougher in areas of medium vegetation	correct, mainly dependent on sector search selection and number of points per sector	high degradation of small terrain details
KRI MAX	good	4	none	correct	at high reduction level, too much loss of detail	in areas of medium vegetation, smoother than with the MIN option	correct, mainly dependent on sector search selection and number of points per sector	high degradation of small terrain details
RBF MIN	good	3 (roughness in the area of medium vegetation)	none	correct	at high reduction level, single loss of continuity, slightly worse than MAX method	rougher in areas of medium vegetation	correct, mainly dependent on sector search selection and number of points per sector	loss of model detail, blurring of land–water lines
RBF MAX	good	4	none	correct	at high reduction rate, single loss of continuity, slightly better than MIN	in areas of medium vegetation, smoother than with the MIN option	correct, mainly dependent on sector search selection and number of points per sector	loss of model detail, blurring of land–water lines

References

1. Genchi, S.A.; Vitale, A.J.; Perillo, G.M.E.; Seitz, C.; Delrieux, C.A. Mapping Topobathymetry in a Shallow Tidal Environment Using Low-Cost Technology. *Remote Sens.* **2020**, *12*, 1394. [CrossRef]
2. Quadros, N.; Collier, P.; Fraser, C. Integration of Bathymetric and Topographic LIDAR: A Preliminary Investigation. *Remote Sens. Spat. Inf. Sci.* **2008**, *36*, 1299–1304.
3. Gesch, D.; Wilson, R. Development of a Seamless Multisource Topographic/Bathymetric Elevation Model of Tampa Bay. *Mar. Technol. Soc. J.* **2001**, *35*, 58–64. [CrossRef]
4. Flener, C.; Vaaja, M.; Jaakkola, A.; Krooks, A.; Kaartinen, H.; Kukko, A.; Kasvi, E.; Hyyppä, H.; Hyyppä, J.; Alho, P. Seamless Mapping of River Channels at High Resolution Using Mobile LiDAR and UAS-Photography. *Remote Sens.* **2013**, *5*, 6382–6407. [CrossRef]
5. Simeone, S.; Palombo, L.; Molinaroli, E.; Brambilla, W.; Conforti, A.; De Falco, G. Shoreline Response to Wave Forcing and Sea Level Rise along a Geomorphological Complex Coastline (Western Sardinia, Mediterranean Sea). *Appl. Sci.* **2021**, *11*, 4009. [CrossRef]
6. Enríquez, A.R.; Marcos, M.; Álvarez-Ellacuría, A.; Orfila, A.; Gomis, D. Changes in beach shoreline due to sea level rise and waves under climate change scenarios: Application to the Balearic Islands (Western Mediterranean). *Nat. Hazards Earth Syst. Sci. Discuss.* **2016**, *17*, 1075–1089. [CrossRef]
7. Pratomo, D.G.; Saputro, I. Comparative analysis of singlebeam and multibeam echosounder bathymetric data. *IOP Conf. Ser. Mater. Sci. Eng.* **2021**, *1052*, 012015.
8. El-Hattab, A.I. Single beam bathymetry data modelling techniques for accurate maintenance dredging. *Egypt. J. Remote Sens. Space Sci.* **2014**, *17*, 189–195.
9. Pike, S.; Traganos, D.; Poursanidis, D.; Williams, J.; Medcalf, K.; Reinartz, P.; Chrysoulakis, N. Leveraging Commercial High-Resolution Multispectral Satellite and Multibeam Sonar Data to Estimate Bathymetry: The Case Study of the Caribbean Sea. *Remote Sens.* **2019**, *11*, 1830. [CrossRef]
10. Randazzo, G.; Barreca, G.; Cascio, M.; Crupi, A.; Fontana, M.; Gregorio, F.; Lanza, S.; Muzirafuti, A. Analysis of Very High Spatial Resolution Images for Automatic Shoreline Extraction and Satellite-Derived Bathymetry Mapping. *Geosciences* **2020**, *10*, 172. [CrossRef]
11. Specht, M.; Specht, C.; Stateczny, A.; Marchel, Ł.; Lewicka, O.; Paliszewska-Mojsiuk, M.; Wiśniewska, M. Determining the Seasonal Variability of the Territorial Sea Baseline in Poland (2018–2020) Using Integrated USV/GNSS/SBES Measurements. *Energies* **2021**, *14*, 2693. [CrossRef]
12. Wlodarczyk-Sielicka, M.; Blaszczak-Bak, W. Processing of Bathymetric Data: The Fusion of New Reduction Methods for Spatial Big Data. *Sensors* **2020**, *20*, 6207. [CrossRef]
13. Wlodarczyk-Sielicka, M.; Lubczonek, J. The Use of an Artificial Neural Network to Process Hydrographic Big Data during Surface Modeling. *Computers* **2019**, *8*, 26. [CrossRef]
14. Blaszczak-Bak, W.; Koppanyi, Z.; Toth, C. Reduction Method for Mobile Laser Scanning Data. *ISPRS Int. J. Geo-Inf.* **2018**, *7*, 285. [CrossRef]
15. Osowski, S. *Metody i Narzędzia Eksploracji Danych*; BTC: Legionowo, Poland, 2013.
16. Ur Rehman, M.H.; Liew, C.S.; Abbas, A.; Jaraman, P.P.; Wah, T.Y.; Khan, S.U. Big Data Reduction Methods: A Survey. *Data Sci. Eng.* **2016**, *1*, 265–284. [CrossRef]
17. Stateczny, A.; Lubczonek, J. Digital Terrains Models. In *Methods of Comparative Navigation*; Stateczny, A., Ed.; Gdańsk Scientific Society: Gdynia, Poland, 2004. (In Polish)
18. Wojciech, M. Kriging Method Optimization for the Process of DTM Creation Based on Huge Datasets Obtained from MBESs. *Geosciences* **2018**, *8*, 433. [CrossRef]
19. Maleika, W. Inverse distance weighting method optimization in the process of digital terrain model creation based on data collected from a multibeam echosounder. *Appl. Geomat.* **2020**, *12*, 397–407. [CrossRef]
20. Căţeanu, M.; Ciubotaru, A. The Effect of LiDAR Sampling Density on DTM Accuracy for Areas with Heavy Forest Cover. *Forests* **2021**, *12*, 265. [CrossRef]
21. Chowdhury, E.H.; Hassan, Q.K.; Achari, G.; Gupta, A. Use of Bathymetric and LiDAR Data in Generating Digital Elevation Model over the Lower Athabasca River Watershed in Alberta, Canada. *Water* **2017**, *9*, 19. [CrossRef]
22. Desmet, P.J.J. Effects of Interpolation Errors on the Analysis of Dems. *J. Br. Geomorphol. Group* **1997**, *22*, 563–580. [CrossRef]
23. Curtarelli, M.; Leão, J.; Ogashawara, I.; Lorenzzetti, J.; Stech, J. Assessment of Spatial Interpolation Methods to Map the Bathymetry of an Amazonian Hydroelectric Reservoir to Aid in Decision Making for Water Management. *ISPRS Int. J. Geo-Inf.* **2015**, *4*, 220–235. [CrossRef]
24. Stereńczak, K.; Ciesielski, M.; Bałazy, R.; Zawiła-Niedźwiecki, T. Comparison of Various Algorithms for DTM Interpolation from LIDAR Data in Dense Mountain Forests. *Eur. J. Remote Sens.* **2016**, *49*, 599–621. [CrossRef]
25. Agüera-Vega, F.; Agüera-Puntas, M.; Martínez-Carricondo, P.; Mancini, F.; Carvajal, F. Effects of Point Cloud Density, Interpolation Method and Grid Size on Derived Digital Terrain Model Accuracy at Micro Topography Level. *Int. J. Remote Sens.* **2020**, *41*, 8281–8299. [CrossRef]

26. Habib, M.; Alzubi, Y.; Malkawi, A.; Awwad, M. Impact of Interpolation Techniques on the Accuracy of Large-Scale Digital Elevation Model. *Open Geosci.* **2020**, *12*, 190–202. [CrossRef]
27. Podobnikar, T. Methods for Visual Quality Assessment of a Digital Terrain Model. *Sapiens* **2009**, *2*, 1–10.
28. Łubczonek, J. *Comparative Analyse of Surface Modelling Methods with Regard to Building a Numerical Sea Bottom Model Annals of Geomatics*; Polish Association for Spatial Information: Warsaw, Poland, 2006; Volume IV/3, pp. 151–163. (In Polish)
29. Uysal, M.; Yilmaz, M.; Tiryakioğlu, I. 'Comparison of Data Reduction Algorithms for Image-Based Point Cloud Derived Digital Terrain Models '. World Academy of Science, Engineering and Technology, Open Science Index 148. *Int. J. Geol. Environ. Eng.* **2019**, *13*, 245–249.
30. Sharma, R.; Xu, Z.; Sugumaran, R.; Oliveira, S. Parallel Landscape Driven Data Reduction & Spatial Interpolation Algorithm for Big LiDAR Data. *ISPRS Int. J. Geo-Inf.* **2016**, *5*, 97. [CrossRef]
31. Lee, D.T.; Schachter, B.J. Two Algorithms for Constructing a Delaunay Triangulation. *Int. J. Comput. Inf. Sci.* **1980**, *9*, 219–242. [CrossRef]
32. Sibson, R. A Brief Description of Natural Neighbor Interpolation. In *Interpreting Multivariate Data*; Barnett, V., Ed.; John Wiley and Sons: New York, NY, USA, 1981; pp. 21–36.
33. Franke, R. Scattered Data Interpolation: Test of Some Methods. *Math. Comput.* **1982**, *33*, 181–200.
34. Cressie, N.A.C. The Origins of Kriging. *Math. Geol.* **1990**, *22*, 239–252. [CrossRef]
35. Powell, M.J.D. The Theory of Radial Basis Function Approximation in 1990. In *Advances in Numerical Analysis II*; Oxford University Press: Oxford, UK, 1992; Volume 2, pp. 105–210.
36. Briggs, I.C. Machine Contouring Using Minimum Curvature. *Geophysics* **1974**, *39*, 39–48. [CrossRef]
37. Varga, M.; Grgić, M.; Bjelotomić Oršulić, O.; Bašić, T. Influence of digital elevation model resolution on gravimetric terrain correction over a study-area of Croatia. *Geofizika* **2019**, *36*, 17–32. [CrossRef]
38. Wolski, W.I. Changes in Dąbie Lake bathymetry in the period 1962–1996. *Limnol. Rev.* **2005**, *5*, 258.
39. IHO—International Hydrographic Organization. *S-44: IHO Standards for Hydrographic Surveys*, 6th ed.; IHO Publication: Monaco, 2020.
40. Li, Z. *Algorithmic Foundation of Multi-Scale Spatial Representation*; CRC Press: Boca Raton, FL, USA, 2007.

remote sensing

MDPI

Communication

Atmospheric Correction of Airborne Hyperspectral CASI Data Using Polymer, 6S and FLAASH

Mengmeng Yang [1], Yong Hu [2], Hongzhen Tian [3], Faisal Ahmed Khan [4], Qinping Liu [3], Joaquim I. Goes [5], Helga do R. Gomes [5] and Wonkook Kim [1,*]

1 Department of Civil and Environmental Engineering, Pusan National University, Pusan 43241, Korea; mengmeng.yang2021@pusan.ac.kr
2 Chongqing Institute of Surveying and Monitoring for Planning and Natural Resources, Chongqing 400120, China; yong@ceode.ac.cn
3 School of Economics and Management, Tiangong University, Tianjin 300387, China; tianhongzhen@tiangong.edu.cn (H.T.); liuqinping@tiangong.edu.cn (Q.L.)
4 Institute of Environmental Studies, University of Karachi, Karachi 75270, Pakistan; tariqmak@uok.edu.pk
5 Lamont-Doherty Earth Observatory, Marine Biology and Paleo Environment, Columbia University, Palisades, NY 10964-8000, USA; jig@ldeo.columbia.edu (J.I.G.); helga@ldeo.columbia.edu (H.d.R.G.)
* Correspondence: wonkook@pusan.ac.kr; Tel.: +82-51-510-2353

Abstract: Airborne hyperspectral data play an important role in remote sensing of coastal waters. However, before their application, atmospheric correction is required to remove or reduce the atmospheric effects caused by molecular and aerosol scattering and absorption. In this study, we first processed airborne hyperspectral CASI-1500 data acquired on 4 May 2019 over the Uljin coast of Korea with Polymer and then compared the performance with the other two widely used atmospheric correction approaches, i.e., 6S and FLAASH, to determine the most appropriate correction technique for CASI-1500 data in coastal waters. Our results show the superiority of Polymer over 6S and FLAASH in deriving the R_{rs} spectral shape and magnitude. The performance of Polymer was further evaluated by comparing CASI-1500 R_{rs} data with those obtained from the MODIS-Aqua sensor on 3 May 2019 and processed using Polymer. The spectral shapes of the derived R_{rs} from CASI-1500 and MODIS-Aqua matched well, but the magnitude of CASI-1500 R_{rs} was approximately 0.8 times lower than MODIS R_{rs}. The possible reasons for this difference were time difference (1 day) between CASI-1500 and MODIS data, higher land adjacency effect for MODIS-Aqua than for CASI-1500, and possible errors in MODIS R_{rs} from Polymer.

Keywords: atmospheric correction; CASI-1500; Polymer; remote sensing reflectance; coastal waters

Citation: Yang, M.; Hu, Y.; Tian, H.; Khan, F.A.; Liu, Q.; Goes, J.I.; Gomes, H.d.R.; Kim, W. Atmospheric Correction of Airborne Hyperspectral CASI Data Using Polymer, 6S and FLAASH. *Remote Sens.* **2021**, *13*, 5062. https://doi.org/10.3390/rs13245062

Academic Editors: Jacek Lubczonek, Paweł Terefenko, Katarzyna Bradtke and Marta Wlodarczyk-Sielicka

Received: 8 November 2021
Accepted: 11 December 2021
Published: 13 December 2021

1. Introduction

Remote sensing is an effective tool for monitoring coastal waters, which are complex and highly variable ecosystems. Airborne hyperspectral sensors have several advantages over spaceborne sensors, and thus they play a vital role in remote sensing. First, imagery from airborne sensors are of a higher spatial resolution. For instance, the Compact Airborne Spectrographic Imager (CASI)-1500 has a spatial resolution that varies with flight altitude and can be 1 m × 1 m if the aircraft is at an altitude of approximately 2 km above the ground, whereas the spaceborne sensors such as SeaWiFS (Sea-viewing Wide Field-of-view Sensor) and MODIS (MODerate resolution Imaging Spectroradiometer) have a spatial resolution of 1 km × 1 km. This higher spatial resolution of airborne hyperspectral sensors makes them more desirable for coastal water applications [1,2].

The second advantage is that most airborne hyperspectral sensors are programmable in terms of the number and locations of spectral channels and their bandwidths. For example, 36 spectral bands may be acquired with 1 m × 1 m spatial resolution and a flight speed of 120 knots, whereas 48 spectral bands are possible at a flight speed of 90 knots.

Therefore, for a given location, it is possible to adjust spectral channels and spectral resolutions as desired [3]. Another advantage is that the timing of airborne data acquisition can be controlled, so adverse weather conditions such as cloud cover, rain, and haze can be avoided, whereas data collected from spaceborne sensors under these conditions are often invalid. Disadvantages of airborne remote sensing compared to satellite and in situ sampling include high cost and lengthy data processing time due to large data volume. Additionally, airborne remote sensing has smaller spatial coverage than satellite sampling. Despite this, airborne remote sensing is particularly useful for mapping coastal waters, as spatial and spectra resolutions are demanding in these areas.

However, atmospheric correction is critical before the quantitative application of airborne hyperspectral data. Atmospheric correction removes or reduces the effects caused by molecular and particular scattering and absorption, thus converting the radiances measured by the sensors to remote sensing reflectances (R_{rs}) of the target surfaces [4,5]. In the past three decades, atmospheric correction algorithms have been developed that range from empirical models to rigorous physically based radiative transfer models [6–13]. Some of the most popular atmospheric correction approaches include Polymer (POLYnomial based algorithm applied to MERIS), 6S (Second Simulation of a Satellite Signal in the Solar Spectrum), FLAASH (Fast Line-of-sight Atmospheric Analysis of Spectral Hypercubes), and ATCOR (ATmospheric CORrection) [14–17]. A few studies were carried that compare several radiative transfer models using medium and high-resolution data. For instance, Eugenio et al. [18] implemented 6S, FLAASH, and ATCOR; and Marcello et al. [19] implemented DOS (Dark Object Subtraction), QUAC (Quick Atmospheric Correction), 6S, FLAASH, and ATCOR for high-resolution WordView-2, a satellite-based imaging system with a spatial resolution of 0.46 m in the panchromatic (PAN) band and 1.84 m in eight multispectral (MS) channels. Nguyen et al. [20] evaluated 6S, FLAASH, and DOS using Landsat ETM+, which has a spatial resolution of 30 m for band 1 to 7 and a resolution of 15 m for the panchromatic band 8. In general, radiative transfer models outperform empirical methods, but no single algorithm performs best in all scenarios [18].

In this study, the performance of three atmospheric correction algorithms, i.e., Polymer, 6S, and FLAASH, were evaluated using a CASI-1500 imagery acquired on 4 May 2019 over the Uljin coast, the southeastern coast of Korea (Figure 1). Polymer was purposed to recovering ocean colour from the top-of-atmosphere signal measured by satellite sensors in the visible spectrum. One of the strengths of this algorithm is the possibility to recover ocean colour in the presence of sun glint. It has been applied to multiple sensors from ESA (MERIS/ENVISAT, MSI/Sentinel-2, and OLCI/Sentinel-3), NASA (SeaWiFS, MODIS/Aqua, VIIRS), and the Korean Geostationary Ocean Colour Imager (GOCI) [14,21–23]. However, to our knowledge, Polymer has not yet been applied to airborne hyperspectral sensors. It is to be expected that the atmospheric correction procedures useful for satellites can be valid for airborne sensors, although satellites and airborne sensors are at different altitudes. The 6S algorithm is one of the most widely used, rigorously validated, and heavily documented radiative transfer codes [24], although it is mainly applied to satellite sensors as well [18,25–27]. FLAASH has been applied to airborne hyperspectral data [28]. However, it does not separate the surface Fresnel reflectance from the target reflectance, which is acceptable for land areas but invalid for water areas [29]. In addition, FLAASH uses homogeneous aerosol properties derived from a dark target for the entire scene [30].

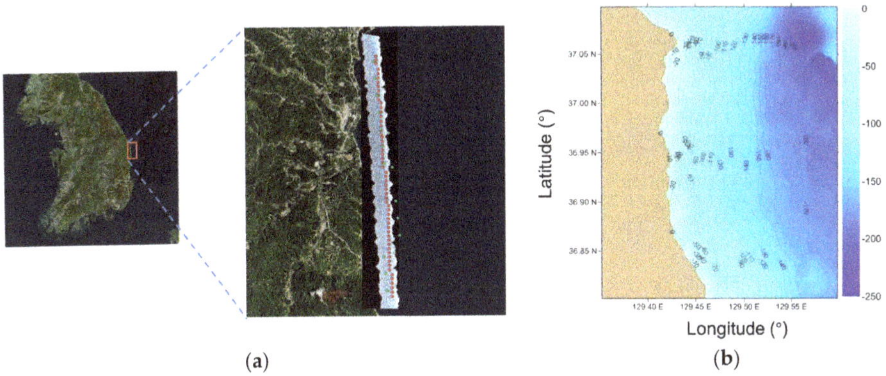

Figure 1. Location of Uljin coast, Korea: (**a**) red symbols represent locations where CASI data were processed using Polymer, 6S, and FLAASH, and green symbols denote locations where MODIS data processed by Polymer were extracted; (**b**) water depth of the coastal waters with labelled contour lines.

The waters along the Uljin coast are shallow with water depths < 20 m (Figure 1b). Seaweed is the dominant primary producer in this ecosystem, but whitening of coralline algae has adversely affected this region. Therefore, whitening detection is necessary to better understand the degradation of the ecosystem and help the local government to develop mitigation measures. In a recent study, Kim et al. [31] investigated the whitening of the Uljin coast via benthic mapping using CASI-1500 data, atmospherically corrected using FLAASH. However, the resulting CASI-1500 R_{rs} data were not validated as no field measurements of R_{rs} were available.

The objective of this study was to find the most suitable atmospheric correction algorithm to apply to CASI-1500 data for future studies in shallow coastal ecosystems. Additionally, for the first time, we applied Polymer to airborne hyperspectral data.

2. Materials and Methods

2.1. Airborne Hyperspectral Data

CASI-1500 hyperspectral data were acquired at 14:17:47 (+9 h GMT) on 4 May 2019, for an area ranging from 36.84°N to 37.05°N latitude and 129.41°E to 129.46°E longitude (Figure 1). The data were collected at an altitude of 2 km with a spatial resolution of 1 m across 48 spectral bands with a full width at half maximum of 7.2 nm (Table 1).

Pre-processing of the CASI-1500 data included radiometric calibration, which converted digital number (DN) into spectral radiance unit (SRU, $\mu w\ cm^{-2}\ sr^{-1}\ nm^{-1}$) with a scale factor of 1000, and geometric calibration using the GEOCORR program, which geo-referenced the data to UTM WGS84 [31]. Use of CASI-1500 also required use of inputs from MODIS-Aqua described in Section 2.2.

Table 1. Spectral bands and width of the CASI-1500 sensor for data acquired on 4 May 2019.

Band Number	Center Wavelength (nm)	Band Number	Center Wavelength (nm)	Band Number	Center Wavelength (nm)	Band Number	Center Wavelength (nm)	Full Width at Half Maximum (nm)
1	370.2	13	542.8	25	715.1	37	887.1	
2	384.6	14	557.2	26	729.4	38	901.5	
3	399	15	571.5	27	743.7	39	915.8	
4	413.4	16	585.9	28	758.1	40	930.2	
5	427.8	17	600.3	29	772.4	41	944.5	
6	442.2	18	614.6	30	786.8	42	958.8	
7	456.6	19	629.0	31	801.1	43	973.2	7.2
8	471	20	643.3	32	815.4	44	987.5	
9	485.3	21	657.7	33	829.8	45	1001.9	
10	499.7	22	672.0	34	844.1	46	1016.2	
11	514.1	23	686.4	35	858.5	47	1030.6	
12	528.5	24	700.7	36	872.8	48	1044.9	

2.2. Satellite and Reanalysis Data

MODIS (Moderate Resolution Imaging Spectroradiometer), a key instrument aboard the Aqua (EOS PM) satellite (MODIS/Aqua), has a spatial and temporal resolution of 1 km and 1 day, respectively. The Level-1A data at 5:00 (GMT) on 3 May 2019 were obtained from the NASA Goddard Space Flight Center. They were converted into Level-1B data using the OCSSW tools in the SeaDAS 7.5.3 software, and finally into Level-1C data that included all the necessary radiometric corrections using the l2gen command in the Polymer codes [14,21]. The Level-1C data were further processed to Level-2 data using the Polymer approach. These Level-2 R_{rs} data had a spatial resolution of 1 km at bands 412, 443, 488, 531, 547, 667, 678, 748, 859, and 869 nm, 500 m at bands 469 and 555 nm, and 250 m at band 645 nm. MODIS R_{rs} data at all bands except for band 678 were compared with the CASI R_{rs} data obtained after applying the Polymer atmospheric correction to radiance data. The R_{rs} at band 678 was not compared because Polymer failed to retrieve CASI R_{rs} at this band.

It should be noted that the MODIS and CASI data were acquired on different dates (1 day difference) because no valid MODIS data were available for the 4 May 2019 when the CASI data were collected. Therefore, when comparing the MODIS and CASI R_{rs} data, we were mindful that this time difference needed to be considered in our analysis. In addition, the spatial resolutions of CASI and MODIS sensors are much different (1 m vs. 1 km). To compare CASI and MODIS R_{rs}, we averaged CASI values within the MODIS 1 km pixel.

Next, the 6S atmospheric correction approach was applied to MODIS atmosphere and chlorophyll-a (Chl-a) products as well as reanalysis Modern-Era Retrospective analysis for Research and Applications, Version 2 (MERRA-2) products provided by the National Aeronautics and Space Administration Goddard Space Flight Center. Specifically, we used two MODIS atmosphere products, including MYD05 (water vapour) and MYD07 (total ozone) at 4:05 (GMT) on 4 May 2019, approximately 1 h before the acquisition of CASI-1500 data. The spatial resolution of MYD05 and MYD07 was 0.05°. As the MODIS aerosol product from MYD04 (aerosol optical thickness at 550 nm, aot_550) was not available for the same day, we used the hourly MERRA-2 aerosol product (Total Aerosol Extinction AOT [550 nm]) with a much coarser spatial resolution of 0.5° × 0.625°. Additionally, we also used MODIS Level-2 Chl-a product on 3 May 2019 (5:00 GMT) and hourly MERRA-2 wind speed product (U2M and V2M) on 4 May 2019. The spatial resolution of these products was 1 km × 1 km and 0.5° × 0.625°, respectively.

2.3. Polymer Atmospheric Correction Approach

Polymer proved advantageous in recovering pixels under sun glint compared to traditional atmospheric correction algorithms. In addition, it uses a polynomial function of wavelength (λ) and an ocean reflectance model to calculate water surface reflectance based on a spectral matching technique. A brief summary of the Polymer procedure is as follows:

First, an initial atmospheric correction is carried out for the top-of-atmosphere reflectance ($\rho_{TOA}(\lambda)$) which is decomposed according to the following equation:

$$\rho_{TOA}(\lambda) = t_{oz}(\lambda)\left[\rho_{mol}(\lambda) + T(\lambda)\rho_{gli} + \rho_{aer}(\lambda) + \rho_{coupl}(\lambda) + t(\lambda)\rho_w{}^+(\lambda)\right] \quad (1)$$

where $t_{oz}(\lambda)$ is the transmittance of ozone; $\rho_{mol}(\lambda)$, ρ_{gli}, $\rho_{aer}(\lambda)$, $\rho_{coupl}(\lambda)$, and $\rho_w{}^+(\lambda)$ are the reflectance of Rayleigh scattering, sun glint, non-absorbing aerosols, and the various couplings between sun glint, molecules, and aerosols, as well as of the water above the water-air interface, respectively; $T(\lambda)$ and $t(\lambda)$ are the direct transmission factors and the total (direct and diffuse) transmission for atmospheric scattering, respectively. The $t_{oz}(\lambda)$, $\rho_{mol}(\lambda)$, and ρ_{gli} were obtained using the Successive Order of Scattering (SOS) radiative transfer model in [14].

After the initial atmospheric correction, the remaining term ($\rho'(\lambda)$) is expressed by the following equation:

$$\rho'(\lambda) = \triangle\rho_{gli}(\lambda) + \rho_{aer}(\lambda) + \rho_{coupl}(\lambda) + t(\lambda)\rho_w{}^+(\lambda) \quad (2)$$

Furthermore, the first three terms and the last term are modelled by $T_0(\lambda)c_0 + c_1\lambda^{-1} + c_2\lambda^{-4}$ and $\rho_{wmod}{}^+([chl], b_{bNC}, \lambda)$, respectively. The $T_0(\lambda)$ is a transmission factor that can be calculated using the formula in [14]. The variables *Chl* and b_{bNC} represent the chlorophyll-*a* concentration and the backscattering coefficient of non-covarying particles, respectively. Therefore, $\rho'(\lambda)$ is also expressed by the following equation:

$$\rho'(\lambda) = T_0(\lambda)c_0 + c_1\lambda^{-1} + c_2\lambda^{-4} + t(\lambda)\rho_{wmod}{}^+([chl], b_{bNC}, \lambda) \quad (3)$$

The spectral matching technique retrieves c_0, c_1, c_2, *chl*, and b_{bNC}, and finally $\rho_w{}^+(\lambda)$ is obtained based on these parameters.

To quality control the MODIS-Aqua R_{rs} after application of the Polymer atmospheric algorithm, several criteria were established to exclude the R_{rs} data that did not meet our strict requirements: (1) High sun glint ($L_{gn} > 0.005$ sr^{-1}), where L_{gn} is a sun glint coefficient calculated with the model by Cox and Munk [32] and wind [33]; (2) Thick clouds ($\rho_{TOA}(869) - t_{oz}(869)\rho_{mol}(869) > 0.027$); (3) high air mass ($1/\cos(\theta_s) + 1/\cos(\theta_v) > 5$), where θ_s and θ_v are solar zenith angle and sensor zenith angle, respectively; and (4) $\theta_s > 88°$. The criteria for quality control of CASI R_{rs} after application of the Polymer atmospheric algorithm were the same as for MODIS-Aqua R_{rs} but the band used for (2) was 873 nm, which is the closest band to the MODIS-Aqua 869 nm band.

The main modification that we undertook when using Polymer for processing CASI data is that we used the 6S output parameters, including $t_{oz}(\lambda)$, $\rho_{mol}(\lambda)$, ρ_{gli}, transmission of Rayleigh scattering ($T_{mol}(\lambda)$), and Rayleigh optical thickness from the top-of-atmosphere to sensor ($\tau_{m\_sensor}(\lambda)$), and total Rayleigh optical thickness ($\tau_{m\_total}(\lambda)$) from the top-of-atmosphere to sea surface, for the initial atmospheric correction and Polymer atmospheric correction model. The difference between $\tau_{m\_total}(\lambda)$ and $\tau_{m\_sensor}(\lambda)$ ($\tau_{m\_total}(\lambda) - \tau_{m\_sensor}(\lambda)$) is the Rayleigh optical thickness from sensor to sea surface.

The CASI radiance data were converted into apparent reflectance using the following equation:

$$\rho^*(\lambda) = L/E_{s\_sensor} \quad (4)$$

where L and E_{s_sensor} represent the CASI radiance and the simulated solar downwelling irradiance at the sensor altitude, respectively.

We calculated E_{s_sensor} using the following equations [14,34]:

$$E_{s\_sensor}(\lambda) = E_{sol}{}^{dir}(\lambda) + E_{sol}{}^{diff}(\lambda) \quad (5)$$

$$E_{sol}{}^{dir}(\lambda) = \mu_s E_{s\_TOA} T_0{}^{dir}(\lambda) \quad (6)$$

$$E_{sol}{}^{diff}(\lambda) = \mu_s E_{s\_TOA} T_0{}^{diff}(\lambda) \tag{7}$$

$$E_{s\_TOA} = \frac{\int_{\lambda_1}^{\lambda_2} E(\lambda)S(\lambda)d\lambda}{\int_{\lambda_1}^{\lambda_2} S(\lambda)d\lambda} \tag{8}$$

$$T_0{}^{dir}(\lambda) = exp[-(\tau_{m\_total}(\lambda)) - \tau_{m\_sensor}(\lambda)) \times \left(\frac{1}{\mu_s} + \frac{1}{\mu_v}\right)] \tag{9}$$

$$T_0{}^{diff}(\lambda) = exp[-0.5 \times (\tau_{m\_total}(\lambda) - \tau_{m\_sensor}(\lambda)) \times \left(\frac{1}{\mu_s} + \frac{1}{\mu_v}\right)] \tag{10}$$

where $E_{sol}{}^{dir}(\lambda)$ and $E_{sol}{}^{diff}(\lambda)$ are the direct and diffuse solar downwelling irradiance at the sensor altitude, respectively; E_{s_TOA} is the solar downwelling irradiance at the top-of-atmosphere; $T_0{}^{dir}(\lambda)$ and $T_0{}^{diff}(\lambda)$ are the direct and diffuse transmittance factors, respectively; $E(\lambda)$ and $S(\lambda)$ are the 1985 Wehrli Standard Extraterrestrial Solar Irradiance Spectrum and spectral response function of CASI, respectively; λ_1 and λ_2 are the start and end wavelength of each CASI band, respectively; and μ_s and μ_v are the cosine of solar zenith angle and sensor zenith angle, respectively.

2.4. 6S Atmospheric Correction Approach

The 6S algorithm is one of the most widely used, rigorously validated, and elaborately documented radiative transfer codes [24]. The vector version of 6S was used in this study and it accounts for radiation polarization, in contrast to the scalar version of 6S. Accuracy of the atmospheric correction results by 6S is affected by the input atmospheric parameters, including the aerosol optical thickness and total amount of water vapour and ozone. However, field measurements of these parameters are often unavailable. Therefore, we used satellite and reanalysis atmospheric products (as described in Section 2.2) in this study. Furthermore, 6S simulated the atmospheric parameters at the airborne sensor altitude that were used as the final 6S input.

The parameter settings of 6S for the CASI-1500 imagery are described as below (Table 2).

Table 2. Description of the 6S input parameters for CASI-1500 data collected on 4 May 2019.

Parameters	Data Source	Spatial Resolution	Temporal Resolution	Value
Solar zenith angle	–	–	–	33.663°
Solar azimuth angle	–	–	–	239.218°
Sensor zenith angle	–	–	–	By pixel
Sensor azimuth angle	–	–	–	270°
Total water vapour	MYD05	1 km × 1 km	Daily	1.77 cm
Total ozone	MYD07	1 km × 1 km	Daily	0.34 cm-atm
Aerosol model	–	–	–	Maritime
aot_550	MERRA-2	0.5° × 0.625°	Hourly	0.32
Wind speed	MERRA-2	0.5° × 0.625°	Hourly	1.91 m s^{-1}
Wind azimuth angle	–	–	–	313.004°
Chl-a concentration	MODIS-Aqua Level-2	1 km × 1 km	Daily	3.76 mg m^{-3}
Sea water salinity	–			34.3 ppt

The solar zenith angle and solar azimuth angle were calculated from one-pixel location in the CASI-1500 imagery using Pysolar codes. Sensor zenith angle and sensor azimuth angle were calculated for each pixel. The former was obtained by dividing the DN values of the CASI-1500 NAD (NADIR Channel) imagery by 1000, whereas the latter using the formula acquired from https://www.omnicalculator.com/other/azimuth#what-is-the-azimuth, accessed on 25 August 2021. In general, the sensor azimuth angles were 0° for the nadir view pixels, 90° for the pixels on the left side of the nadir view pixels, and 270° for the pixels on the right side of the nadir view pixels. As the location points for validation (red symbols in Figure 1) were generally on the right side of the nadir view pixels, the sensor azimuth angle was set to be 270°. For total water vapour, total ozone, and Chl-*a* concentration, averaged values from the pixels covering the CASI-1500 imagery were used. For aot_550 and wind speed, a single value from the pixel covering the whole CASI-1500 imagery was used. In addition, the aerosol model was chosen to be Maritime based on the location and climate of the study area. Finally, sea water salinity was set to be 34.3 ppt for each pixel.

The 6S model is expressed by the following equation [18,35]. Reference to wavelength (λ) is omitted for clarity of the equation:

$$\rho_{TOA}(\theta_s, \theta_v, \Delta\phi) = t_g(\theta_s, \theta_v)\left\{\rho_a(\theta_s, \theta_v, \Delta\phi) + \left[e^{\frac{-\tau}{\mu_s}} + td(\theta_s)\right]\frac{\rho_{su}e^{\frac{-\tau}{\mu_v}} + \rho_e td(\theta_v)}{1 - \rho_e S}\right\} \quad (11)$$

where $\rho_{TOA}, \rho_a, \rho_{su},$ *and* ρ_e represent the top-of-atmosphere reflectance, atmospheric reflectance, surface reflectance, and a homogenous environment of reflectance, respectively; $\Delta\phi$ represents the difference between solar and sensor azimuth; t_g represents the total transmissivity of the gases, considering the absorption of different gases; τ represents the atmospheric thickness; $td(\theta_s)$ and $td(\theta_v)$ represent the diffuse transmittance of the atmosphere; S represents the spherical albedo of the atmosphere; and the $1 - \rho_e S$ term considers the multiple scatterings between the surface and the atmosphere.

2.5. FLAASH Atmospheric Correction Approach

FLAASH is available in ENVI and it incorporates MODTRAN4 radiative transfer code [34]. The FLAASH model is expressed by the following equation. Again, reference wavelength (λ) is also omitted for simplicity:

$$L = (\frac{A\rho}{1 - \rho_e S}) + (\frac{B\rho_e}{1 - \rho_e S}) + L_a \quad (12)$$

where L represents the radiance at the sensor pixel; ρ represents the pixel surface reflectance; ρ_e represents an average surface reflectance for the pixel and the surrounding region; S represents the spherical albedo of the atmosphere; L_a represents the radiance backscattered by the atmosphere; and A and B are coefficients that depend on atmospheric and geometric conditions.

For the FLAASH parameter setting, any of the standard MODTRAN model atmosphere and aerosol types can be chosen to represent a scene and a unique MODTRAN solution is computed for each image. In this study, mid-latitude summer and maritime were selected for the atmospheric and aerosol models, respectively. The initial visibility value was obtained from the 6S simulation, which used the reanalysis aot_550 as the input data. Water vapour was retrieved using the water absorption feature at 940 nm.

The parameter settings of FLAASH applied to CASI-1500 imagery are described in Table 3.

Table 3. Description of the FLAASH input parameters applied to CASI-1500 data collected on 4 May 2019.

Parameter	Value
Image center location	37.05057344°N, 129.41963701°E.
Sensor altitude	2 km
Ground elevation	0.01 km
Pixel size	1 m
Flight date	4-May-19
Flight time GMT	5:17:47
Atmospheric model	Mid-Latitude Summer
Aerosol model	Maritime
Aerosol retrieval	None
Initial visibility	14.85 km
Water retrieval	Yes
Water absorption feature	940 nm
Modtran resolution	5 cm^{-1}
Modtran multiscatter model	Scaled DISORT
DISCORT streams number	8

2.6. Statistical Analysis

To quantitatively measure the spectral similarity between the CASI-1500 and MODIS R_{rs} spectra obtained by Polymer, we calculated the cosine distance using the following equation [23,36]:

$$\cos(\alpha) = \frac{\sum_{i}^{N}(Rrs_{CASI,i}Rrs_{MODIS,i})}{\sqrt{\sum_{i}^{N}(Rrs_{CASI,i})^2}\sqrt{\sum_{i}^{N}(Rrs_{MODIS,i})^2}} \tag{13}$$

where α represents the angle between the *CASI* R_{rs} spectrum ($R_{rsCASI,i}$) and MODIS R_{rs} spectrum ($R_{rsMODIS,i}$). A closer match of the two spectra is indicated by a smaller α and a bigger $\cos(\alpha)$.

3. Results

The results from applying the atmospheric correction approaches—Polymer, 6S, and FLAASH—to CASI-1500 data were compared. MODIS data were processed with Polymer as the atmospheric algorithm was used to evaluate the performance of Polymer for CASI-1500 data.

3.1. CASI-1500 Radiance and Apparent Reflectance

CASI-1500 radiance data (Figure 1a) were first converted into apparent reflectance data and then processed using Polymer to obtain the CASI-1500 R_{rs} data.

Radiance data were collected from eight locations, i.e., No. 1, No. 9, No. 17, No. 25, No. 33, No. 41, No. 49, and No. 57 shown as evenly distributed (from top to bottom) red symbols on the CASI radiance image in Figure 1. The radiance spectra (Figure 2a), had similar spectral shapes but different magnitudes. Specifically, the spectra peaked at the green band near 500 nm, which is typical for coastal waters. Several spectral valleys seen near 720 nm, 760 nm, 820 nm, and 940 nm are due to absorption by water vapour and oxygen. The apparent radiance data had similar spectral shapes as the radiance data.

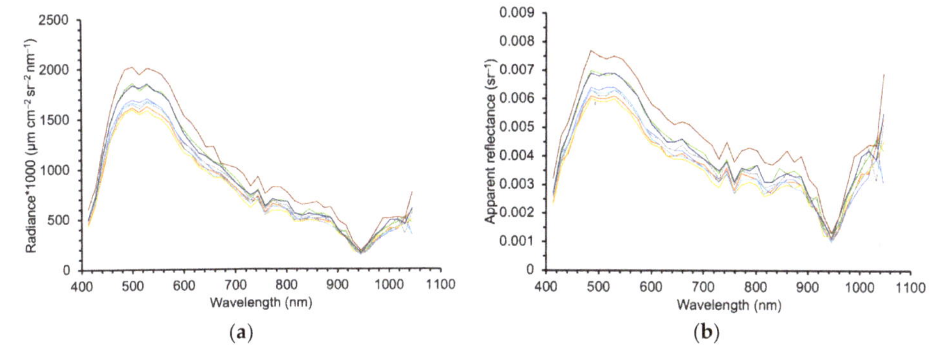

Figure 2. (**a**) CASI-1500 radiance data; (**b**) the calculated apparent reflectance data. The data were acquired on 4 May 2019 over the Uljin coast. Point1′ to Point8′ (Nos. 1, 9, 17, 25, 33, 41, 49, and 57) correspond from top to bottom to the evenly distributed line of red symbols in Figure 1.

3.2. Polymer, 6S, and FLAASH Results for CASI-1500

To compare the atmospheric correction approaches, the entire CASI-1500 radiance imagery (Figure 1a) was corrected using Polymer, 6S, and FLAASH, after which R_{rs} data were extracted from 57 pixels, i.e., the red symbols in Figure 1a. All R_{rs} spectra from the 57 pixels were averaged and plotted with error bars of one standard deviation (Figure 3). The R_{rs} spectra from wavelengths above 900 nm are not shown because the CASI-1500 data were usually unstable and not informative in that wavelength range.

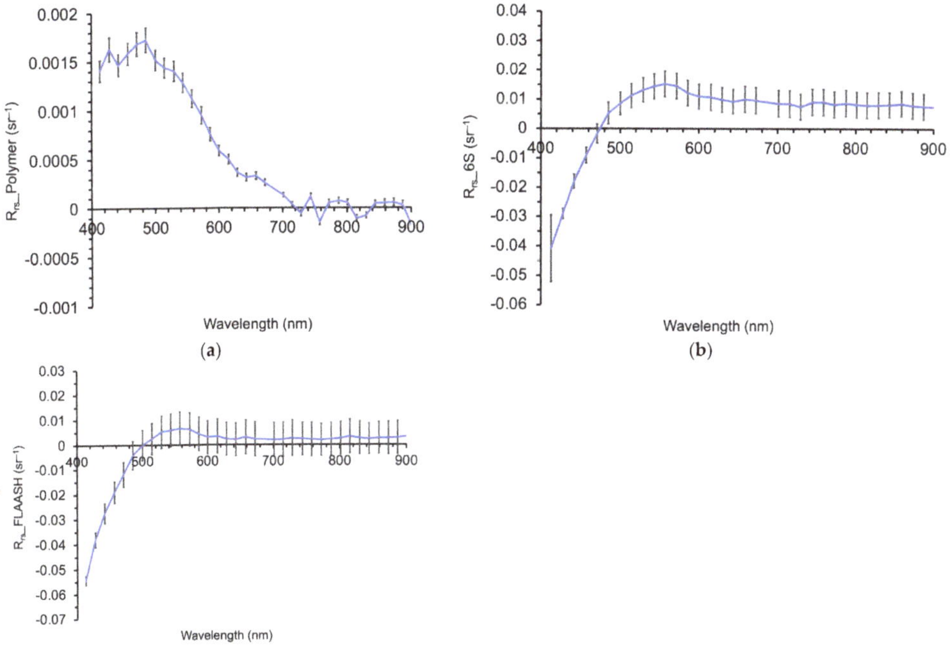

Figure 3. Averaged R_{rs} spectra from CASI-1500 atmospherically corrected using Polymer (**a**), 6S (**b**), FLAASH (**c**). The error bars in each plot represent one standard deviation.

The spectral shape of R_{rs} after applying the Polymer algorithm (Figure 3a) was generally consistent with that of CASI-1500 radiance and apparent reflectance (Figure 2), but the magnitude of R_{rs} was smaller at all spectral bands. In contrast, although R_{rs} from both 6S and FLAASH (Figure 3b,c) had similar spectral shape and magnitude, they differed greatly from R_{rs} obtained using Polymer both in spectral shape and magnitude. In addition, R_{rs} using 6S and FLAASH were negative in the blue bands (400–500 nm) and higher than apparent reflectance values in other bands.

CASI R_{rs} values using 6S and FLAASH were overcorrected in the blue bands. This was probably caused by inaccurate input parameters, particularly the atmospheric parameters. Therefore, we examined a variation of CASI R_{rs} with aot_550 and water vapour for 6S (Figure 4a,b) and visibility for FLAASH (Figure 4c) at a test point (129.435°E, 36.985°N). For 6S atmospheric correction, the referenced aot_550 and water vapour values were 0.316 and 1.64 cm, respectively; for FLAASH atmospheric correction, the referenced visibility value was 14.85 km. The other two atmospheric values in each plot were 0.5 times and 1.5 times of the referenced value, respectively.

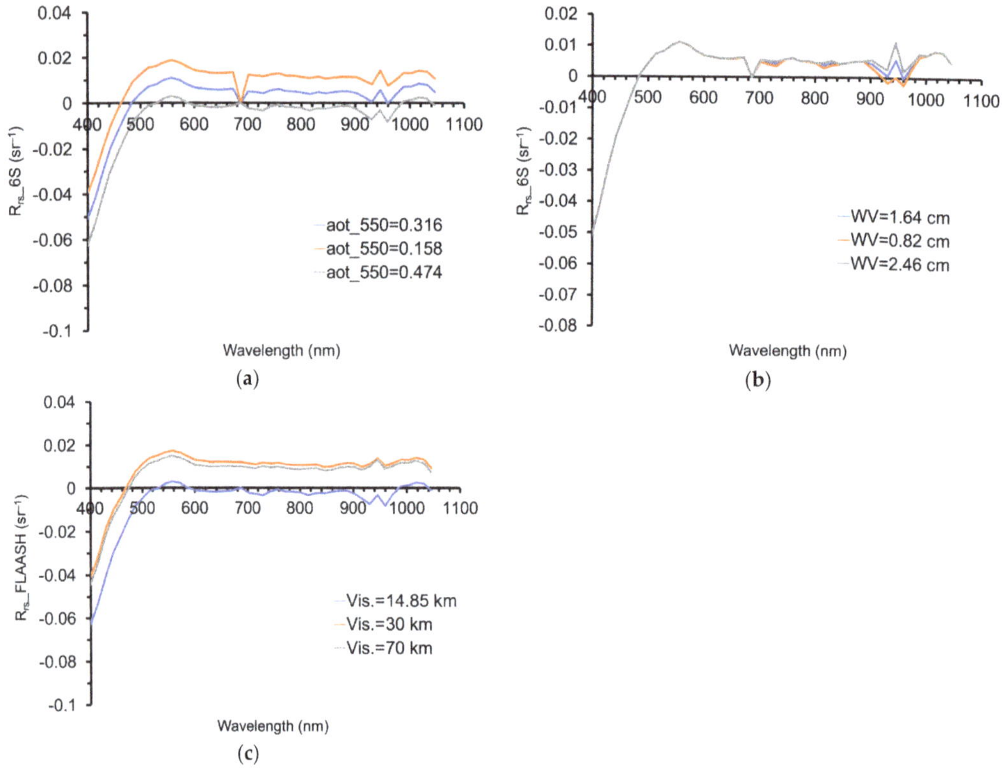

Figure 4. CASI-1500 R_{rs} spectra at various aot_550 values (**a**) and various water vapour values (**b**) after application of the 6S atmospheric correction, and at various visibility values (**c**) after application of the FLAASH atmospheric correction.

For 6S atmospheric correction, the magnitude of the entire R_{rs} spectrum increased and decreased significantly with the decrease and increase in aot_550 (Figure 4a), respectively, whereas the magnitude of R_{rs} changed little with variation of water vapour. Similarly, for FLAASH atmospheric correction, the magnitude of the entire R_{rs} spectrum increased significantly with the increase in visibility.

3.3. Comparison of the Polymer Results for CASI-1500 and MODIS

Previously, Zhang et al. [22] applied Polymer to MODIS data and compared its performance with in situ R_{rs} to conclude that it did not perform better than the standard NASA atmospheric correction algorithm (traditional near-infrared (NIR) approach [37]), at blue bands but was comparable at green and red bands. Therefore, they proposed using Polymer as a surrogate for the NASA NIR approach when MODIS R_{rs} data were limited.

In this study, due to lack of in situ R_{rs} data, we evaluated the performance of Polymer by comparing the R_{rs} results from CASI-1500 and MODIS. MODIS bands selected were 412, 443, 488, 531, 547, 667, 748, and 869 nm, and the corresponding CASI-1500 bands were 413, 442, 485, 529, 543, 672, 744, and 873 nm. Regarding the different pixel resolution of the two sensors, spatial distribution of CASI R_W (water-leaving reflectance) using Polymer was examined. As the data volume of whole CASI radiance image (Figure 1) was too big to process at one time, the image was divided into eight parts. Images of CASI R_W at 413, 442, 485, 529, 543, and 672 nm were made for one part of the whole area (129.428°–129.438°E, 37.002°–37.027°N) (Figure 5).

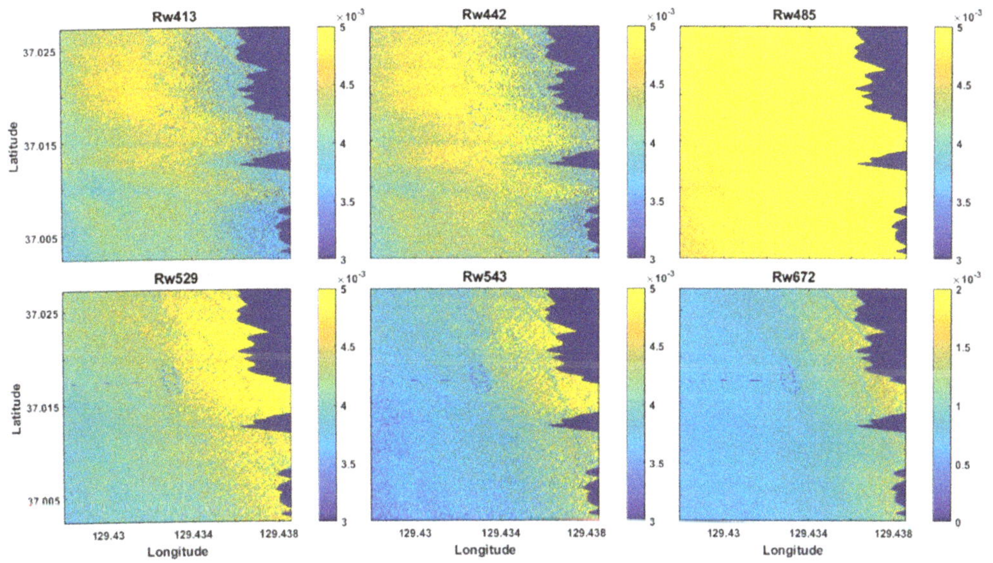

Figure 5. CASI-1500 R_w images at 413, 442, 485, 529, 543, and 672 nm.

Spatial variability of R_W at each band was observed. To quantify the variability, pixel and averaged R_{rs} were compared for eight points (No. 4, No. 5, No. 6, No. 7, No. 8, No. 9, No. 10, and No. 11) in each R_w image (Figure 6). In general, the pixel and averaged CASI R_{rs} showed similar spectral shape and magnitude for each point. Therefore, pixel CASI R_{rs} was compared with pixel MODIS R_{rs} for simplifying data processing (Figure 7).

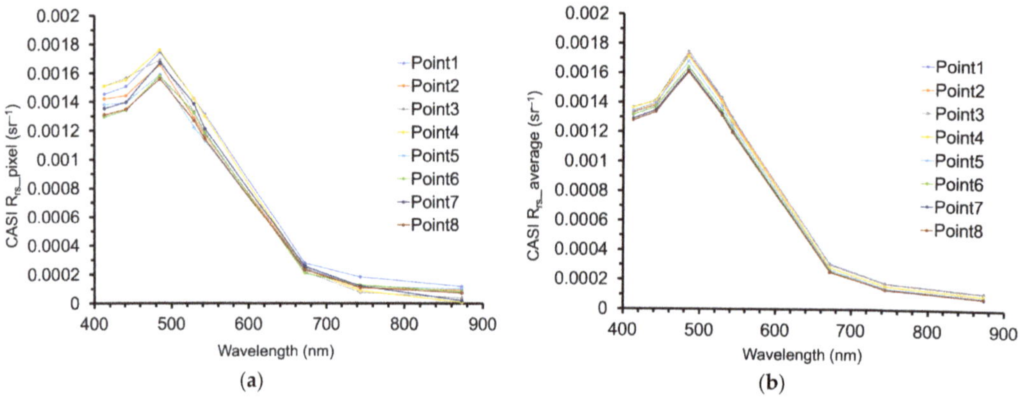

Figure 6. Comparison of pixel (**a**) and averaged (**b**) CASI-1500 R_w spectra after applying the Polymer atmospheric correction.

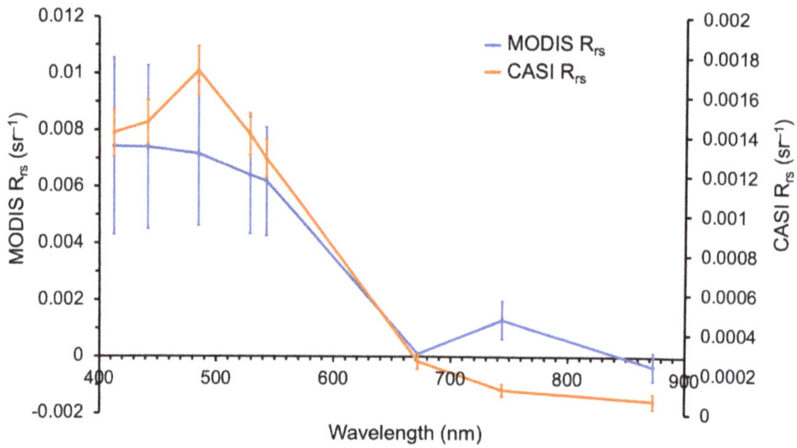

Figure 7. Comparison of averaged CASI-1500 and MODIS R_{rs} spectra after applying the Polymer atmospheric corrected algorithm. CASI-1500 and MODIS R_{rs} were averaged for R_{rs} from all locations denoted by red and green symbols in Figure 1a, respectively. The error bars represent one standard deviation.

Despite being measured on different days, R_{rs} spectra from CASI-1500 and MODIS generally matched except for inconsistencies at 485 and 744 nm. In addition, across the entire spectrum, MODIS R_{rs} values were much higher than CASI-1500 R_{rs} values. At 488 nm for instance, MODIS R_{rs} was approximately 0.007 sr^{-1}, whereas the CASI-1500 R_{rs} at 485 nm was 0.002 sr^{-1}.

The correlation between the CASI-1500 and MODIS R_{rs} spectra (Figure 8) was statistically significant (R^2 = 0.98), but the slope (0.21) was low, suggesting that the CASI-1500 R_{rs} were generally 0.8 times lower than MODIS R_{rs}. We further calculated the cosine distance between the two spectra and the cos(α) was approximately 0.99, which suggested that the two spectra were closely matched.

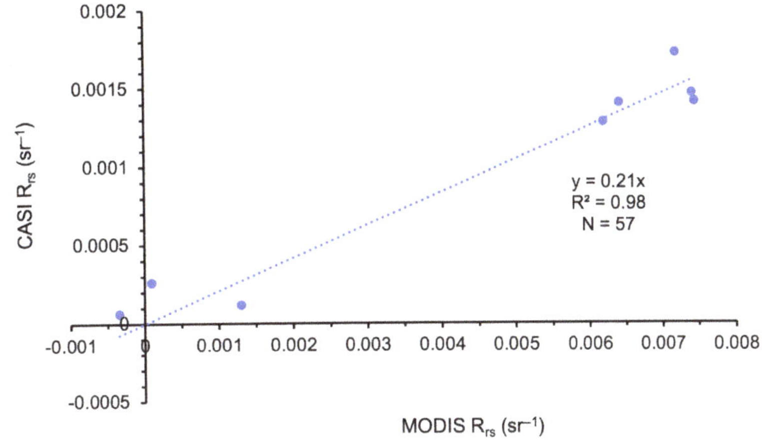

Figure 8. Scatter plot of the averaged CASI-1500 R_{rs} at bands 413, 442, 485, 529, 543, 672, 744, and 873 nm versus the averaged MODIS R_{rs} at bands 412, 443, 488, 531, 547, 667, 748, and 869 nm; N represents the data number for each band.

3.4. *Variation of Polymer Results with Aerosol and Water Vapor for CASI-1500*

For understanding the discrepancy between CASI-1500 and MODIS R_{rs} at each band, the variation in CASI-1500 R_{rs} with aot_550 and water vapour was investigated (Figure 9). The processing of CASI-1500 radiance data (as described in Section 2.3) was undertaken again with various aot_550 values, including 0, 0.01, 0.05, 0.1, 0.2, 0.3, and 0.4, and various water vapour values, including 0, 1, 1.5, 2, 3, and 4 cm, for the 6S input. The R_{rs} spectra obtained by applying Polymer to CASI-1500 R_{rs} for test locations showed that variation in CASI-1500 derived R_{rs} spectra due to aerosols and water vapour was small.

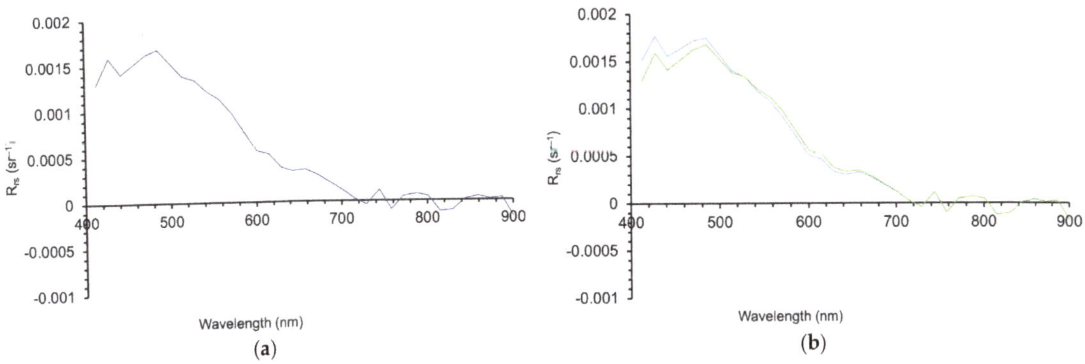

Figure 9. CASI-1500 R_{rs} spectra at various aot_550 values (**a**) and various water vapour values (**b**) after application of the Polymer atmospheric correction.

4. Discussion

4.1. *Comparison among Polymer, 6S, and FLAASH*

In this study, we applied Polymer to airborne hyperspectral CASI-1500 data and compared the resulting R_{rs} to two other widely used atmospheric correction approaches, i.e., 6S and FLAASH. The spectral shape of CASI-1500 R_{rs} using Polymer (Figure 3a) was similar to that of CASI-1500 apparent reflectance (Figure 2b), suggesting that the atmospheric effects influencing CASI-1500 radiance data on the acquisition day might be small. Moreover, the spectral shapes of CASI-1500 R_{rs} and MODIS R_{rs} using Polymer were highly correlated statistically, suggesting that the R_{rs} derived using Polymer were accurate.

In contrast, both 6S and FLAASH underestimated CASI-1500 R_{rs}, with negative values in the blue bands (400–500 nm). The R_{rs} values at other wavelengths were even larger than the CASI-1500 apparent reflectances, a clear indication that these two atmospheric correction approaches were estimating R_{rs} erroneously. Therefore, we conclude that the performance of Polymer is better than the performance of 6S and FLAASH. Despite the superior performance of Polymer, rigorous validation with in situ R_{rs} is necessary, which we plan to carry out when in situ R_{rs} data become available.

Polymer was originally designed to retrieve ocean colour in the presence of sun glint using MERIS data. The author concluded that Polymer greatly increased the spatial coverage of MERIS measurements for ocean colour and that the accuracy of retrieved data was not significantly reduced in high glint areas and remained the same as the standard algorithm outside sun glint areas [14]. The performance of Polymer was further evaluated by applying it to other satellite sensors. A recent study [22] compared Polymer with the traditional near-infrared (NIR) approach for MODIS-Aqua data and reported that Polymer did not perform better at blue bands but was comparable at green and red bands. Therefore, MODIS R_{rs} from Polymer was considered as reliable, at least not worse. The performance of Polymer to CASI data was then evaluated by comparing CASI R_{rs} with MODIS R_{rs}. However, CASI and MODIS R_{rs} showed big difference in their magnitude though their spectral shapes were similar. One possible cause for this difference may be errors in the MODIS R_{rs}.

Several studies [19,20] have compared R_{rs} after application of 6S and FLAASH to medium and high-resolution data. Recently, Eugenio et al. [18] applied 6S and FLAASH to WorldView-2 high-resolution satellite data and validated the atmospheric correction results with fields above water R_{rs}. They reported that both approaches achieved excellent results, although 6S produced slightly superior results. Our sub-optimal results using 6S and FLAASH for CASI-1500 hyperspectral imagery could be because the atmospheric products used were not accurate and the algorithms did not apply to these waters. In the future, to obtain more realistic atmospheric products, we need to consider two approaches. One is to make field measurements during CASI-1500 overflights, and the other is to develop algorithms to retrieve atmospheric products from CASI-1500 data.

4.2. Discrepancy of the CASI-1500 and MODIS R_{rs} with Polymer

The magnitude of CASI-1500 R_{rs} was approximately 0.8 times lower than that of MODIS R_{rs} (Figure 8). The possible reasons include: (1) dfference in the acquisition time of CASI-1500 versus MODIS data, the CASI-1500 data used was one day later than the MODIS data; (2) MODIS pixel size (1 km $\times$ 1 km) is much coarser than that of CASI-1500 (1 m $\times$ 1 m); (3) atmospheric products, such as MODIS water vapour and reanalysis aot_550, were also of much coarser spatial resolution than CASI-1500 data, which might affect the 6S output parameters and thus the atmospheric correction of CASI-1500 data with Polymer; (4) satellite measurements of coastal waters near land/water surfaces suffer from land adjacency effects [38], thus the MODIS 1 km pixels will most likely be affected by the adjacency and contamination of land signals much more than the low altitude and high resolution CASI data; and (5) errors in MODIS R_{rs} from Polymer. However, as we investigated the pixel CASI R_{rs} and averaged CASI R_{rs} within MODIS 1 km pixel were similar in spectral shape and magnitude, and the variation in R_{rs} using Polymer and CASI-1500 data with various values of aot_550 and water vapour and found them to be very small (Figure 6), we conclude that the discrepancies in R_{rs} between CASI-1500 and MODIS R_{rs} may be due to (1), (4), and (5).

5. Conclusions

Atmospheric correction may be required before application of airborne hyperspectral data analysis. The traditional atmospheric correction approach for CASI data has been FLAASH. However, the performance of FLAASH depends on the accuracy of input atmospheric parameters, including aerosol, water vapour, and ozone, which are often unavailable during the acquisition of airborne data.

Therefore, in this study, we evaluated three of the most widely used atmospheric correction approaches, i.e., Polymer, 6S, and FLAASH, and take the position that Polymer was the most suitable for application to CASI-1500 data. We further evaluated Polymer by comparing the R_{rs} derived from CASI-1500 and MODIS after using this atmospheric correction. Spectral shapes of R_{rs} from both sensors correlated significantly, although the magnitude of CASI-1500 R_{rs} was much lower than that of MODIS R_{rs}. We ascribe this difference to the one day difference in data acquisition between the two sensors, higher land adjacency effect for MODIS than for CASI-1500, and possible errors in MODIS R_{rs} from Polymer.

Our future plans include collecting in situ R_{rs} data to validate R_{rs} from CASI-1500 after application of the Polymer atmospheric correction, after which we will continue to modify the Polymer algorithm to improve its atmospheric correction and ocean reflectance models. In case of the atmospheric correction model, we plan to remove or reduce the effect of sky radiance, whereas for the ocean reflectance model, we plan to address the bottom reflectance effect in shallow waters.

Author Contributions: Conceptualization, M.Y. and W.K.; methodology, M.Y. and W.K.; software, M.Y., Y.H., F.A.K. and H.T.; validation, M.Y.; formal analysis, M.Y.; investigation, M.Y.; resources, W.K.; data curation, M.Y. and W.K.; writing—original draft preparation, M.Y.; writing—review and editing, M.Y., W.K., Y.H., H.T., F.A.K., Q.L., J.I.G. and H.d.R.G.; visualization, M.Y.; supervision, W.K.; project administration, W.K.; funding acquisition, W.K. All authors have read and agreed to the published version of the manuscript.

Funding: This research was supported by (1) basic research program through the National Research Foundation of Korea (NRF) funded by the Ministry of Science, ICT and Future Planning, grant number: NRF-2019R1F1A1062585.

Institutional Review Board Statement: Not applicable.

Informed Consent Statement: Not applicable.

Acknowledgments: We are thankful to the NRF and FIRA for providing the funding and to the Ocean Ecology Laboratory and Ocean Biology Processing Group of NASA Goddard Space Flight Center for providing daily Moderate-resolution Imaging Spectroradiometer (MODIS) Aqua L2 data, MODIS atmospheric products (MYD04, MYD05, and MYDO7), and reanalysis MERRA-2 data of Total Aerosol Extinction AOT [550 nm] (TOTEXTTAU), 2-m eastward wind (U2M), and 2-m northward wind (V2M). We are also thankful to the Korea Fisheries Resources Agency (FIRA) funded by the Ministry of Oceans and Fisheries, Republic of Korea. JIG and HRG are supported by NASA 80LARC21DA002-GLIMR AABO8078 and NOAA GST SA18-CUNY01.

Conflicts of Interest: The authors declare no conflict of interest. The funders had no role in the design of the study; in the collection, analyses, or interpretation of data; in the writing of the manuscript, or in the decision to publish the results.

References

1. Moses, W.J.; Gitelson, A.A.; Perk, R.L.; Gurlin, D.; Rundquist, D.C.; Leavitt, B.C.; Barrow, T.M.; Brakhage, P. Estimation of chlorophyll-a concentration in turbid productive waters using airborne hyperspectral data. *Water Res.* **2012**, *46*, 993–1004. [CrossRef]
2. Gould, R.W., Jr.; Arnone, R.A. Remote sensing estimates of inherent optical properties in a coastal environment. *Remote Sens. Environ.* **1997**, *61*, 290–301. [CrossRef]
3. Kallio, K.; Kutser, T.; Hannonen, T.; Koponen, S.; Pulliainen, J.; Vepsaälaäinen, J.; Pyhaälahti, T. Retrieval of water quality from airborne imaging spectrometry of various lake types in different seasons. *Sci. Total Environ.* **2001**, *268*, 59–77. [CrossRef]
4. Ibrahim, A.; Franz, B.; Ahmad, Z.; Healy, R.; Knobelspiesse, K.; Gao, B.; Proctor, C.; Zhai, P.-W. Atmospheric correction for hyperspectral ocean color retrieval with application to the Hyperspectral Imager for the Coastal Ocean (HICO). *Remote Sens. Environ.* **2018**, *204*, 60–75. [CrossRef]
5. Gao, B.C.; Montes, M.J.; Davis, C.O.; Goetz, A.F.H. Atmospheric correction algorithms for hyperspectral remote sensing data of land and ocean. *Remote Sens. Environ.* **2009**, *113*, S17–S24. [CrossRef]

6. Gao, B.C.; Davis Curtiss, O.; Goetz, A.F.H. A review of atmospheric correction techniques for hyperspectral remote sensing of land surfaces and ocean colour. In Proceedings of the 2006 IEEE International Symposium on Geoscience and Remote Sensing, Denver, CO, USA, 31 July–4 August 2006; pp. 1979–1981.
7. Gao, B.-C.; Montes, M.J.; Ahmad, Z.; Davis, C.O. Atmospheric correction algorithm for hyperspectral remote sensing of ocean color from space. *Appl. Opt.* **2000**, *39*, 887–896. [CrossRef] [PubMed]
8. Qu, Z.; Goetz, A.F.H.; Heidebrecht, K.B. High accuracy atmospheric correction for hyperspectral data (HATCH). In *Proceedings of the Ninth JPL Airborne Earth Science Workshop, 23–25 February 2000*; Green, R.O., Ed.; Jet Propulsion Laboratory: Pasadena, CA, USA, 2000.
9. Gao, B.-C.; Heidebrecht, K.H.; Goetz, A.F.H. Derivation of scaled surface reflectances from AVIRIS data. *Remote Sens. Environ.* **1993**, *44*, 165–178. [CrossRef]
10. Kruse, F.A. Use of airborne imaging spectrometer data to map minerals associated with hydrothermally altered rocks in the northern Grapevine Mountains, Nevada and California. *Remote Sens. Environ.* **1988**, *24*, 31–51. [CrossRef]
11. Clark, R.N.; King, T.V.V. Causes of spurious features in spectral reflectance data. In *Proceedings of the 3rd Airborne Imaging Spectrometer Data Analysis Workshop*; Vane, G., Ed.; Jet Propulsion Laboratory: Pasadena, CA, USA, 1987; pp. 49–61.
12. Conel, J.E.; Green, R.O.; Vane, G.; Bruegge, C.J.; Alley, R.E. AIS-2 radiometry and a comparison of methods for the recovery of ground reflectance. In *Proceedings of the 3rd Airborne Imaging Spectrometer Data Analysis Workshop*; Vane, G., Ed.; Jet Propulsion Laboratory: Pasadena, CA, USA, 1987; pp. 18–47.
13. Roberts, D.A.; Yamaguchi, Y.; Lyon, R. Comparison of various techniques for calibration of AIS data. In *Proceedings of the 2nd Airborne Imaging Spectrometer Data Analysis Workshop*; Vane, G., Goetz, A.F.H., Eds.; Jet Propulsion Laboratory: Pasadena, CA, USA, 1986; pp. 21–30.
14. Steinmetz, F.; Deschamps, P.-Y.; Ramon, D. Atmospheric correction in presence of sun glint: Application to MERIS. *Opt. Express* **2011**, *19*, 9783–9800. [CrossRef]
15. Pu, R.; Landry, S.; Zhang, J. Evaluation of Atmospheric Correction Methods in Identifying Urban Tree Species with WorldView-2 Imagery. *IEEE J. Sel. Top. Appl. Earth Obs. Remote Sens.* **2015**, *8*, 1886–1897. [CrossRef]
16. Vanonckelen, S.; Lhermitte, S.; Van Rompaey, A. The effect of atmospheric and topographic correction methods on land cover classification accuracy. *Int. J. Appl. Earth Obs. Geoinf.* **2013**, *24*, 9–21. [CrossRef]
17. El Hajj, M.; Bégué, A.; Lafrance, B.; Hagolle, O.; Dedieu, G.; Rumeau, M. Relative Radiometric Normalization and Atmospheric Correction of a SPOT 5 Time Series. *Sensors* **2008**, *8*, 2774–2791. [CrossRef] [PubMed]
18. Eugenio, F.; Marcello, J.; Martin, J.; Esparragón, D.R. Benthic habitat mapping using multispectral high-resolution imagery: Evaluation of shallow water atmospheric correction techniques. *Sensors* **2017**, *17*, 2639. [CrossRef] [PubMed]
19. Marcello, J.; Eugenio, F.; Perdomo, U.; Medina, A. Assessment of Atmospheric Algorithms to Retrieve Vegetation in Natural Protected Areas Using Multispectral High-resolution Imagery. *Sensors* **2016**, *16*, 1624. [CrossRef]
20. Nguyen, H.C.; Jung, J.; Lee, J.; Choi, S.U.; Hong, S.Y.; Heo, J. Optimal Atmospheric Correction for Above-Ground Forest Biomass Estimation with the ETM+ Remote Sensor. *Sensors* **2015**, *15*, 18865–18886. [CrossRef]
21. Steinmetz, F.; Ramon, D. Sentinel-2 MSI and Sentinel-3 OLCI consistent ocean colour products using POLYMER. In *Remote Sensing of the Open and Coastal Ocean and Inland Waters*; International Society for Optics and Photonics: Bellingham, WA, USA, 2018; p. 107780E.
22. Zhang, M.; Hu, C.; Cannizzaro, J.; English, D.; Barnes, B.B.; Carlson, P.; Yarbro, L. Comparison of two atmospheric correction approaches applied to MODIS measurements over north American waters. *Remote Sens. Environ.* **2018**, *216*, 442–455. [CrossRef]
23. Wang, J.W.; Lee, Z.P.; Wei, J.W.; Du, K.P. A Scheme of Atmospheric Correction with A Revised POLYMER model Using Same-day Observations. *Opt. Express* **2020**, *28*, 26953–26976. [CrossRef]
24. Vermote, E.; Tanré, D.; Deuzé, J.; Herman, M.; Morcrette, J.; Kotchenova, S. Second simulation of a satellite signal in the solar spectrum-vector (6SV) 6S User Guide Version 3. In *Remote Sensing: Models and Methods for Image Processing*; Schowengerdt, R.A., Ed.; Academic Press: Cambridge, MA, USA, 2006.
25. Kotchenova, S.Y.; Vermote, E.F. Validation of a vector version of the 6S radiative transfer code for atmospheric correction of satellite data, Part II: Homogeneous Lambertian and anisotropic surfaces. *Appl. Opt.* **2007**, *46*, 4455–4464. [CrossRef]
26. Kotchenova, S.Y.; Vermote, E.F.; Matarrese, R.; Klemm, F.J., Jr. Validation of a vector version of the 6S radiative transfer code for atmospheric correction of satellite data, Part I: Path radiance. *Appl. Opt.* **2006**, *45*, 6762–6774. [CrossRef]
27. Hu, Y.; Liu, L.Y.; Liu, L.L.; Peng, D.L.; Jiao, Q.J.; Zhang, H. A Landsat-5 atmospheric correction based on MODIS atmosphere products and 6S model. *IEEE J. Sel. Top. Appl. Earth Obs. Remote Sens.* **2014**, *7*, 1609–1615. [CrossRef]
28. Kruse, F.A. Comparison of ATREM, ACORN, and FLAASH atmospheric corrections using low-altitude AVIRIS data of Boulder, Colorado. In *Proceedings of the 13th JPL Airborne Geosci. Workshop*; International Society for Optics and Photonics: Pasadena, CA, USA, 2004.
29. Hu, C.; Carder, K. Atmospheric correction for airborne sensors: Comment on a scheme used for CASI. *Remote Sens. Environ.* **2002**, *79*, 134–137. [CrossRef]

30. Adler-Golden, S.; Berk, A.; Bernstein, L.S.; Richtsmeier, S.; Acharya, P.K.; Matthew, M.W.; Anderson, G.P.; Allred, C.L.; Jeong, L.S.; Chetwynd, J.H. FLAASH, a modtran4 atmospheric correction package for hyperspectral data retrievals and simulations. In Proceedings of the 7th JPL Airborne Earth Sci Workshop, Pasadena, CA, USA, 12–16 January 1998; Jet Propulsion Laboratory: Pasadena, CA, USA, 1998.
31. Kim, W.; Kim, S.H.; Baek, S.; Kim, H.; Oh, J. A Preliminary Study on Benthic Mapping of Uljin Coast Based on Airborne Hyperspectral Imagery Towards the Whitening Detection. *J. Coast. Res.* **2020**, *114*, 464–468. [CrossRef]
32. Cox, C.; Munk, W. Measurement of the roughness of the sea surface from photographs of the sun's glitter. *J. Opt. Soc. Am.* **1954**, *44*, 838–850. [CrossRef]
33. Wang, M.; Bailey, S.W. Correction of sun glint contamination on the SeaWiFS ocean and atmosphere products. *Appl. Opt.* **2001**, *40*, 4790–4798. [CrossRef]
34. Cooley, T.; Anderson, G.P.; Felde, G.W.; Hoke, M.L.; Ratkowski, A.J.; Chetwynd, J.H.; Gardner, J.A.; Adler-Golden, S.M.; Matthew, M.W.; Berk, A. FLAASH, a MODTRAN4-based atmospheric correction algorithm, its application and validation. *IEEE Int. Geosci. Remote Sens. Symp.* **2002**, *3*, 1414–1418.
35. Schowengerdt, R.A. *Remote Sensing: Models and Methods for Image Processing*; Academic Press: Cambridge, MA, USA, 2006.
36. Wei, J.; Lee, Z.-P.; Shang, S. A system to measure the data quality of spectral remote sensing reflectance of aquatic environments. *J. Geophys. Res. Ocean.* **2016**, *121*, 8189–8207. [CrossRef]
37. Gordon, H.R.; Wang, M. Retrieval of water-leaving radiance and aerosol optical thickness over the oceans with SeaWiFS: A preliminary algorithm. *Appl. Opt.* **1994**, *33*, 443–452. [CrossRef]
38. Feng, L.; Hu, C. Land adjacency effects on MODIS Aqua top-of-atmosphere radiance in the shortwave infrared: Statistical assessment and correction. *J. Geophys. Res. Ocean.* **2017**, *122*, 4802–4818. [CrossRef]

remote sensing

MDPI

Article

Open-Source Analysis of Submerged Aquatic Vegetation Cover in Complex Waters Using High-Resolution Satellite Remote Sensing: An Adaptable Framework

Arthur de Grandpré [1,2,*], Christophe Kinnard [1,3] and Andrea Bertolo [1,2]

1 Centre de recherche sur les interactions bassins versants-écosystèmes aquatiques (RIVE), Université du Québec à Trois-Rivières, 3351 Boul. des Forges, C.P. 500, Trois-Rivieres, QC G9A 5H7, Canada; christophe.kinnard@uqtr.ca (C.K.); andrea.bertolo@uqtr.ca (A.B.)

2 Interuniversity Research Group in Limnology (GRIL), Université de Montréal, C.P. 6128, Succursale Centre-Ville, Montreal, QC H3C 3J7, Canada

3 Centre d'études nordiques (CEN), Université Laval, A.P. 1202, Quebec, QC G1V 0A6, Canada

* Correspondence: arthur.de.grandpre@uqtr.ca

Abstract: Despite being recognized as a key component of shallow-water ecosystems, submerged aquatic vegetation (SAV) remains difficult to monitor over large spatial scales. Because of SAV's structuring capabilities, high-resolution monitoring of submerged landscapes could generate highly valuable ecological data. Until now, high-resolution remote sensing of SAV has been largely limited to applications within costly image analysis software. In this paper, we propose an example of an adaptable open-sourced object-based image analysis (OBIA) workflow to generate SAV cover maps in complex aquatic environments. Using the R software, QGIS and Orfeo Toolbox, we apply radiometric calibration, atmospheric correction, a de-striping correction, and a hierarchical iterative OBIA random forest classification to generate SAV cover maps based on raw DigitalGlobe multispectral imagery. The workflow is applied to images taken over two spatially complex fluvial lakes in Quebec, Canada, using Quickbird-02 and Worldview-03 satellites. Classification performance based on training sets reveals conservative SAV cover estimates with less than 10% error across all classes except for lower SAV growth forms in the most turbid waters. In light of these results, we conclude that it is possible to monitor SAV distribution using high-resolution remote sensing within an open-sourced environment with a flexible and functional workflow.

Keywords: submerged aquatic vegetation; macrophytes; OBIA; GeOBIA; high-resolution

Citation: de Grandpré, A.; Kinnard, C.; Bertolo, A. Open-Source Analysis of Submerged Aquatic Vegetation Cover in Complex Waters Using High-Resolution Satellite Remote Sensing: An Adaptable Framework. *Remote Sens.* **2022**, *14*, 267. https://doi.org/10.3390/rs14020267

Academic Editor: Jacek Lubczonek

Received: 12 November 2021
Accepted: 3 January 2022
Published: 7 January 2022

1. Introduction

The ecological importance of submerged aquatic vegetation (SAV; syn. hydrophytes or macrophytes) has been recognized for a long time in both marine and freshwater sciences. It is acknowledged that SAV provides diverse and important functions at the local, ecosystem, and global scales. By generating productive habitats able to sustain multiple trophic levels [1–3] and playing major roles in the biogeochemical cycles [4–9], SAV represents a major asset for both conservation and climate change research [10,11]. Because of the interactions between their engineering capabilities and their sensitivity to environmental stress, SAV has been identified as a good functional group of candidates for sentinel species [10]. On one hand, their role as ecological engineers is related to their ability to regulate water quality through density-dependent feedback mechanisms acting as sediment traps, nutrient sinks, and biodiversity hotspots, all of which modulate the availability of resources to other species in the ecosystem [6,12–15]. On the other hand, at lower densities of SAV, individuals become more exposed to the impact of multiple ecological stressors and their distribution starts to depend on environmental factors related to ecological integrity, including nutrient and light availability, hydrology, and climate [10,16–18].

Despite the recognition of SAV's importance, large-scale observations of their ecological integrity are lacking, and their monitoring is generally limited to qualitative point sampling [19]. While some efforts have been made to model SAV distribution at multiple scales [20–23], it remains difficult to generalize, transfer and validate these results because calibration data is rare, costly, and difficult to obtain. This lack of data can be attributed to the inherent difficulty to observe and quantify submerged landscapes. This is especially true in complex and dynamic systems such as fluvial lakes and estuaries. As a complement to classical field observations, medium to high-resolution satellite remote sensing can be used to efficiently monitor SAV cover and canopies, given the right atmospheric and hydrological conditions [19,24]. Optical satellite remote sensing should be able to provide environmental scientists with some level of information regarding large-scale SAV landscapes. However, shallow, complex, and dynamic inland waters are usually excluded from most remote sensing products because their high variability generates large radiometry changes that can lead to modeling errors and biases. The inherent optical properties (IOPs) of water (turbidity, color, and dissolved organic matter), water depth, atmospheric and meteorological conditions, phytoplankton biomass and species, adjacency to land as well as sensor noise all play a role in modulating the quantity and quality of the remote sensing signal of SAV [24,25]. Those issues are vastly documented in the remote sensing literature and impose limits to current aquatic remote sensing applications. For these reasons, classical remote sensing methods that would require precise quantification of SAV's expected and observed reflectance spectra currently represent a very complex endeavor.

Regardless of these limitations, some of SAV's ecological sentinel abilities do not require precise radiometric measurements to be monitored. Indeed, large parts of SAV's engineering capabilities can be attributed to its structuring effects within aquatic landscapes, which increases their spatial complexity [26–28]. While radiometric accuracy is of high importance for tasks such as species distribution or biomass studies, shifting the focus towards the quantification of the spatial structure of SAV landscapes using vegetation cover maps could reveal precious information about ecosystem integrity. Some authors have even suggested that SAV landscape patterns can be linked with ecological status, and even act as early warning signals towards non-linear ecological transitions [29–31]. In this sense, methods capable of detecting local reflectance gradients and contrasts generated by SAV landscapes would represent strong candidates to generate ecologically relevant SAV maps.

As an alternative to pixel-based modeling, object-based image analysis (OBIA) offers a framework that can be used to integrate more complex data than the raw remote sensing signal by including different levels of spatial information related to scale, shape, and texture [32–36]. This higher level of information offers a more robust comparative basis than pixel-based approaches when radiometric accuracy is uncertain, such as in complex water cases. Unfortunately, OBIA workflows in remote sensing frequently rely on proprietary software that diminishes analysis replicability, passing through "black box" functions that limit the interpretation of the results [37]. This is especially critical considering the delicate nature of remote sensing over complex inland waters where calibration and validation data are often limited. We suggest that, with sufficient understanding of the radiometric challenges and acknowledging some limitations, large-scale SAV distribution data can be obtained from optical remote sensing using OBIA in an open-source environment.

Because calibration and validation data are difficult to obtain (and often non-existent, especially for older satellite imagery), evaluating the accuracy of remotely sensed SAV metrics can be difficult. In some cases, it might be necessary to rely upon expert knowledge to train classifiers and determine whether the results depict an acceptable representation of reality [32–34]. This decision relies on factors such as appropriate knowledge of the study system, the sensitivity of the application of the results to type I and II statistical errors, and the capacity to interpret and correctly communicate the limitations of the data generating process (from sensing to classification). Since this process can be subjective, complete transparency is necessary when describing methods, discussing results, and transferring results. Based on this principle, open data and open-source solutions should be preferred when possible.

Here, we propose an open-source high-resolution optical remote sensing workflow to detect SAV cover in high-resolution satellite remote sensing imagery, applicable to optically complex waters and going from raw data processing to vegetation cover map. The workflow is presented in the form of general and adaptable guidelines based on the use of the R, QGIS, and Orfeo Toolbox (OTB) software [38–40]. This is followed by real-world applications of the workflow on high-resolution DigitalGlobe imagery products in two large shallow fluvial lakes of the Saint-Lawrence River system, Canada, with the specific objective of generating binary vegetation cover maps from detectable SAV. The steps that will be addressed here are pre-processing, including radiometric calibration, atmospheric correction, and image de-striping, followed by image segmentation, feature extraction, training set building, and finally, the application of a two-level hierarchical random forest image classification.

2. Materials and Methods

2.1. Workflow Overview

The suggested image processing workflow is based on the concept of geographic object-based image analysis (OBIA, or GEOBIA) where objects within an image are delineated by segmentation, described by feature extraction, zonal statistics and landscape indices, and identified by object classification [41] (Figure 1). As opposed to pixel-based approaches, objects can be described not only by zonal statistics generated by pixel values, but also by their shape, texture, and spatial relationship to one another. This allows for the generation of highly diverse descriptors which can then be used as inputs into machine learning methods to produce complex classification rules.

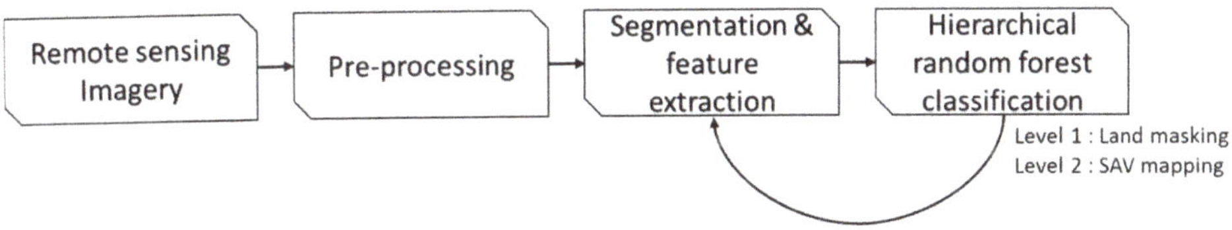

Figure 1. Typical remote sensing object-based image analysis (OBIA) workflow, where a sequence of classifications is performed on image segments (objects) based on training sets and large arrays of features.

In the case of SAV remote sensing using high-resolution satellite imagery such as DigitalGlobe satellite products, the pre-processing should at least include steps such as radiometric calibration, atmospheric corrections, and de-striping of push-broom sensor artifacts. Following pre-processing, image segmentation, feature extraction, and a hierarchical random forest image classification are used to generate the OBIA results (Figure 2). This workflow has been scripted within the R software [39] in the R Markdown format and is available on Github (URL: https://github.com/arthurdegrandpre/Open_HRRS_W2 (accessed on 1 November 2021)). While the following workflow could be used as is, it should nonetheless be adapted to fit the user's needs and resources, such as specific objectives, available imagery, available calibration and validation data, and access to computing power.

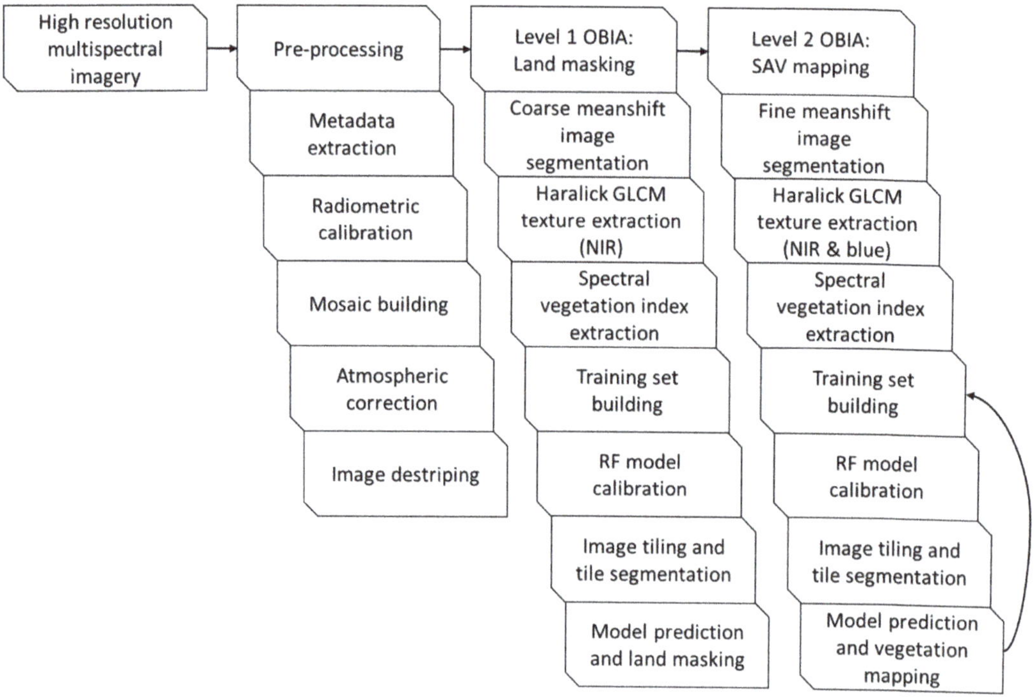

Figure 2. Detailed schematic representation of the high-resolution multispectral remote sensing workflow of submerged aquatic vegetation (SAV) proposed in this study. The associated scripts are available online, on Github, URL: https://github.com/arthurdegrandpre/Open_HRRS_W2 (accessed on 1 November 2021).

2.2. *Pre-Processing*

In this section, the process of preparing high-resolution multispectral imagery for OBIA will be detailed, going from raw digital numbers data to surface reflectance values using DigitalGlobe multispectral products. While OBIA is robust to radiometric errors, absolute radiometric calibration and atmospheric correction allows us to make images comparable and favors better interpretability of the data.

2.2.1. Metadata Extraction

In order to perform the subsequent steps, it is recommended to parse through image and satellite metadata to extract or generate the required information for radiometric calibration and atmospheric correction. In the case of an image collection with multiple data sources, this includes for each image their date and time of acquisition in the appropriate format, their product identification numbers and parts (in case it was split) for mosaic building, the identity of the satellite used for sensing and its orientation, including the solar zenith angle and the viewing angle. For every band, their name, effective bandwidth, absolute calibration factor and time delay integration value (TDI) are used for radiometric calibration, and image correction. Finally, for every satellite, the altitude, gain and offset for every band, as well as estimated exoatmospheric radiance values, are extracted (available from DigitalGlobe's technical notes [42–44]).

2.2.2. Radiometric Calibration

Since the raw sensor products are typically delivered with relative digital numbers (DN) as radiance units, radiometric calibration is necessary to generate absolute reflectance values. Doing so is a first step towards obtaining radiometrically comparable values across different remote sensing products. DigitalGlobe's technical notes [42,43] recommend the use of the following equation for radiometric calibration:

$$L = GAIN \times DN \times (abscalfactor / effectivebandwidth) + OFFSET \quad (1)$$

where L is the top-of-atmosphere radiance ($W\mu m^{-1}\ m^{-2}\ sr^{-1}$), DN is the digital number raw pixel value, and GAIN, OFFSET, abscal factor, and effective bandwidth are band-specific radiometric calibration for image and sensor parameters available within the metadata.

2.2.3. Atmospheric Correction and Mosaic Building

Atmospheric correction is a controversial topic over complex inland waters, even more so when shallow waters and submerged aquatic vegetation are expected. Since the state-of-the-art methods are very context-dependent, heavily debated, and often rely on large amounts of external cal/val inputs [45,46], an image-based empirical atmospheric correction has been implemented, based on the dark object subtraction method (DOS) and the cosine of the solar zenith angle (COST) [47]. This correction is based on the assumptions that (i) a dark object exists within an image, and that its remote sensing signal can be attributed to atmospheric effects, (ii) the atmospheric signal is relatively homogeneous across the image, and (iii) that the cosine of the solar zenith angle represents an acceptable approximation of the atmospheric transmittance. Because of those assumptions, it is recommended to build mosaics of compatible images, i.e., by grouping (tiling) and correcting images from the same satellite overpass together, thus providing more dark pixel candidates to smaller tiles. Once the images are grouped accordingly, the following equation is used to apply the correction:

$$\rho\ (BOA)_\lambda = ((L_\lambda - \min(L_\lambda)) \times d^2 \times \pi)/(E_\lambda \times \cos(\theta_S) \times TAU_Z) \quad (2)$$

where $\rho(BOA)_\lambda$ is the bottom of atmosphere reflectance (ρ) at a given wavelength (λ), L_λ is the at sensor radiance as previously calculated, d is the distance between the earth and the sun given in astronomical units, E_λ is the exoatmospheric radiance (here the mean of the three values estimated by DigitalGlobe was used), θ_S is the solar zenith angle and TAUz is the estimated atmospheric transmittance along the path from the sun to the surface, which is roughly equal to $\cos(\theta_S)$ according to the COST model. More recent and promising methods such as dark spectrum fitting (DSF) are currently under development but have yet to be tested on DigitalGlobe products [48].

2.2.4. Image De-Striping

While push-broom optical scanners are very efficient tools for generating high-resolution multispectral imagery, they also bring some challenges. Over areas with low remote sensing signals such as water, even small sensor noise and irregularities can have great effects on the quality of the data. The push broom architecture is composed of multiple side-by-side scanners that together generate a larger image, and while the image is corrected and calibrated as a whole, it is not perfectly calibrated for the radiometric differences that can occur between each parallel sensor, sometimes generating striping artifacts along the orbital line that can be quite significant. Correction of such artifacts over complex waters is difficult because homogeneity between stripes cannot be assumed. For this reason, we suggest the application of an empirical correction based on Marmorino & Chen's work [49] where "jumps" between contiguous stripes are identified manually, and their height estimated by comparing homogeneous objects on both sides of the jumps. Using QGIS, jumps were identified as line vectors, and for every line, 3 to 5 homogeneous polygon objects were drawn on each side. In R, a function was built to retrieve the values on both sides of every jump and to apply an offset to fix the height difference, effectively correcting most of the striping effect.

2.3. Object-Based Image Analysis

Pixel-based classification of submerged features being a very complex task across complex water environments, alternative object-based approaches offer a wide variety of spatial descriptors that can inform on landscape composition. Object-oriented workflows depend on three major steps: Image segmentation, feature extraction, and object classification. A wide variety of methods exist for each of these steps, and they should be adapted to fit the end-users needs. OBIA is often applied in a hierarchical workflow, defining broader regions, and then working towards more precise identifications. To apply it efficiently over SAV, two levels of OBIA classification have been performed: land masking and vegetation mapping.

2.3.1. Image Segmentation

Segmentation separates the image into a mosaic of smaller objects based on various homogeneity criteria that varies greatly between algorithms. A great variety of methods exists, the broader categories being edge- or region-based algorithms, respectively detecting steep gradients or homogeneous regions [50]. For its ease of implementation, speed, scalability, support of multiband rasters, and open code, we used OTB's region-based mean-shift segmentation algorithm [38] to perform image segmentation, calling it from R software's shell function. Segmentation parameters were tuned to fit the spatial and radiometric range of expected objects, with the spatial range set to 5 pixels (~10 to 12 m, depending on the sensor resolution), the radiometric range set between 0.001 and 0.003 depending on the intensity of the water-leaving reflectance, a spectral threshold of 0.0005, and a minimum size of 5 pixels.

2.3.2. Feature Extraction

Feature extraction can be described as the generation of information describing the objects. Features can include radiometric indices such as the normalized difference vegetation index (NDVI), local statistical moments, zonal statistics, shape and size of the object, and its spatial relationships to its neighbors, textural information such as grey-level co-occurrence matrix (GLCM) based metrics, etc. Due to the high diversity of possible features, many OBIA applications end up using very large arrays of features that could be useful when generating end results, relying on machine learning and computing capabilities to prune the less important candidates [32,33]. In this paper, we used a conservative array of features to favor the interpretability of the results, preventing a black-box effect (Tables 1 and 2).

2.3.3. Image Classification

Image classification needs to be able to deal with a high number of predictors (hundreds) over a high number of objects (millions). For those reasons, supervised machine learning algorithms are preferred, such as support vector machines (SVM), random-forest classifiers (RF), or deep learning-based methods such as convoluted neural networks (CNN). The main difficulty of using such methods is the construction of an appropriate training set, which is often built upon expert knowledge and image interpretation. While the use of ground validation data is to be preferred, especially for validation purposes, such data is often non-existent or unreliable. Because of its ease of use, both high interpretability and performance, and strong robustness to large amounts of predictors, we have used R package Caret's parallel random forest (parRF) algorithm. The training sets were built manually by expert image interpretation using the segmentation output and QGIS with classes determined prior to the process. To reduce the number of classes and generate more relevant out-of-bag accuracy estimates, two levels of classification were performed. The first one is for masking terrestrial objects, including wetlands, forests, agriculture, and human structures, and the second is for discriminating objects within the aquatic part of the scene, such as canopy-forming or submerged macrophytes, shallow or deep water, and water masses. An additional 10 folds cross-validation was performed on the final training set for each image to better explore risks of overfitting.

Table 1. Parameters, features and classes used for object-based image analysis level 1 classification (land masking).

Meanshift Segmentation and Haralick Texture Extraction Parameters.	Features per Objects	Object Classes
Meanshift • Spatial range = 5 • Spectral range = 0.0025 • Treshold = 0.001 • Max. iterations = 100 • Min. size = 5 Haralick (NIR band) • Min. = 0 • Max. = 1 • X radius = 5 • Y radius = 5 • X offset = 10 • Y offset = 10 • N bins = 64	Zonal statistics for all spectral bands (4 or 8) • Min/Max • Mean/Standard deviation Simple Haralick textures for NIR band • Energy • Entropy • Correlation • Inverse difference moment • Inertia • Cluster shade • Cluster prominence • Haralick correlation Vegetation indices [1] • Normalized difference vegetation index (NDVI) • Submerged aquatic vegetation index (SAVI) • Enhanced vegetation index (EVI) • Normalized difference aquatic vegetation index (NDAVI) • Water adjusted vegetation index (WAVI)	Land • Anthropological (buildings/roads) • Boats • Wetlands (high NDVI, attached to land) • Shadows • Agriculture/grasslands • Forest • Sand/soil Water • Blue/green-deep • Blue/green-shallow • Brown • Dark Aquatic vegetation • Floating (debris in deep water) • Emergent (high NIR signal) • Submerged—high (NIR signal) • Submerged—low (no NIR signal) Other • Boat waves • Floating foam

Table 2. Parameters, features and classes used for object-based image analysis level 2 classification (vegetation cover mapping).

Meanshift Segmentation and Haralick Texture Extraction Parameters (Image 1/Image 2)	Features per Objects	Object Classes, Image 1	Object Classes, Image 2
Meanshift • Spatial range = 5 • Spectral range = (0.0025/0.001) • Threshold = 0.001 • Max. iterations = 100 • Min. size = 5 Haralick (NIR, blue and NDVI bands) • Min. = 0 • Max. = 0.3 • X radius = 5 • Y radius = 5 • X offset = 10 • Y offset = 10 • N bins = 64	Zonal statistics for all spectral bands (4 or 8) • Min/Max • Mean/Standard deviation • Simple Haralick textures for NIR, blue and NDVI bands • Energy • Entropy • Correlation • Inverse difference moment • Inertia • Cluster shade • Cluster prominence • Haralick correlation Vegetation indexes [1] • Normalized difference vegetation index (NDVI) • Submerged aquatic vegetation index (SAVI) • Enhanced vegetation index (EVI) • Normalized difference aquatic vegetation index (NDAVI) • Water adjusted vegetation index (WAVI)	Water • Blue/green-deep • Shallow • Brown • Dark • Dark—adjacency effect Aquatic vegetation • Emergent (high NIR signal) • Emergent—adjacency effect (very high NIR signal due to land adjacency) • Submerged—high (NIR signal) • Submerged—low (no NIR signal)	Water • Deep • Shallow • Aquatic vegetation • Floating (debris in deep water) • Submerged—high (NIR signal) • Submerged—low (no NIR signal)

[1] See [51].

To construct the training sets, object candidates were selected one class at a time, by manual photointerpretation by an observer familiar with the SAV distribution at the study sites. Objects were identified by overlapping the segmentation results to the color-composite image, alternating between true-color and false-color using blue, green, and NIR, and then assigning a class to the selected objects. A representative sample of high confidence identifications was collected across the scene, looking for candidates of multiple sizes, shapes, and positions. To favor the independence of training objects, non-adjacent objects were preferred. While a single classification was usually satisfactory for level-1 classification, level-2 training sets were built iteratively, adding objects into the training sets in areas where obvious classification errors were observed. The sampling effort for every class should be proportional to their importance in the image and their importance relative to the classification objectives. In cases where adjacency effects were observed such as certain types of emergent vegetation, classes were split into two sub-classes, identifying "regular" object candidates and "adjacency affected" candidates. When relevant, distinct water masses were also separated in the training set to facilitate discrimination of SAV, especially in cases where high turbidity differences were expected.

2.4. Application to Real Ecosystems

The workflow was applied to produce vegetation cover maps for two fluvial lakes of the Saint-Lawrence River system, Quebec, Canada (Figure 3). The first scene classified is a Quickbird-2 four bands multispectral image of the lake Saint-Pierre (lat. 46.2°, lon. −72.8°), a shallow enlargement of the Saint-Lawrence River fed by multiple watersheds, creating a complex and dynamic mosaic of water masses varying in color and turbidity (Figure 3). The image was taken on 5 September 2009T15:50:00 GMT with a mean off-nadir view angle of 12.3° and a ground sampling distance of 2.4 m. Because of its complexity and high NIR signal near the shores, additional classes were created in the training set to favor narrower classes and diminish classification errors. The second scene is a WorldView-03 eight bands multispectral image of the lake Saint-François (lat. 45.15°, lon. −74.4°) a shallow fluvial lake mostly fed by clear water coming from the Great Lakes (Figure 3). It was taken on 2 August 2019T16:12:16 GTM with a mean off-nadir view angle of 7.3°, a ground sampling distance of 2 m, and shows traces of wave-generated sun glint and push-broom striping artifacts.

2.4.1. Level 1 Classification (Land Masking)

Land masking, while being in principle optional, highly facilitates subsequent classifications. The goal of this step is to exclude terrestrial objects, reducing the expected reflectance range and the global number of objects to classify for computation purposes. Because of the high reflectance differences between emerged and submerged objects, level 1 classification is relatively robust to poor segmentation optimization and training set calibration. Performance is evaluated by an expert observer based on random forest out-of-bag accuracy and visual validation of the results. For both test images, the same parameters have been used for segmentation, feature selection, and object classes (see Table 1 for details).

Following visual validation of the results, land classes are excluded from the dataset, a land mask is built from the polygons, and a new raster image is extracted to perform the level 2 classification.

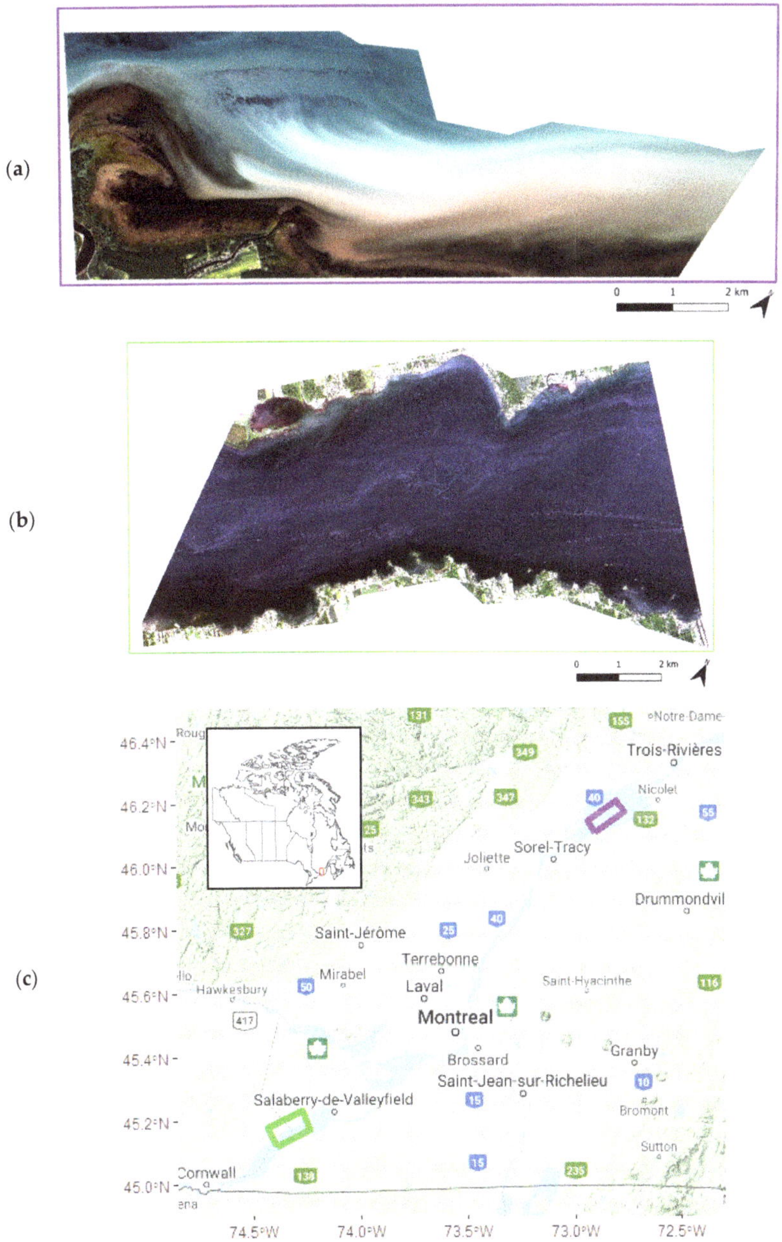

Figure 3. Images used in this study and their location. (**a**) Image 1: Quickbird-2 four band imagery of lake Saint-Pierre (© 2022 Maxar Technologies) (**b**) Image 2: Worlview-03 eight band imagery of lake Saint-François (© 2022 Maxar Technologies) (**c**) Location of the two fluvial lakes in southern Quebec, Canada (map data © 2022 Google). Purple is (**a**), green is (**b**) and the red square in the insert shows the location within Canada.

2.4.2. Level 2 Classification (Vegetation Cover Mapping)

Vegetation mapping is a very sensitive step because reflectance spectra overlap between different classes of objects. In cases where direct validation data is non-existent or insufficient, performance is evaluated by an expert observer based on random forest out-of-bag accuracy and visual validation of the results. For this reason, we suggest an iterative modeling approach where the training set is manually adjusted three to six times or until the end results are satisfactory. When adjusting the training set, the focus should be put on the most obvious type 1 and 2 errors concerning the classes of higher interest. Such obvious error cases include water objects wrongly classified as vegetation because of the presence of optical edges within the image (laminar water masses, steep bathymetry changes, etc.). To facilitate accuracy assessment, simple metrics are computed from the confusion matrices, called meaningful false positives (mFP) and meaningful false negatives (mFN). Those metrics exclude the same group classes from the error rate calculations, thus giving a better representation of the actual performances relative to the objectives. As an example, confusion between submerged and emerged vegetation would not be considered meaningful errors in this case, since we are ultimately seeking to produce a vegetation cover map. In contrast to level 1 classification, segmentation parameters are important to capture the right spatial scale for the end-user. Hence, object classes should be defined in accordance with the context of the scene and their potential for differentiation, and special attention should be given to training set calibration, reducing bias in out-of-bag accuracy estimation. This may mean creating additional classes to balance for steep turbidity gradients, different vegetation types, or in some cases zones of high environmental noise caused by wind. The parameters, features, classes, and training sets used are defined in Table 2.

2.5. R Libraries

Scripts used in the suggested R workflow are highly dependent on external libraries published by open source developers. At the moment of submitting this paper, the workflow is based on the following packages and their own dependencies: Caret, data.table, doParallel, foreach, gdalUtils, GeoLight, GSIF, kableExtra, landscapemetrics, plyr, raster, rgdal, satellite, sf, snowfall, sp, spatialEco, tidyverse, randomForest and velox [52–72].

3. Results

3.1. Radiometric Calibration and Atmospheric Correction

Concerning radiometric corrections and calibration, the radiometric spectra of deep-water pixels across all products (raw, absolute radiometric correction, top and bottom-of-atmosphere reflectances) of both images behaved as expected across all pre-processing, resulting in similar bottom-of-atmosphere reflectance (Figure 4). The raw level 1 product displays high DN values optimized for product visualization. Absolute radiometric calibration performed with Equation (1) generates radiance values very similar for both images, which translates into top-of-atmosphere (TOA) reflectance that are high in the lower wavelengths due to aerosol scattering and lower in the higher wavelengths due to aerosol absorption. The COST method (Equation (2)) for atmospheric correction used to generate bottom of atmosphere reflectances appears to efficiently correct those biases, effectively flattening the signal of both images into a comparable range close to 0.05 reflectance units for deep-water pixels.

3.2. Empirical Image De-Striping

Because of the architecture of push-broom sensors, noticeable striping artifacts can sometimes appear over low reflectance areas (i.e., water). While no noticeable reflectance jumps were observed in the 2009 Quickbird-02 image, severe reflectance jumps were observed in the 2019 Worldview-03 image (Figure 5a). Those jumps happened approximatively every 1110 m, and their height varied between 0.00001 and 0.00389 reflectance units, the worst offenders being in the green, red, and near-infrared wavelengths with mean jumps of 0.00127, 0.00092, and 0.00076, respectively. The method adapted from Marmorino

& Chen (2019) generated a product with noticeably fewer striping artifacts, equalizing the image to the mean of all objects used for jump height quantification (Figure 5b).

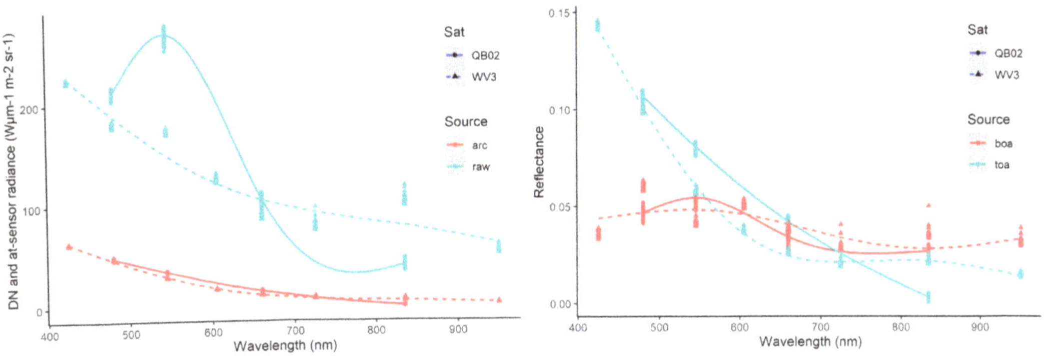

(**a**) (**b**)

Figure 4. Reflectance spectra of deep-water pixels from both images for all 4 product types. The full line and circles refer to image 1 (QB02), and the dotted lines and triangles refer to image 2 (WV3). (**a**) Shows the raw image in digital numbers (DN) and the absolute radiometric calibrated (arc) at-sensor radiance product. (**b**) Shows the top and bottom of atmosphere reflectance spectra (toa and boa, respectively).

(**a**)

(**b**)

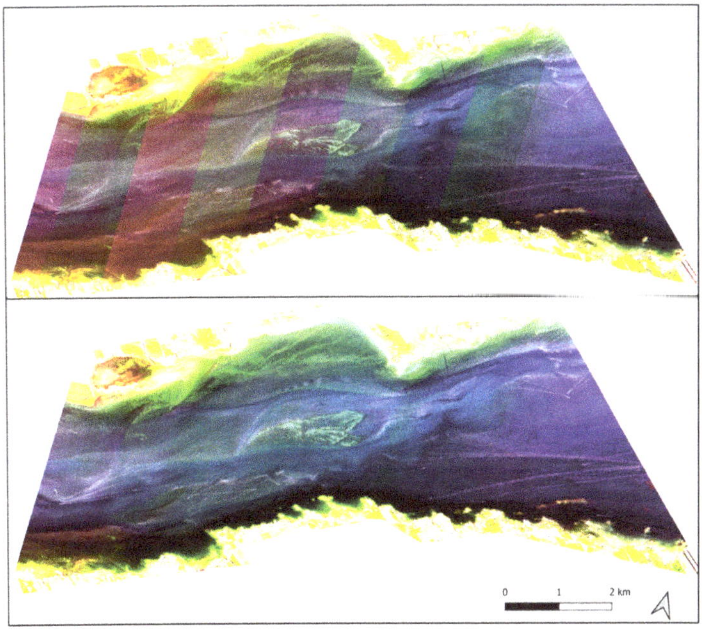

Figure 5. RGB composite of the 2019 Worldview-03 image of lake Saint-François (© 2022 Maxar Technologies) before (**a**) and after (**b**) empirical de-striping.

3.3. Level 1 OBIA, Land Masking

The first level of OBIA is a coarse classification of land and aquatic objects to narrow the scope of subsequent classifications. Using OTB's Meanshift segmentation with the parameters described in Table 1, image 1, the 2009 Quickbird-02 image generated 206,890 segments of (mean $\pm$ 1 S.D.) 83.3 $\pm$ 75.6 m^2 in size while image 2, the 2019 Worldview-03 image, generated 450,941 segments of 39.3 $\pm$ 39.9 m^2. This difference is associated with the different image spatial resolutions, but also different segmentation parameters. Because land and water reflectance are very different, a single random forest classification based run on a simple training set using the features from Table 1 was sufficient to generate a robust land mask for both images (Figure 6).

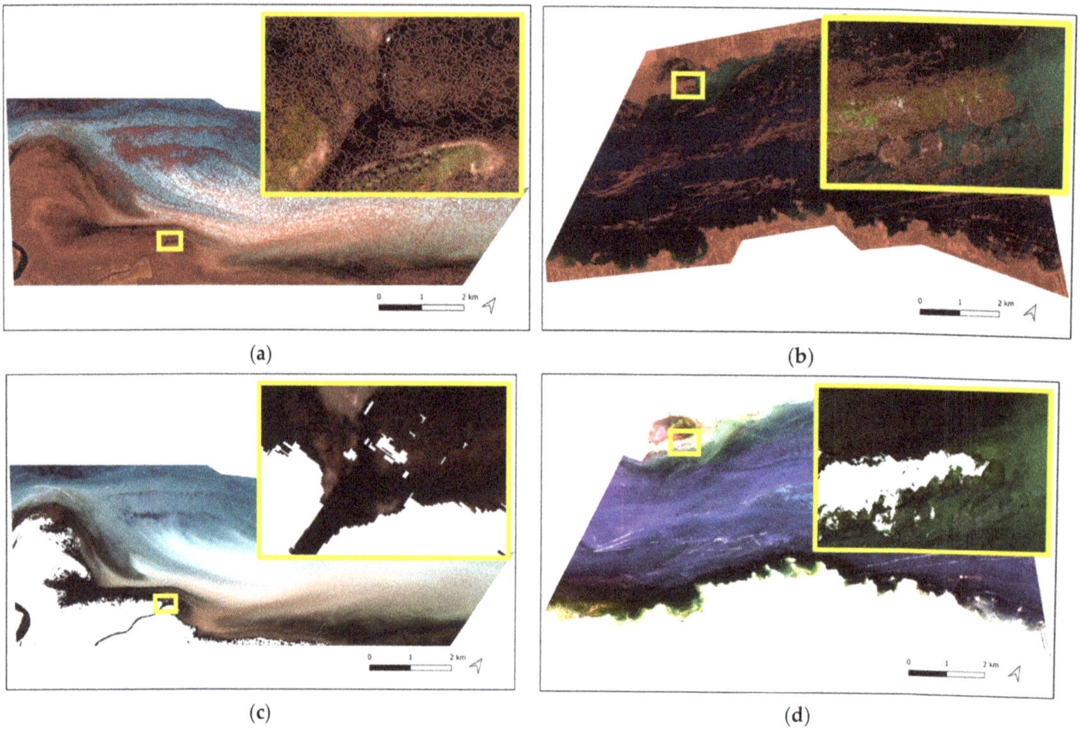

Figure 6. Visualisation of the coarse initial segmentation of the 2009 (**a**) and 2019 (**b**) images and their corresponding land masked products (**c**,**d**) (© 2022, 2019 Maxar Technologies).

For image 1, the most important features used in the land masking random forest models were largely based on the NIR signal and vegetation indexes zonal statistics, with the most important features being mean EVI, NIR, NDAVI, and maximum WAVI (Figure S1a). On the other hand, image 2 used more contrasting features, led by mean red reflectance and including NIR-based Haralick textures such as energy and entropy, green, NIR, blue, yellow, and red-edge region zonal statistics, as well as the NDVI indice (Figure S1b).

The random forest model for the 2009 image predicted the training set with an average accuracy of 98.17%, with the most confused classes being within the land and water-based classes, which is a non-issue for land masking purposes (Table S1). Boats (l_boats) were wrongly classified as emerged vegetation (v_em), but this was mainly linked to their low representation in the training set and the image. The 2019 training set was predicted with an average accuracy of 98.29%, again with low confusion between land and water classes (Table S2). The most confused classes were shallow water (w_shallow), deep water (w_deep), and boat waves (o_boatwave), all of which could be dealt with in the level 2 OBIA.

3.4. Level 2 OBIA, Vegetation Cover Mapping

The second level of OBIA is an iterative process of object classification targeted towards the identification of submerged aquatic vegetation. Using OTB's Meanshift segmentation with the parameters described in Table 2, the 2009 Quickbird-02 image generated (mean ± 1 S.D.), 112,615 segments of 94 ± 98.2 m^2 size while the 2019 Worldview-03 image generated 1,152,826 segments of 42.8 ± 32.5 m^2 size. Since water objects tend to have low reflectance across all wavelengths, differentiating between classes can represent a challenge both for the expert building the training set, and the algorithm building classification rules. For this reason, the training and predictions of object classes are performed multiple times, until the results are satisfactory (Figure 7). For images 1 and 2, four and six runs were required, respectively, to obtain results satisfactory to the expert observer (Figure 8).

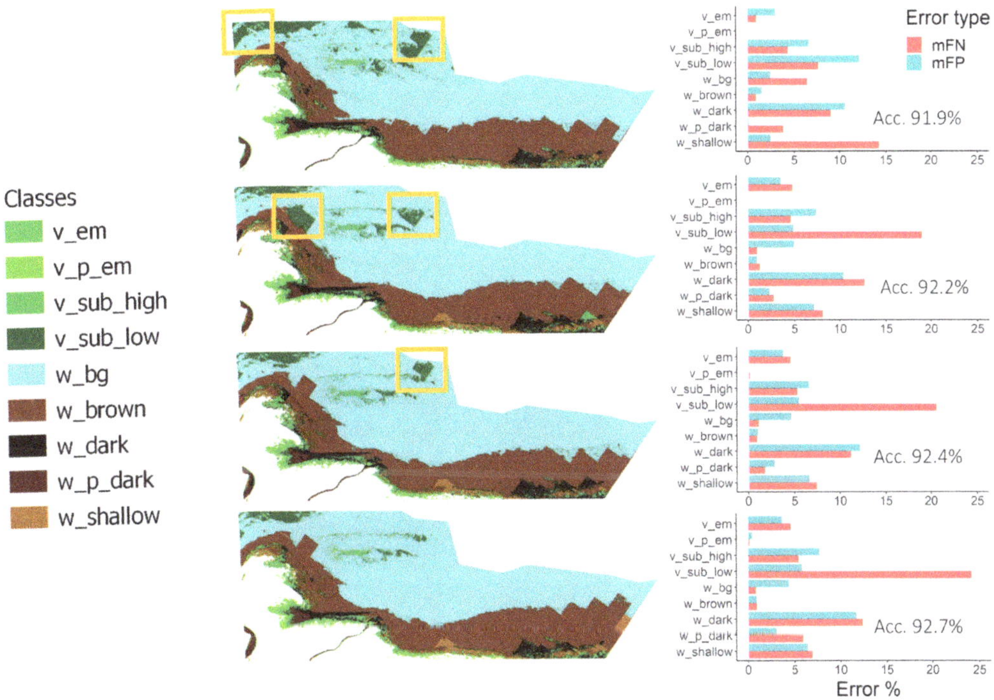

Figure 7. Visualization of the iterative classification process over image 1, the 2009 lake Saint-Pierre Quickbird-2 image (© 2022 Maxar Technologies). From top to bottom, we see the first to the last classification attempt, where the main difficulty was to prevent overestimation of vegetated areas in low contrast areas. On the left we see the predicted classes for all objects using the random forest model, where the orange boxes represent areas to be corrected in the next iteration by adjusting the training set. On the right we see the proportions of meaningful classification errors (false negatives and false positives) for every class with the total OOB model accuracy. The classes prefixes stand for vegetation (v) and water (w), and the suffixes for emergent (em), submerged (sub), blue-green (bg), and proximity to land (p).

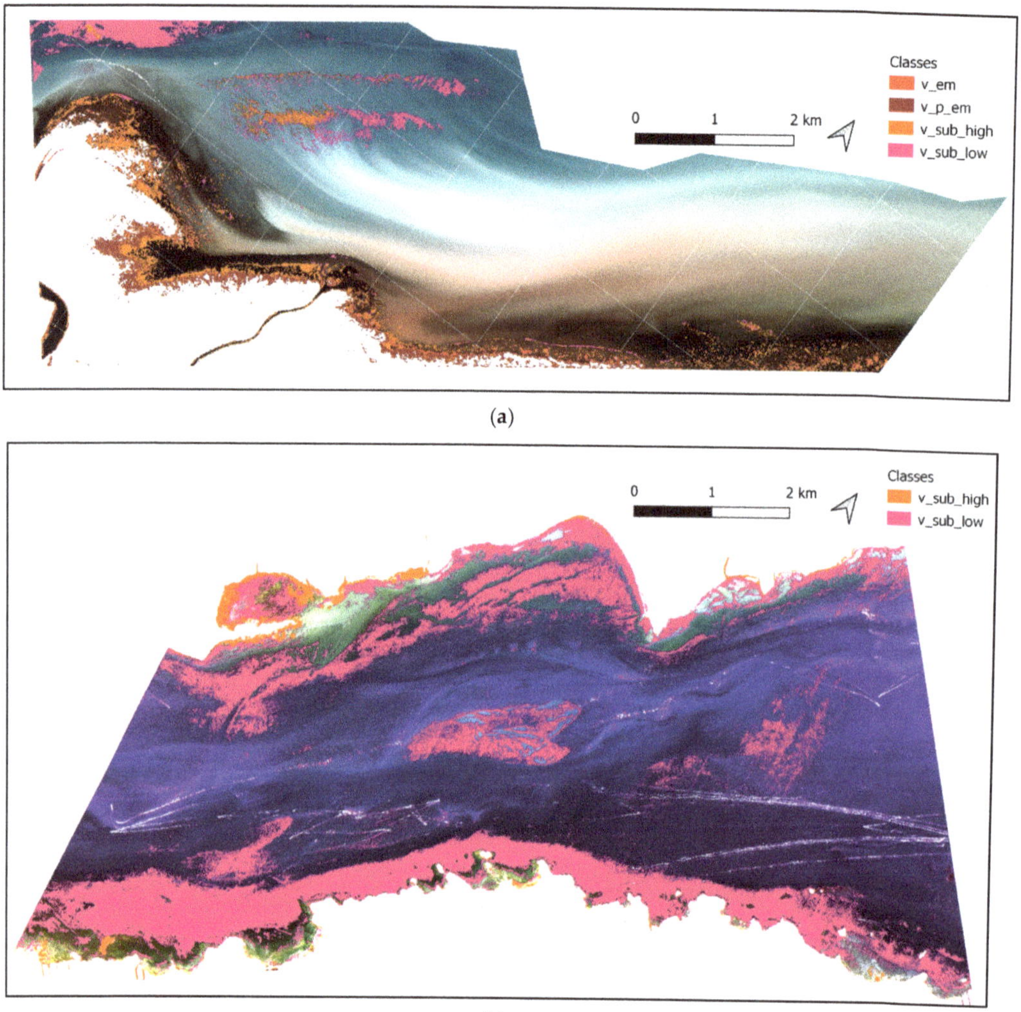

(a)

(b)

Figure 8. Final vegetation cover map as produced by the proposed workflow. Panel (**a**) shows the 2009 lake Saint-Pierre Quickbird-2 image (© 2022 Maxar Technologies) and panel (**b**) shows the 2019 lake Saint-François image (© 2022 Maxar Technologies).

The most important features used in the random forest for SAV cover mapping for the 2009 image were again largely based on the NIR signal and vegetation indexes, but with the addition of blue-based Haralick correlation (Figure S2a). The 2019 image again used more contrasting features, led by minimum yellow, green and blue reflectance, followed by some vegetation indexes and blue-based Haralick inertia (Figure S1b).

The random forest model for the first image predicted the training set with an average OOB accuracy of 92.74% and a 10 folds accuracy of 93.6%, which suggests a low probability of overfitting. While the global performance is very high, some classes of interest had large errors, with meaningful false negatives and positives up to 23.46% and 10.62%, respectively, for the low submerged vegetation class (v_sub_low), which was mainly confused with blue-green water (w_bg) (Figure 7, Table S3), suggesting an underestimation of the total vegetation cover. The second most confused classes were high submerged vegetation (v_sub_high) and dark water (w_dark), with comparable amounts of false negatives and positives (Table S3), suggesting a trade-off between both classes. As for the second image, the average random forest model OOB accuracy for SAV cover mapping was 96.87% and 93.8% for the k-fold cross-validation. Again, the highest error rates were associated with the low submerged vegetation class (v_sub_low), which was often confused with shallow (w_shallow) and deep water (w_deep) (Table S4). Nevertheless, meaningful false negatives were more important than meaningful false positives (8.86% vs. 5.88%) for low submerged vegetation, indicating conservative SAV cover estimates.

4. Discussion

The proposed workflow allowed the production of image-based submerged aquatic vegetation cover maps for two different high-resolution multispectral sensors. By using open-source OBIA methods, the need for precise radiometric measurements was circumvented, shifting the modeling effort towards the quantification of local gradients and textural information.

Pre-processing generated radiometrically comparable images, as shown by the bottom of atmosphere reflectance spectra (Figure 4b). Although the lack of field radiometric measurements prevented the validation of the observed reflectance values, the magnitude and general shape of the deep-water reflectance spectra are in accordance with what has been found elsewhere for inland waters when compared with the literature. Our deep-water spectra show fewer variations and a higher baseline, but the peak values appear at the expected wavelengths, with around 0.05 surface reflectance units around the 500–600 nm region [24,25,73]. The high reflectance baseline and the flatness of the reflectance spectra could be attributed to a combination of the simple atmospheric correction, adjacency effects, traces of bottom reflection, and the IOPs of the target water bodies, which are subject to high loads of suspended matter, especially in the case of the lake Saint-Pierre [74].

De-striping offered a visually satisfying result, despite minor hiccups at closer inspection. Because the distance between every stripe jump, and their orientation is estimated and assumed constant, some pixels located very close to the jumps can remain problematic and be under/overcorrected. Also, the jumps do not occur over a single pixel, but rather over a range of about 7 pixels [49], so that the ideal correction would take into account this transition.

Level-1 OBIA random forest classification generated robust land masks despite low object counts within the training set together with low attention put into classes equalization (large variations between the number of objects per class). As expected, NIR reflectance-based features were of great importance for distinguishing between land and water objects, although a more complex array of features was used in the lake Saint-François (Figure S1), including lower wavelength bands such as red, green, yellow, and blue. This might be related to higher amounts of noise within the 2019 lake Saint-François image, partly due to boat waves and higher water transparency. In other studies, land-masking is often not discussed and land is roughly excluded [33,75]. On the other hand, in a knowledge-based OBIA classification of SAV, Visser et al. [34] used NDVI, mean NIR, and red reflectance thresholds as classification rules to classify exposed bank and bank vegetation, generating similar rules to our random forest. The confusion matrices for both images show low confusion between land and water-based classes (respectively identified by the *l*_ and *w*_ or *v*_ prefixes; Tables S2 and S3). Looking at meaningful misclassifications, classes with low object count are the most prone to errors, but they are also the less important in terms of cover, such as foam formations, boats, and boat waves, which could be compensated on level-2 OBIA.

Level-2 OBIA random forest classification performed relatively well across the classes of interest, with errors distributing as expected among classes and deeper growth forms displaying relatively high confusion with optically deep-water objects. The most important features for classification were quite different between the two images, being heavily reliant upon vegetation indices and NIR reflectance for the lake Saint-Pierre image, and again much more diverse for the lake Saint-François, including zonal statistics from the yellow, green, and blue bands (Figure S2). This is probably related to the higher turbidity of the lake Saint-Pierre, which limits the growth of SAV species with short height and favors canopy-forming species that can reach the top of the water column. Accordingly, image exploration reveals a much higher proportion of SAV showing traces of NIR reflectance (higher when SAV is close or at the surface) in the lake Saint-Pierre, especially within the brown and mixed water masses (see Figure 3a), as opposed to the blue-green waters from the main channel which is similar to the lake Saint-François, where most SAV is truly submerged, reflecting almost no NIR signal. This suggests that the features fed to the random forest account for enough variation to be robust across optically different fluvial ecosystems. Although SAV cover mapping is more sensitive to error than land masking, the confusion matrices show satisfactory classification results, with less than 10% meaningful false positives and negatives for all vegetation classes, except low submerged vegetation in the lake Saint-Pierre. Understandably, low submerged vegetation has the weakest remote sensing signal and is expected to be the least performing class. Ecologically, this weakness is partially compensated by the fact that shorter SAV would be expected to perform less as an engineer (i.e., generating smaller environmental feedbacks) than taller SAV growth forms [76], making them less important as sentinels, but nonetheless valuable as indicators. This also justifies that it is acceptable in this context for parts of the SAV landscape to be completely undetectable because of high turbidity. For the lake Saint-François image, floating/drifting vegetation generates large errors, is often confused with optically deep water. This is acceptable to us because drifting vegetation debris are not an object of interest to this classification despite their observed abundance in the field. Their large classification errors are associated with the size of the detached debris which is often smaller than the minimum segmentation object size, behaving similarly to white caps or NIR remote sensing noise. Finally, and more importantly, false negatives appear to be more common than false positives for SAV classes (Figure 7, Tables S3 and S4), suggesting overall conservative estimates of SAV cover. Further examination of Figure 7 also shows that visual validation criteria can be of higher importance than prediction accuracy over the training set, where an actual decrease in accuracy estimates can result in more satisfactory cover maps, especially for submerged aquatic vegetation in low contrast regions. To further validate the performance of the classification for a specific application, a multi-observer approach should be used to either estimate the error of the expert training set object identification, or the variability of resulting vegetation cover maps from multiple observers.

Despite acceptable classification results for routine monitoring of SAV, this paper represents a basic application of a very adaptable workflow, which could be improved and optimized in multiple ways. Furthermore, for the sake of the exercise, we used only image-based information and expert knowledge to calibrate and validate the model, which represents a worst-case scenario in terms of data availability. From calibration to validation, field observations should be capitalized upon whenever possible, although their availability and relevancy rapidly decrease with time passed since image capture. For those reasons, the rest of this section will serve as application guidelines, discussing best practices and limits regarding the state-of-the-art methods in the context of highly limited data availability.

Starting with data acquisition, multiple factors can affect classification results and require further consideration. These factors include sensor noise, spatial resolution, spectral resolution, and off-nadir angle, as well as environmental factors such as sun elevation angle, atmospheric conditions, wind, waves, water level, turbidity, the progress of macrophytes growing season, and terrain. Concerning sensor choice, Mouw et al. [45] offer an in-depth review of how and why most current water sensors are ill-fitted to study inland complex waters, leading researchers to use terrestrial earth observation missions such as Landsat, or in this case, Quickbird-2 and Worldview-3. The main issues with this approach are that those satellites were not built for low reflectance targets such as submerged objects, generating high signal-to-noise ratios (SNR) [25], and that their slow revisit time can be problematic for dynamic systems such as temperate fluvial lakes. Water level variations, water mass mixing, turbidity spikes, and growth phenology all affect how much plant canopy can be detected by satellite remote sensing. Nonetheless, methods such as multi-temporal, high-resolution OBIA based remote sensing of SAV have been successfully performed in the past in coastal ecosystems using appropriate field validation and proprietary software [35]. Understanding the effects of such environmental conditions on the temporal variability of the remote sensing signal is crucial for the correct use of remotely sensed SAV cover maps and represents one of the major challenges of this method.

Remote sensing noise can originate from multiple sources, including sensor architecture (as seen in Figure 5), wind and waves generating localized glint, white caps, water depth, and surface angle variations, as well as atmospheric and adjacency effects [77]. In addition, different bands can have slightly different sensing times, as indicated by the time delay integration (TDI) in DigitalGlobe's products, thus visualizing different waves at different wavelengths. An example of high noise can be seen by observing deep-water areas of the lake Saint-François Worldview image, where a sub-scene can generate reflectance values ranging from 0.02 to 0.1 just in the 725 nm band due to what appears to be dominantly wave action and despite the absence of identifiable submerged features (Figure 9). While glint can be corrected over optically deep waters, shallow, and turbid waters glint correction applications are very complex and challenging, and it is usually preferred to select images where the sun and off-nadir angles minimize glinting [78]. Because of low SNR, atmospheric correction becomes disproportionately important over water pixels [25,46]. While this remains valid, the narrow swath width of high-resolution sensors allows the assumption that atmospheric conditions are homogeneous across a single scene, arguably allowing whole-image atmospheric corrections as opposed to pixel-based methods [48]. One last source of noise that should be considered is the presence of adjacency effects caused by nearby landmasses that can have strong effects on the radiometry of close-by water objects because of the atmospheric scattering of upwelling signals. While some methods, such as the dark spectrum fitting, tend to limit those effects [48], no generalized solution currently exists for adjacency effect correction. For the lake Saint-Pierre image, additional object classes were thus included such as "v_p_em" and "w_p_dark", to represent emerging vegetation and dark water objects in proximity to land and diminish errors associated with adjacency effects and emergent vegetation.

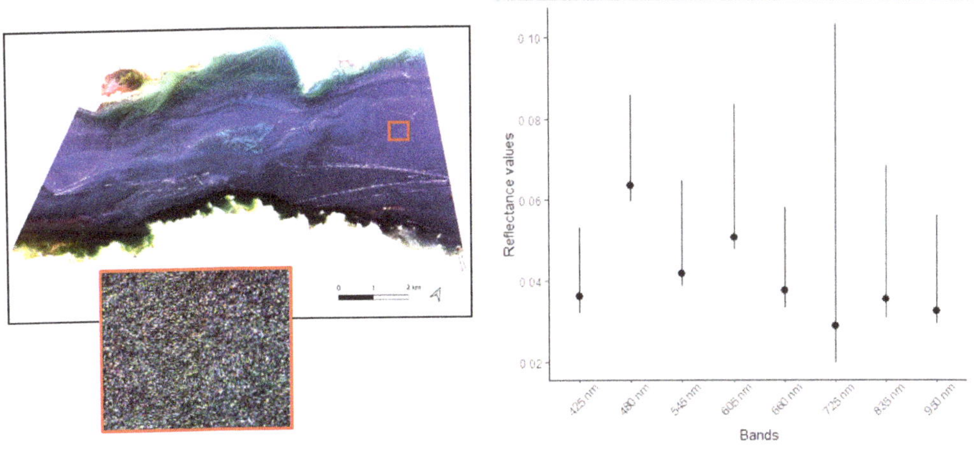

(a) (b)

Figure 9. Value range for all bands in a deep-water area affected by surface waves. (**a**) Shows a seemingly homogenous area of deep water in the eastern part of lake St-François. (**b**) Shows minimum, mean (dot) and maximum pixel values at every wavelength within this area.

Regarding the OBIA workflow, segmentation tuning, feature selection/engineering, training set building, and interpretation of results all add considerable flexibility to its applications, but also make it more vulnerable to inconsistent usage. Despite the growing interest towards OBIA in remote sensing, guidelines for the correct application of these steps are scarce and rarely generalizable across studies. We believe that further research is required to truly understand the importance of segmentation methods, spatial and spectral resolution, the choice of spectral, textural, and contextual indices included in the predictors, and the construction of robust and standardized training sets. In previous vegetation studies, different segmentation methods were used, including ENVI's watershed segmentation [33] and eCognition's multi-resolution segmentation [34,36,79], with little justification to their choice of segmentation solution despite their licensing fees. Following the same pattern, little information is available to guide feature selection in building the modeling data sets, with similar sets of features being widely used (zonal statistics plus vegetation and textural indexes), despite little apparent research to understand the effects of the feature array, especially regarding their specific tuning (such as scale parameters when calculating Haralick texture indexes) [33,36,79,80]. While machine learning-based models allow for some robustness despite wild arrays of inputs, widespread routine monitoring of SAV would highly benefit from a stronger understanding of feature engineering for its detection. In this paper, relative feature importance as estimated by the random forest algorithm provides some insight regarding feature selection in contrasted systems (Figure S4), where NIR-based metrics performed better in SAV-rich, turbid water, while shorter wavelengths and texture-based metrics were more appropriate in clear waters where SAV rarely reaches the surface.

In terms of modeling, the building of the training sets represents a critical task that would ideally require large amounts of field observations designed for training and validation of such methods, which is non-existent for most water bodies across time. While this paper suggests that it is possible to use Earth observation tools to estimate SAV coverage using only expert visual observation, this exercise would be hardly transposable to ecosystems unknown to the observer or the study of more complex metrics such as community composition. Additionally, little is known about the importance of the number of end members to the random forest classifier, the balancing of the training sets, or the error associated with the expert training approach. Further investigations regarding the transferability of models between dates and ecosystems could give us access to proper tools for large-scale monitoring of SAV landscapes across time and space. Finally, attention should also be given to quantify the impact of different observers on the classification results. Many of the issues inherent to expert or knowledge-based approaches are well discussed in the study by Visser et al. [34] about the knowledge-based mapping of SAV, which further reinforces the arguments for establishing open standards for SAV monitoring

Given the many challenges and knowledge gaps yet to be addressed, we believe high-resolution remote sensing of SAV would highly benefit from more open-source-based research, allowing true testing and validation of methods. The workflow proposed in this study represents a useful step in this direction. By adapting generic methods to the specific needs of SAV remote sensing, not only highlights the potential of OBIA for SAV mapping, but also the large amount of work remaining before achieving a truly generalizable high-resolution remote sensing framework for SAV. Until then, this workflow offers a foundation for further open-source research, and acts as a testimony for better SAV field monitoring programs in complex waters, built to accommodate calibration and validation of high-resolution remote sensing products.

5. Conclusions

In this paper, we propose an adaptable framework and an open-sourced workflow to map SAV cover in complex waters using OBIA. By shifting the focus of the classification efforts from precise radiometric measurements to local gradient quantification, we were able to conduct satisfactory SAV mapping in two contrasted fluvial lakes, using two different high-resolution multi-spectral sensors. By using an iterative expert approach, conservative estimates of SAV cover area could be extracted both for a recent image acquisition and an older image from over 10-years back, despite the absence of ground-truthing data, hinting towards the large potential of similar methods for long-term monitoring of aquatic ecosystems. To better calibrate and validate this process, we argue for the importance of implementing and maintaining remote-sensing appropriate field observations and call for more transparent research in aquatic remote sensing OBIA applications. By proposing a workflow for high-resolution remote sensing of SAV covers in complex systems, we hope to enable better monitoring of sensible ecosystems and facilitate landscape-level research about SAV dynamics in a changing world.

Supplementary Materials: The following are available online at https://www.mdpi.com/article/10.3390/rs14020267/s1, Table S1: level 1 confusion matrix for image 1, Table S2: level 1 confusion matrix for image 2, Table S3: level 2 confusion matrix for image 1, Table S4: level 2 confusion matrix for image 2, Figure S1: most important features for level 1 classification, Figure S2: most important features for level 2 classification.

Author Contributions: Conceptualization, A.d.G., A.B. and C.K.; methodology, A.d.G. and C.K.; software, A.d.G.; validation, A.d.G.; formal analysis, A.d.G.; investigation, A.d.G., A.B. and C.K.; resources, A.B. and C.K.; data curation, A.d.G.; writing—original draft preparation, A.d.G.; writing—review and editing, A.d.G., A.B. and C.K.; visualization, A.d.G.; supervision, A.B. and C.K.; project administration, A.B. and C.K.; funding acquisition, A.d.G., A.B. and C.K. All authors have read and agreed to the published version of the manuscript.

Funding: This study was financially supported by the Canada Research Chair program (grant number 231380, Christophe Kinnard), the Natural Sciences and Engineering Research Council of Canada (NSERC discovery grant CRSNG-RGPIN-2017-05451, Andrea Bertolo), the Réseau Québec Maritime (UQAR-RQM-FRQNT 307819), the Interuniversity Research Group in Limnology (GRIL, ÉcoLac—NSERC CREATE training program, doctoral scholarship), the Centre de Recherche sur les Interactions Bassins Versants—Écosytèmes Aquatiques (RIVE, doctoral scholarships) and the Ministère des Forêts, de la Faune et des Parcs (MFFP, contracts 2019-1-SGHAPP-251 & 2020-1-SGHAPP-328).

Institutional Review Board Statement: Not applicable.

Data Availability Statement: The satellite imagery used in this paper were obtained in collaboration with the MFFP and licensed by Digital Globe. Inc. The imagery products are found in the Digital Globe catalog under the DigitalGlobe Catalog IDs A0100104D7CB2A00 and A0100104CD914100 for the 2009 Quickbird and 2019 Worldview images respectively. The scripts to run the workflow used to map the vegetation cover is available on Github under the cc-by-nc license under the following URL: https://github.com/arthurdegrandpre/Open_HRRS_W2 (accessed on 1 November 2021).

Acknowledgments: We gratefully acknowledge all of our funders for supporting our work. We would also like to acknowledge the MFFP for providing the DigitalGlobe satellite imagery with the appropriate licenses for us to be able to conduct this research. Also, we thank our anonymous reviewers for their knowledgeable and relevant comments that made this work stronger. Finally, we would like to thank all the open-source software and library developers that made this workflow possible.

Conflicts of Interest: The authors declare no conflict of interest.

References

1. Heck, K.L.; Able, K.W.; Roman, C.T.; Fahay, M.P. Composition, abundance, biomass, and production of macrofauna in a New England estuary: Comparisons among eelgrass meadows and other nursery habitats. *Estuaries* **1995**, *18*, 379–389. [CrossRef]
2. Pang, S.; Zhang, S.; Lv, X.Y.; Han, B.; Liu, K.; Qiu, C.; Wang, C.; Wang, P.; Toland, H.; He, Z. Characterization of bacterial community in biofilm and sediments of wetlands dominated by aquatic macrophytes. *Ecol. Eng.* **2016**, *97*, 242–250. [CrossRef]
3. Law, A.; Baker, A.; Sayer, C.; Foster, G.; Gunn, I.D.M.; Taylor, P.; Pattison, Z.; Blaikie, J.; Willby, N.J. The effectiveness of aquatic plants as surrogates for wider biodiversity in standing fresh waters. *Freshw. Biol.* **2019**, *64*, 1664–1675. [CrossRef]
4. Caffrey, J.; Kemp, W. Nitrogen cycling in sediments with estuarine populations of Potamogeton perfoliatus and Zostera marina. *Mar. Ecol. Prog. Ser.* **1990**, *66*, 147–160. [CrossRef]

5. Jeppesen, E.; Søndergaard, M.; Søndergaard, M.; Christoffersen, K. *The Structuring Role of Submerged Macrophytes in Lakes*; Ecological Studies; Springer: New York, NY, USA, 1998; Volume 131, ISBN 978-1-4612-6871-0.
6. Marion, L.; Paillisson, J.M. A mass balance assessment of the contribution of floating-leaved macrophytes in nutrient stocks in an eutrophic macrophyte-dominated lake. *Aquat. Bot.* **2003**, *75*, 249–260. [CrossRef]
7. Fourqurean, J.W.; Duarte, C.M.; Kennedy, H.; Marbà, N.; Holmer, M.; Mateo, M.A.; Apostolaki, E.T.; Kendrick, G.A.; Krause-Jensen, D.; McGlathery, K.J.; et al. Seagrass ecosystems as a globally significant carbon stock. *Nat. Geosci.* **2012**, *5*, 505–509. [CrossRef]
8. Duarte, C.M.; Krause-Jensen, D. Export from Seagrass Meadows Contributes to Marine Carbon Sequestration. *Front. Mar. Sci.* **2017**, *4*, 13. [CrossRef]
9. Xing, Y.; Xie, P.; Yang, H.; Wu, A.; Ni, L. The change of gaseous carbon fluxes following the switch of dominant producers from macrophytes to algae in a shallow subtropical lake of China. *Atmos. Environ.* **2006**, *40*, 8034–8043. [CrossRef]
10. Orth, R.J.; Dennison, W.C.; Lefcheck, J.S.; Gurbisz, C.; Hannam, M.; Keisman, J.; Landry, J.B.; Moore, K.A.; Murphy, R.R.; Patrick, C.J.; et al. Submersed aquatic vegetation in chesapeake bay: Sentinel species in a changing world. *Bioscience* **2017**, *67*, 698–712. [CrossRef]
11. Macreadie, P.I.; Anton, A.; Raven, J.A.; Beaumont, N.; Connolly, R.M.; Friess, D.A.; Kelleway, J.J.; Kennedy, H.; Kuwae, T.; Lavery, P.S.; et al. The future of Blue Carbon science. *Nat. Commun.* **2019**, *10*, 3998. [CrossRef]
12. De Boer, W.F. Seagrass-sediment interactions, positive feedbacks and critical thresholds for occurrence: A review. *Hydrobiologia* **2007**, *591*, 5–24. [CrossRef]
13. Barko, J.W.; James, W.F. Effects of Submerged Aquatic Macrophytes on Nutrient Dynamics, Sedimentation, and Resuspension. In *The Structuring Role of Submerged Macrophytes in Lakes*; Ecological Studies; Springer: New York, NY, USA, 1998; pp. 197–214. [CrossRef]
14. Challen Hyman, A.; Frazer, T.K.; Jacoby, C.A.; Frost, J.R.; Kowalewski, M. Long-term persistence of structured habitats: Seagrass meadows as enduring hotspots of biodiversity and faunal stability. *Proc. R. Soc. B Biol. Sci.* **2019**, *286*, 20191861. [CrossRef] [PubMed]
15. Jones, C.G.; Lawton, J.H.; Shachak, M. Organisms as Ecosystem Engineers. *Oikos* **1994**, *69*, 373. [CrossRef]
16. Koch, E.W. Beyond light: Physical, geological, and geochemical parameters as possible submersed aquatic vegetation habitat requirements. *Estuaries* **2001**, *24*, 1–17. [CrossRef]
17. Lacoul, P.; Freedman, B. Environmental influences on aquatic plants in freshwater ecosystems. *Environ. Rev.* **2006**, *14*, 89–136. [CrossRef]
18. Bornette, G.; Puijalon, S. Response of aquatic plants to abiotic factors: A review. *Aquat. Sci.* **2011**, *73*, 1–14. [CrossRef]
19. Madsen, J.D.; Wersal, R.M. A review of aquatic plant monitoring and assessment methods. *J. Aquat. Plant Manag.* **2017**, *55*, 1–12.
20. Kemp, W.M.; Batiuk, R.; Bartleson, R.; Bergstrom, P.; Carter, V.; Gallegos, C.L.; Hunley, W.; Karrh, L.; Koch, E.W.; Landwehr, J.M.; et al. Habitat requirements for submerged aquatic vegetation in Chesapeake bay: Water quality, light regime, and physical-chemical factors. *Estuaries* **2004**, *27*, 363–377. [CrossRef]
21. Jayathilake, D.R.M.; Costello, M.J. A modelled global distribution of the seagrass biome. *Biol. Conserv.* **2018**, *226*, 120–126. [CrossRef]
22. McKenzie, L.J.; Nordlund, L.M.; Jones, B.L.; Cullen-Unsworth, L.C.; Roelfsema, C.; Unsworth, R.K.F. The global distribution of seagrass meadows. *Environ. Res. Lett.* **2020**, *15*, 74041. [CrossRef]
23. Jin, K.R.; Ji, Z.G. A long term calibration and verification of a Submerged aquatic vegetation model for Lake Okeechobee. *Ecol. Process.* **2013**, *2*, 23. [CrossRef]
24. Rowan, G.S.L.; Kalacska, M. A review of remote sensing of submerged aquatic vegetation for non-specialists. *Remote Sens.* **2021**, *13*, 623. [CrossRef]
25. Giardino, C.; Brando, V.E.; Gege, P.; Pinnel, N.; Hochberg, E.; Knaeps, E.; Reusen, I.; Doerffer, R.; Bresciani, M.; Braga, F.; et al. Imaging Spectrometry of Inland and Coastal Waters: State of the Art, Achievements and Perspectives. *Surv. Geophys.* **2019**, *40*, 401–429. [CrossRef]
26. Bolduc, P.; Bertolo, A.; Pinel-Alloul, B. Does submerged aquatic vegetation shape zooplankton community structure and functional diversity? A test with a shallow fluvial lake system. *Hydrobiologia* **2016**, *778*, 151–165. [CrossRef]
27. Madsen, J.D.; Chambers, P.A.; James, W.F.; Koch, E.W.; Westlake, D.F. The interaction between water movement, sediment dynamics and submersed macrophytes. *Hydrobiologia* **2001**, *444*, 71–84. [CrossRef]
28. Koch, E.W.; Ackerman, J.D.; Verduin, J.; Keulen, M. Van Fluid dynamics in seagrass ecology-from molecules to ecosystems. In *Seagrasses: Biology, Ecology and Conservation*; Springer: Dordrecht, The Netherlands, 2006; pp. 193–225. ISBN 140202942X.
29. Ruiz-Reynés, D.; Gomila, D.; Sintes, T.; Hernández-García, E.; Marbà, N.; Duarte, C.M. Fairy circle landscapes under the sea. *Sci. Adv.* **2017**, *3*, e1603262. [CrossRef] [PubMed]
30. Rietkerk, M.; van de Koppel, J. Regular pattern formation in real ecosystems. *Trends Ecol. Evol.* **2008**, *23*, 169–175. [CrossRef] [PubMed]
31. Wang, R.; Dearing, J.A.; Langdon, P.G.; Zhang, E.; Yang, X.; Dakos, V.; Scheffer, M. Flickering gives early warning signals of a critical transition to a eutrophic lake state. *Nature* **2012**, *492*, 419–422. [CrossRef] [PubMed]
32. Husson, E.; Ecke, F.; Reese, H. Comparison of manual mapping and automated object-based image analysis of non-submerged aquatic vegetation from very-high-resolution UAS images. *Remote Sens.* **2016**, *8*, 724. [CrossRef]

33. Chabot, D.; Dillon, C.; Shemrock, A.; Weissflog, N.; Sager, E.P.S. An object-based image analysis workflow for monitoring shallow-water aquatic vegetation in multispectral drone imagery. *ISPRS Int. J. Geo-Inf.* **2018**, *7*, 294. [CrossRef]
34. Visser, F.; Buis, K.; Verschoren, V.; Schoelynck, J. Mapping of submerged aquatic vegetation in rivers from very high-resolution image data, using object-based image analysis combined with expert knowledge. *Hydrobiologia* **2018**, *812*, 157–175. [CrossRef]
35. Roelfsema, C.M.; Lyons, M.; Kovacs, E.M.; Maxwell, P.; Saunders, M.I.; Samper-Villarreal, J.; Phinn, S.R. Multi-temporal mapping of seagrass cover, species and biomass: A semi-automated object based image analysis approach. *Remote Sens. Environ.* **2014**, *150*. [CrossRef]
36. Mishra, N.B.; Crews, K.A. Mapping vegetation morphology types in a dry savanna ecosystem: Integrating hierarchical object-based image analysis with Random Forest. *Int. J. Remote Sens.* **2014**, *35*, 1175–1198. [CrossRef]
37. Hay, G.J.; Castilla, G. Geographic object-based image analysis (GEOBIA): A new name for a new discipline. In *Object-Based Image Analysis; Spatial Concepts for Knowledge-Driven Remote Sensing Applications*; Blaschke, T., Lang, S., Hay, G.J., Eds.; Springer Science and Business Media Deutschland GmbH: Berlin/Heidelberg, Germany, 2008; pp. 75–89. ISBN 9783540770589.
38. Grizonnet, M.; Michel, J.; Poughon, V.; Inglada, J.; Savinaud, M.; Cresson, R. Orfeo ToolBox: Open source processing of remote sensing images. *Open Geospatial Data, Softw. Stand.* **2017**, *2*, 15. [CrossRef]
39. R Core Team. *R: A Language and Environment for Statistical Computing*; R Core Team: Vienna, Austria, 2021.
40. QGIS Development Team. QGIS Geographic Information System. 2009. Available online: https://en.wikipedia.org/wiki/QGIS (accessed on 1 November 2021).
41. Blaschke, T.; Hay, G.J.; Kelly, M.; Lang, S.; Hofmann, P.; Addink, E.; Queiroz Feitosa, R.; van der Meer, F.; van der Werff, H.; van Coillie, F.; et al. Geographic Object-Based Image Analysis—Towards a new paradigm. *ISPRS J. Photogramm. Remote Sens.* **2014**, *87*, 180–191. [CrossRef] [PubMed]
42. Updike, T.; Comp, C. *Radiometric Use of WorldView-2 Imagery*; DigitalGlobe: Westminster, CO, USA, 2010; pp. 1–17.
43. Kuester, M. *Radiometric Use of WorldView-3 Imagery*; DigitalGlobe: Westminster, CO, USA, 2016; pp. 1–12.
44. Kuester, M. *Absolute Radiometric Calibration: 2016v0*; DigitalGlobe: Westminster, CO, USA, 2017; pp. 1–8.
45. Mouw, C.B.; Greb, S.; Aurin, D.; DiGiacomo, P.M.; Lee, Z.; Twardowski, M.; Binding, C.; Hu, C.; Ma, R.; Moore, T.; et al. Aquatic color radiometry remote sensing of coastal and inland waters: Challenges and recommendations for future satellite missions. *Remote Sens. Environ.* **2015**, *160*, 15–30. [CrossRef]
46. Moses, W.J.; Sterckx, S.; Montes, M.J.; De Keukelaere, L.; Knaeps, E. Atmospheric Correction for Inland Waters. In *Bio-Optical Modeling and Remote Sensing of Inland Waters*; Elsevier Inc.: Amsterdam, The Netherlands, 2017; pp. 69–100. ISBN 9780128046548.
47. Chavez, P.S. Image-based atmospheric corrections—Revisited and improved. *Photogramm. Eng. Remote Sens.* **1996**, *62*, 1025–1036.
48. Vanhellemont, Q.; Ruddick, K. Atmospheric correction of metre-scale optical satellite data for inland and coastal water applications. *Remote Sens. Environ.* **2018**, *216*, 586–597. [CrossRef]
49. Marmorino, G.; Chen, W. Use of WorldView-2 along-track stereo imagery to probe a Baltic Sea algal spiral. *Remote Sens.* **2019**, *11*, 865. [CrossRef]
50. Hossain, M.D.; Chen, D. Segmentation for Object-Based Image Analysis (OBIA): A review of algorithms and challenges from remote sensing perspective. *ISPRS J. Photogramm. Remote Sens.* **2019**, *150*, 115–134. [CrossRef]
51. Villa, P.; Mousivand, A.; Bresciani, M. Aquatic vegetation indices assessment through radiative transfer modeling and linear mixture simulation. *Int. J. Appl. Earth Obs. Geoinf.* **2014**, *30*, 113–127. [CrossRef]
52. Kuhn, M. Caret: Classification and Regression Training. 2020. Available online: https://cran.r-project.org/web/packages/caret/caret.pdf (accessed on 1 November 2021).
53. Dowle, M.; Srinivasan, A. Data.Table: Extension of 'Data.Frame'. 2021. Available online: https://rdrr.io/cran/data.table/ (accessed on 1 November 2021).
54. Corporation, M.; Weston, S. DoParallel: Foreach Parallel Adaptor for the "Parallel" Package. 2020. Available online: https://cran.r-project.org/web/packages/doParallel/index.html (accessed on 1 November 2021).
55. Microsoft; Weston, S. Foreach: Provides Foreach Looping Construct. 2020. Available online: https://cran.r-project.org/web/packages/foreach/index.html (accessed on 1 November 2021).
56. Greenberg, J.A.; Mattiuzzi, M. gdalUtils: Wrappers for the Geospatial Data Abstraction Library (GDAL) Utilities. 2020. Available online: https://rdrr.io/cran/gdalUtils/ (accessed on 1 November 2021).
57. Lisovski, S.; Hahn, S. GeoLight—Processing and analysing light-based geolocation in R. *Methods Ecol. Evol.* **2012**. [CrossRef]
58. Hengl, T. GSIF: Global Soil Information Facilities. 2020. Available online: https://rdrr.io/rforge/GSIF/ (accessed on 1 November 2021).
59. Zhu, H. kableExtra: Construct Complex Table with "kable" and Pipe Syntax. 2020. Available online: https://mran.microsoft.com/package/kableExtra (accessed on 1 November 2021).
60. Hesselbarth, M.H.K.; Sciaini, M.; With, K.A.; Wiegand, K.; Nowosad, J. Landscapemetrics: An open-source R tool to calculate landscape metrics. *Ecography* **2019**, *42*, 1648–1657. [CrossRef]
61. Wickham, H. The Split-Apply-Combine Strategy for Data Analysis. *J. Stat. Softw.* **2011**, *40*, 1–29. [CrossRef]
62. Liaw, A.; Wiener, M. Classification and Regression by randomForest. *R News* **2002**, *2*, 18–22.
63. Hijmans, R.J. Raster: Geographic Data Analysis and Modeling. 2020. Available online: https://cran.r-project.org/web/packages/raster/raster.pdf (accessed on 1 November 2021).

64. Bivand, R.; Keitt, T.; Rowlingson, B. Rgdal: Bindings for the "Geospatial" Data Abstraction Library. 2020. Available online: https://cran.r-project.org/web/packages/rgdal/index.html (accessed on 1 November 2021).
65. Nauss, T.; Meyer, H.; Detsch, F.; Appelhans, T. Manipulating Satellite Data with {Satellite}. 2019. Available online: https://cran.r-project.org/web/packages/satellite/satellite.pdf (accessed on 1 November 2021).
66. Pebesma, E. Simple Features for R: Standardized Support for Spatial Vector Data. *R J.* **2018**, *10*, 439–446. [CrossRef]
67. Knaus, J. Snowfall: Easier Cluster Computing (Based On Snow). 2015. Available online: https://rdrr.io/cran/snowfall/ (accessed on 1 November 2021).
68. Evans, J.S. spatialEco. 2020. Available online: https://cran.r-project.org/web/packages/spatialEco/index.html (accessed on 1 November 2021).
69. Wickham, H.; Averick, M.; Bryan, J.; Chang, W.; McGowan, L.D.; François, R.; Grolemund, G.; Hayes, A.; Henry, L.; Hester, J.; et al. Welcome to the {tidyverse}. *J. Open Source Softw.* **2019**, *4*, 1686. [CrossRef]
70. Hunziker, P. velox: Fast Raster Manipulation and Extraction. 2018. Available online: https://www.r-bloggers.com/2016/09/velox-fast-raster-manipulation-and-extraction-in-r/ (accessed on 1 November 2021).
71. Pebesma, E.J.; Bivand, R.S. Classes and methods for spatial data in {R}. *R News* **2005**, *5*, 9–13.
72. Bivand, R.S.; Pebesma, E.; Gomez-Rubio, V. *Applied Spatial Data Analysis with {R}*, 2nd ed.; Springer: New York, NY, USA, 2013.
73. Ma, S.; Zhou, Y.; Gowda, P.H.; Dong, J.; Zhang, G.; Kakani, V.G.; Wagle, P.; Chen, L.; Flynn, K.C.; Jiang, W. Application of the water-related spectral reflectance indices: A review. *Ecol. Indic.* **2019**, *98*, 68–79. [CrossRef]
74. Rondeau, B.; Cossa, D.; Gagnon, P.; Bilodeau, L. Budget and sources of suspended sediment transported in the St. Lawrence River, Canada. *Hydrol. Process.* **2000**, *14*, 21–36. [CrossRef]
75. Reshitnyk, L.; Costa, M.; Robinson, C.; Dearden, P. Evaluation of WorldView-2 and acoustic remote sensing for mapping benthic habitats in temperate coastal Pacific waters. *Remote Sens. Environ.* **2014**, *153*, 7–23. [CrossRef]
76. Vilas, M.P.; Marti, C.L.; Adams, M.P.; Oldham, C.E.; Hipsey, M.R. Invasive Macrophytes Control the Spatial and Temporal Patterns of Temperature and Dissolved Oxygen in a Shallow Lake: A Proposed Feedback Mechanism of Macrophyte Loss. *Front. Plant Sci.* **2017**, *8*, 2097. [CrossRef]
77. Ruddick, K.; Vanhellemont, Q.; Dogliotti, A.I.; Nechad, B.; Pringle, N.; Van der Zande, D. New opportunities and challenges for high resolution remote sensing of water colour. In Proceedings of the Ocean Optics 2016, Victoria, BC, Canada, 23–28 October 2016.
78. Kay, S.; Hedley, J.; Lavender, S.; Kay, S.; Hedley, J.D.; Lavender, S. Sun Glint Correction of High and Low Spatial Resolution Images of Aquatic Scenes: A Review of Methods for Visible and Near-Infrared Wavelengths. *Remote Sens.* **2009**, *1*, 697–730. [CrossRef]
79. Husson, E.; Reese, H.; Ecke, F. Combining spectral data and a DSM from UAS-images for improved classification of non-submerged aquatic vegetation. *Remote Sens.* **2017**, *9*, 247. [CrossRef]
80. Peña-Barragán, J.M.; Ngugi, M.K.; Plant, R.E.; Six, J. Object-based crop identification using multiple vegetation indices, textural features and crop phenology. *Remote Sens. Environ.* **2011**, *115*, 1301–1316. [CrossRef]

remote sensing

MDPI

Article

Methodology for Combining Data Acquired by Unmanned Surface and Aerial Vehicles to Create Digital Bathymetric Models in Shallow and Ultra-Shallow Waters

Jacek Lubczonek *, Witold Kazimierski, Grzegorz Zaniewicz and Malgorzata Lacka

Faculty of Navigation, Maritime University of Szczecin, ul. Waly Chrobrego 1-2, 70-500 Szczecin, Poland; w.kazimierski@am.szczecin.pl (W.K.); g.zaniewicz@am.szczecin.pl (G.Z.); m.lacka@am.szczecin.pl (M.L.)

* Correspondence: j.lubczonek@am.szczecin.pl

Abstract: This paper presents a method for integrating data acquired by unmanned surface vehicles and unmanned aerial vehicles. The aim of this work was to create a uniform bathymetric surface extending to the shoreline. Such a body of water is usually characterized by ultra-shallow depths, which makes measurement impossible even with hydrographic autonomous vessels. Bathymetric data acquired by the photogrammetric method are, however, characterized by large errors with increasing depth. The presented method is based on processing of two data sets using a bathymetric reference surface and selection of points on the basis of generated masks. Numerical bathymetric models created by interpolation methods confirmed the usefulness of the concept adopted.

Keywords: USV; UAV; digital bathymetric model; bathymetry; spatial interpolation; data fusion; hydrography; surface modeling

Citation: Lubczonek, J.; Kazimierski, W.; Zaniewicz, G.; Lacka, M. Methodology for Combining Data Acquired by Unmanned Surface and Aerial Vehicles to Create Digital Bathymetric Models in Shallow and Ultra-Shallow Waters. *Remote Sens.* **2022**, *14*, 105. https://doi.org/10.3390/rs14010105

Academic Editor: Dimitris Poursanidis

Received: 23 November 2021
Accepted: 24 December 2021
Published: 27 December 2021

1. Introduction

Shallow waters are among the most interesting and exploited areas on the Earth. They are accessible and have multiple uses, for example, for unique environmental observation and tourist activities, but also pose a number of technological and research challenges. They cover not only areas in the coastal zones of the seas and oceans but also inland waters, such as lakes, rivers, and artificial reservoirs. Taking into account the management and exploitation of coast zones, as well as ecology issues, it is important to know the shape and structure of the terrain, of both the sea and the shore. Thus, many techniques have been developed over the years to present topography and bathymetry in the coastal zones and shallow waters. However, these issues are still a research challenge, and as a result numerous publications with various approaches to solve them are published. Exemplary approaches have been provided in [1] or [2]. In this paper, we are undertaking the problem of building a digital bathymetric model (DBM) [3] based on integrated data from novel measurement techniques on unmanned vehicles.

The traditional method of data gathering in shallow water was manual measurements with a pole. In time, with development of engineering, it was replaced by acoustic methods, first with a single-beam echosounder and then with a multi-beam echosounder. These hydroacoustic methods are presently the basis of most hydrographic works, as mentioned, for example, in [4,5]. They provide the most accurate results [6], and in many research works, they became a reference for assessing other methods' accuracy, e.g., [1,7]. In some works, such as [8], the key disadvantages of these solutions were indicated, which are mostly their labor-intensive nature, relatively small coverage (in shallow waters), and relatively complicated processing of data. Thus, alternative methods of bathymetry measurements in coastal zones have been proposed, such as the photogrammetric (mostly multispectral) method and light detection and ranging (LIDAR), in both aerial and satellite approaches. Remote sensing of river depths has traditionally relied upon passive optical image data [9].

Many studies have proven that passive optical spectrally based approaches to depth retrieval perform well in shallow waters. Examples of use of these techniques for coastal area bathymetry can be found, especially when the mapped area is large and a classical survey would be economically ineffective. They can perform well in many applications; however, there are also some disadvantages, related mostly to technological and environmental issues. Optical methods are highly vulnerable to environmental conditions, and they are able to penetrate only relatively shallow waters. Factors such as water turbidity and color or atmospheric disturbances affect the quality and availability of measurements.

In the traditional approach, based on passive optical image data, the key issue is to estimate depth from multi- or hyperspectral images, which typically involves establishing a relationship between depth and reflectance at one or more wavelengths. The most common means of calibrating such a relation is to link georeferenced image pixels to field-based depth measurements [9]. One of the most interesting approaches presented in literature is the use of optimal band ratio analysis (OBRA) for this purpose, presented in [10]. Other approaches are extraction of pixel values and regression of depth measurements against them [11], some of them also taking into account light attenuation in a water column [12]. These methods generally perform well. However, as mentioned in [13], as the use of remote sensing techniques for mapping shallow river waters continues to expand, so must awareness of the inherent limitations of this approach. This paper shows the need for a hybrid approach combining remote sensing with field-based methods, such as multibeam echosounder surveys, to obtain thorough, complete maps of the bathymetry of large rivers.

Among other photogrammetric approaches, the most important technique is the so-called structure from motion (SfM), which allows the elaboration of a 3D model based on computation that includes camera motion. The guidelines of the National Oceanic and Atmospheric Administration (NOAA) for using this methodology for bathymetric mapping of coastal areas are included in [14]. It was indicated in this work that the accuracy and reliability of the measurements depend on many environmental issues, such as water clarity, seafloor surface texture, and active wave breaking. Water clarity is also crucial for determining the effective measurement depth, which is usually about a few meters. In [15], accuracy better than 30 cm for the digital terrain model (DTM) was achieved, which in the case of shallow waters fulfills IHO special order requirements. However, simultaneously, it was stated that the results show that SfM through water can be a complex proposition. It was also proved in [6] that the quality of the bathymetric SfM is highly sensitive to flow, turbidity, and color. Other factors can also affect the development of any photogrammetric products, such as specific atmospheric conditions affecting the quality of the photo radiometry [16]; blurs appearing on the images [17]; water surface reflections [18]; and sea state, sunglint, and solar elevation angles [19]. In [20], refraction issues are additionally raised. It is claimed that using the SfM approach for submerged areas faces additional challenges, posed by the presence of water and, in particular, the effects of refraction. In this work, a direct comparison of two methods [21,22] of tackling this issue is included, showing that there is minimal difference in results produced by different refraction correction procedures. Other examples of using photogrammetrical SfM for river bathymetry can be found in [23–25]. Studies considering refraction can also be found in [26–28].

Another example of modern systems for bathymetry measurements is bathymetric LIDAR, usually making use of green laser for bathymetric survey. However, as given in [29], the availability of these devices is still relatively low, mostly due to high prices. Nevertheless, in some publications, this approach is presented as functional and in [30], it is presented as an important and developmental data source for coastal areas. In [31,32], it is said that although LIDAR is only feasible for relatively shallow and clear waters, due to the significance of such regions there are many airborne LIDAR systems specifically developed for bathymetry. In [33], based on LIDAR bathymetry measurements on two lakes in Poland, it was concluded that this sensor can be used for measurements in the littoral zone (up to 1.6 m), while above this depth, data can be acquired by hydroacoustic sensors. In the study case of rivers, the maximum obtained depth was 2 m [25]. Topobathymetric LIDAR, which can simultaneously survey land and water areas, is presented as a prospective solution in [34,35].

Satellite technology can also be used for bathymetric survey in coastal areas. Descriptions of various techniques (optical, radar, and laser) can be found in some publications. A fine survey on the technology is provided in [36]; the methods presented are interesting; however, they are mostly used for large areas on a world scale. Another non-obvious example is the method of indirect bathymetry calculation based on gravimetry measurements, given, for example, in [1,37].

Each of the methods above has its advantages and disadvantages and is suitable for various conditions and in different areas. Thus, in many studies, data from various sources are combined in a fusion process to obtain an integrated model. In some works, described, for example, in [30], the wider concept of topobathymetry is included. A recent example is given in [38] in the form of an updated methodology for the topobathymetry survey by United States Geological Survey (USGS). The core idea is to combine geospatial products derived from single sources in raster form. The measurements have to be preprocessed before they are combined based on prioritizing and filling gaps in data sets. In most of the research, data from acoustic measurements are combined with other data sources. In [39], aerial photos and multispectral images are combined with a digital elevation model (DEM) from echosounder measurements based on a pixel- and object-level fusion strategy. Another approach is used in [40], in which a lot of time was spent on preprocessing of data from various sources (echosounder, cartographic data, manual survey), which were then combined into one data set, for which a joint DEM was provided. In [41], a raster from echosounder data is combined with a model elaborated based on IKONOS images. In [42], geoswath bathymetry is combined with LIDAR as spatially complementary data sets, with the analysis of various interpolation methods. In [8], acoustic measurements are combined with data from a photogrammetric survey (SfM) by an unmanned aerial vehicle (UAV). The final product is an interpolated raster with a spatial resolution that depends on the point spacing from the sampling process. For achieving a smooth transition between SfM and echo sounding measurements, each bathymetric raster is classified by its corresponding range of high-precision scanning depths. A similar approach is also provided in [2] for combining data for a topobathymetric model. Another approach for combining acoustic and photogrammetric survey is given in [43], where a photogrammetric survey is used for determining the coastline, while the DEM itself is prepared based on single-beam acoustic measurements.

A complex approach to the integration of data from various sources is presented in [44], in which many databases and sources are used. The basis is the integration of LIDAR and hydroacoustic data, any gaps therein are supplemented by data from various databases, such as Electronic Navigational Chart (ENC), General Bathymetric Chart of the Oceans (GEBCO), and legacy systems. A common database is prepared, and a common DEM is created. As the input data for the model are characterized by high density, the binning minimum value (BMV) method is used for smoothing and denoising.

Unmanned vehicles are being increasingly used in many survey techniques. The main reason for and advantage of this is reducing the workload of surveyors. UAVs as the platform for photogrammetric equipment have been used, for example, in [2,6,14,45] and also for rivers, for example, in [20,46,47], while a survey with LIDAR mounted on a UAV is presented, for example, in [29]. A review of such an application can be found in [48]. An interesting approach is also proposed in [8,49], where a bathymetric survey was performed with an echosounder towed by a low-level UAV. Various echosounders mounted directly on unmanned surface vehicles (USVs) are presented in research given in [50]. In works [2,6,51], the use of UAVs for photogrammetric measurements and USVs for hydroacoustic measurements has been presented. A novel and interesting approach for planning and performing an integrated mission by UAVs and USVs is explained in [52].

The method of combining data is another component in the development of an integrated digital bathymetric model, apart from measurement techniques. In most works, the areas for various data are defined. Usually, some kind of priority is proposed for them, indicating the most reliable source (echosounder in most cases), and then the gaps are filled by data from other sources. Such an approach has been used, for example, in [8,40,42,44]. The main problem with this approach is the ambiguity in the neighboring areas, where models of different qualities (resolution and accuracy of measurement) are fused. Other problems are the necessity of cutting of overlapping areas and the ambiguity of point classification. For example, in [8], for achieving a smooth transition between SfM and echo sounding measurements, each bathymetric raster is classified by its corresponding range of high-precision scanning depths. In some works, after the identification of areas jointly covered, a combination of input data in them is proposed. This approach is used, for example, in [34] for LIDAR data or in [2] for photogrammetric data. The main issue then is the way of combining data of different types. The works involving this approach are in minority, showing research potential in this area. A joined DEM can then be elaborated as one model for integrated input data set [44] or as the integration of a few raster surface models [38].

The development of a digital bathymetric model also involves choosing the method for interpolation, apart from the input data set. In [38], the geostatistical empirical Bayesian kriging method is proposed for this purpose. Kriging has also been used in [6,40]. Triangulation, after local smoothing, is used in [44], while the nearest neighbor method is proposed in [8]. In [2], three methods are analyzed, inverse distance weighted (IDW), kriging, and natural neighbor, and IDW is indicated to give the best results. In [53], the proposal of modification dedicated to bathymetric modeling of IDW is proposed, based on bathymetric data, and in [54], the need for big data set reduction prior to interpolation is emphasized. As seen, a variety of methods can be used for interpolation in modeling, with satisfactory results.

Most of the research focusing on issues with modeling a shallow water surface refers to sea coastal areas, leaving a void in inland areas. Additionally, modern methods should use unmanned vehicles as the most effective survey solution and data should be integrated from various sources. The most reliable is usually derived from hydroacoustic measurements, while the most popular sensor for UAVs is a multispectral camera. In this case, the SfM methodology is usually used, allowing the most effective modeling. Data sets derived from various sources should be integrated taking into account their quality, and then the final digital surface model should be developed. Following these conclusions from literature analysis, we decided to elaborate the method of developing an integrated digital bathymetric model for shallow waters, based on hydroacoustic measurements done with a USV and photogrammetric measurements made with a UAV. In both large-hydrographic-vessel and USV surveys, there is always some part of the area inaccessible for an acoustic survey. As shown in [45], the development of the bathymetric surface in areas with no data, based on interpolation techniques, has low effectiveness and has much bigger errors compared with the surface modeled with measured data. The results from this work were the basis for performing the research on the use of UAVs to gather the data remotely in areas inaccessible for survey surface vehicles and to integrate these data with USV measurements.

The main contribution of this study is the development of a method to combine bathymetric data acquired by hydroacoustic and optical sensors to create a digital bathymetric model for shallow water areas up to the shoreline boundary. Depending on the type of water body, these two methods have limitations related to the acquisition of bathymetric data in shallow water areas. In the case of USVs, the main restrictions are the vehicle's draught, maneuvering capabilities in close vicinity to shoreline, submerged aquatic vegetation, and other dangerous objects (fishing nets, trees, etc.). An identified problem for UAVs is the limited possibility of using optical sensors in bathymetric measurements, which is caused by factors such as depth, water transparency, and others previously mentioned. The main research problem related to the use of photogrammetric data is the answer to the question, to what depth can these data be used to develop a digital bathymetric model? For this purpose, we propose a solution based on the creation of a bathymetric reference surface, which was developed from data acquired by a single-beam echo sounder mounted on a USV. With respect to this surface, UAV measurement points were selected based on the proposed data geoprocessing method, which were then combined with data acquired by hydroacoustic sensors. The UAV bathymetric data were developed based on a photogrammetric method using an underwater control network. In our work, we assumed that it is possible to obtain accurate bathymetric data based on SfM and an underwater control network without taking refraction into account. The final product was qualitatively and quantitatively analyzed. Based on the results, we confirmed that the assumptions (bathymetric data generation using an underwater photogrammetric network, data fusion based on a bathymetric reference surface, and dataset processing based on developed masks) make it possible to create a digital bathymetric model from hybrid data using the proposed methodology. The method was widely assessed with multivariate analysis. Various combinations of input data sets and interpolation methods were considered in the research. The measurements themselves were performed in inland waters covering shallow and ultra-shallow areas. Such a location is also unique in terms of analyzed works. However, the achieved results and the proposed method can be used for any coastal and shallow waters.

2. Materials and Methods

2.1. Study Area

The study area is located within the village of Czarna Łąka (Poland), situated in Western Pomeranian Voivodeship, Goleniów District (Figure 1). The study area is a part of Dąbie Lake, which is a water body with an average depth of 2.61 m [55]. It includes a small bay with the Bystra beach. The water area from the northern and southern side is densely covered with aquatic vegetation and has a sandy beach on the eastern side. The area of the investigated water covers 0.0271 km^2 (2.71 ha). The bottom of the water area is flat for the most part, with a larger depression at the entrance to the bay. The average depth of the water area is about 1 m, and the maximum depth is 3.95 m. During the survey, the sky was slightly overcast; during the UAV flight, the sun was shining. The water surface was slightly wavy. The measurement campaign took place on 4 August 2021, between 7:30 a.m. and 1:00 p.m. UTC.

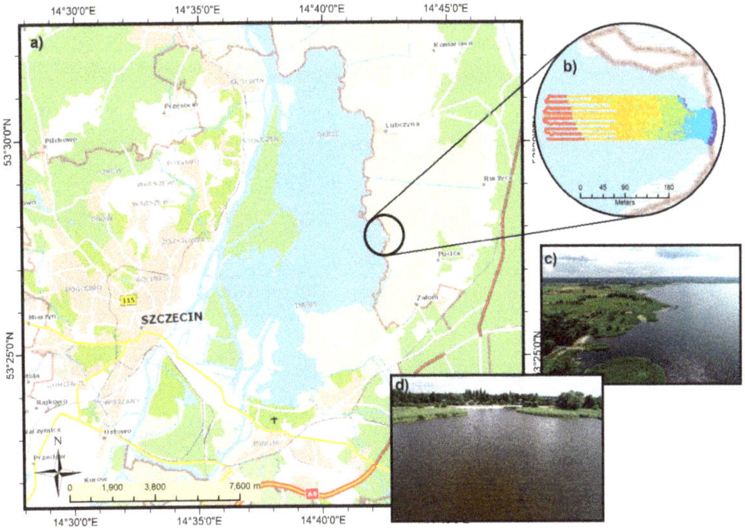

Figure 1. Study area: (**a**) location of Dąbie Lake, (**b**) extent of bathymetric model development, (**c**) top view of the study area, and (**d**) view of the bay with a beach.

2.2. UAV Photogrammetric Data Acquisition

The photogrammetric data were acquired using an unmanned aerial system, a DJI Phantom 4 Pro quadrocopter. One of the advantages of this system is a camera integrated with a three-axis gimbal that allows recording of camera angles while taking pictures. The camera has a CMOS sensor that can take 20-megapixel images. The focal length of the camera is 8.8 mm (35 mm equivalent: 24 mm), and the field of view of the camera is 84 degrees. Images are saved in JPEG format on a microSD card. The system allows autonomous missions to be performed using dedicated software. The flight was taken from an altitude of 120 m using a basic ultraviolet (UV) filter. The UAV uses a dual positioning system: Global Positioning System and Global Navigation Satellite System (GPS-GLONASS); the positioning accuracy specified by the manufacturer for the hover is ± 0.5 m vertically and ± 1.5 m horizontally [56].

Before the flight, a temporary, signalized photogrammetric network was stabilized, consisting of 10 ground control points (GCPs): 6 on land and 4 underwater. Two types of survey GCPs were used: round white discs 27 cm in diameter (fixed with a centered surveyor's nail) to stabilize the land survey network and orange-black survey discs 91 cm in diameter as an underwater network, stabilized with surveying pins to the lake bottom. The distribution of the network was designed during an investigation based on a field inspection. The network points were distributed evenly and on the boundaries of the survey area. The photogrammetric network was measured with a Sokkia GRX-1 receiver operating in the real-time kinematic (RTK) positioning mode.

In addition to the photogrammetric network, check points (CPs) were measured with an RTK receiver and used for height control of the DTM and the DBM. The check points for the land area were distributed evenly over the survey area, where 29 check points were measured. The underwater control points were measured as an underwater profile, where the distance of each successive point was approximately 8–9 m from the previous point; 23 such points were measured. During the flight, which lasted 18 min, 439 images in 12 stripes were acquired for an area with a coverage of 0.3207 km^2. Overlap and sidelap image coverage was 85%.

Figure 2 illustrates the locations of the underwater and land GCPs, check points, USV soundings, the developed UAV orthomosaic, and the study area. In addition, a photograph of the underwater GCP is included.

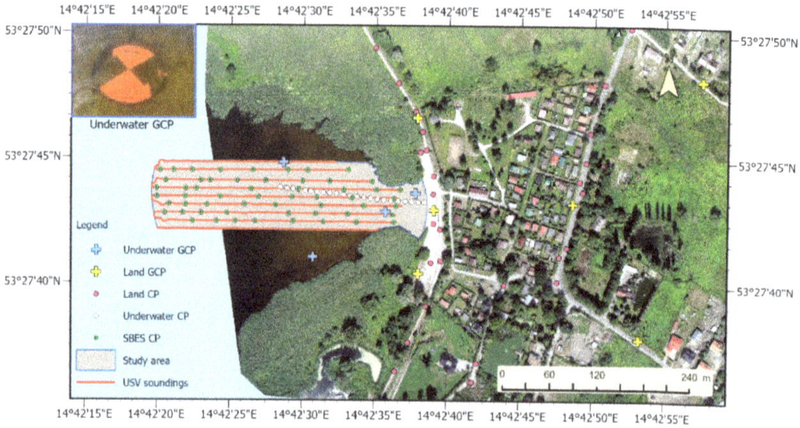

Figure 2. Map of the distribution of underwater and land-based GCPs/CPs and the USV soundings in the background of the generated UAV orthomosaic; in the upper-left corner is a photo of one underwater GCP used in data elaboration.

2.3. USV Hydroacoustic Data Acquisition

USVs are remotely operated floating platforms usually designed for surveys in extremely shallow water areas [57,58], where the use of standard survey boats is not viable or is risky. The Gerris ASV used within this study is dedicated to single-beam echosounder surveys. It can operate in manual and autonomous control mode on designed survey lines.

The vehicle is equipped with two electric motors driving the vehicle and a wireless radio link for control and telemetry data. The measurement system based on the Echologger EU400 single-beam echosounder and the Emlid Reach M2 RTK positioning receiver is integrated in the USV platform industrial computer. HYPACK 2021 hydrographic software is responsible for the integration of bathymetric and position data.

The echosounder transducer is mounted to the pole and connected to the industrial computer. The echosounder is equipped with an inertial measurement unit (IMU) for platform motion correction [59]. The most important parameters of the echosounder are a high acoustic frequency and a small acoustic beam size, which affects the precision of depth measurements in shallow water areas. The GNSS receiver antenna is mounted on top of the echosounder pole. RTK corrections are provided by the Internet module via Global System for Mobile Communications (GSM). Basic parameters of the measurement system used are shown in Table 1. Picture of USV used in the study is in Figure 3.

Table 1. USV survey system parameters.

Single-Beam Echosounder		Unmanned Survey Vehicle		Positioning System	
Frequency	450 kHz	Length	1200 mm	Static accuracy	Static horizontal 4 mm + 0.5 ppm Static vertical 8 mm + 1 ppm
Beam width	5° Conical (−3 dB)	Width	1000 mm	Kinematic accuracy	Kinematic vertical 14 mm + 1 ppm Kinematic horizontal 7 mm + 1 ppm
Transmit pulse width	10~200 µs (10 µs step)	Height	360 mm (without mast)	Signals tracked simultaneously	GPS/QZSS, GLONASS, BeiDou, Galileo
Ranges	0.15~100 m	Survey speed	1.2 m/s		
Data output format	ASCII TXT, NMEA0183	Thrusters	2× T200 Blue Robotics		

Figure 3. Unmanned Survey Vehicle.

2.4. Research Methodology

The research concerns full bathymetric imaging, up to the shoreline boundary, including shallow and ultra-shallow depths. Considering the operational capabilities of a USV, its range is limited by the depth of the body of water, which in the case of shallow waters, often occurring in coastal areas, makes a complete survey impossible.

In practice, other barriers, often natural, limiting the hydrographic survey can also be encountered. These may include partially submerged vegetation that creates barriers preventing a full survey, underwater obstacles, and floating vegetation on the water surface. Bathymetric data for an area inaccessible to USVs can be acquired using UAVs. Although there is some overlap between the two methods, UAV data are more heterogeneous and correct only for a certain depth, below which the values are generated with increasing errors. At a certain depth threshold value, bathymetric data cannot be acquired using the photogrammetric method due to attenuation of the electromagnetic radiation, which dissipates completely in the water, preventing it from reaching the bottom and reflecting. The magnitude of this value will be variable for a given body of water and will depend on many factors, such as water transparency, surface ripples, and bottom type and structure [60,61].

An identified problem related to the development of a unified bathymetric model is the correct integration of the data. For this purpose, research has been carried out in this paper, resulting in the development of a method to conduct this process. The main idea of the method is to create a bathymetric reference surface (BRS) from echo sounder measurements, on the basis of which, UAV cloud points are selected within an established tolerance (maximum vertical deviation from the reference surface). Such a surface was created using the triangulated irregular network (TIN) method. This surface was created to carry out multi-variate studies to create a proper UAV data set for the development of the final digital bathymetric model. The process of creating data sets from the UAV point cloud is described in Section 2.9.

In the conducted research, a UAV point cloud (created in PIX4D Mapper) developed from all classes (PCA) and a point cloud with a ground class (PCG) were tested. Three experimental surfaces (ES) were created from each cloud in raster form. The surfaces were created using statistical map algebra operations based on individual raster cells [62]. Each surface was created in a different way, based on the mean, the highest, and the lowest height values of the points lying within each pixel boundary. These surfaces were further used to create masks to select points in the UAV cloud based on the maximum allowable deviation from the reference surface, which we call tolerance. A mask is a polygon covering an area of the raster where the deviations from the BRS surface are contained within the tolerance limits. Four types of masks were included in the research process.

In the next stage of the research, four sets of UAV measurement points were prepared for the PCA and PCG clouds. Masks were used to create these sets, on the basis of which point selections were made in the point clouds. The result was the creation of four test sets for the PCA cloud and an analogous four sets for the PCG cloud. After analyzing the height range and finding significant deviations above the surface, additional filtering was performed on points located at heights above 0.25 m. Positive depth values were used in the final model mainly to preserve points that, due to errors in the development of the numerical bathymetric model, could be above the water surface in areas of ultra-shallow depth, on the order of a few centimeters.

The data sets prepared in this way were combined with the USV data in a single LAS format file. In the next step, digital surfaces were created for each data set using different interpolation methods. These surfaces allowed a qualitative and quantitative assessment of the created bathymetric surfaces, on the basis of which it was possible to select dedicated UAV point cloud types for merging with USV data, masks for processing point clouds, and interpolation methods in terms of creating a digital bathymetric model. The final quantitative and qualitative analysis formed the basis for discussion and conclusions. Next, the research is described in detail according to the research process adopted in the study, illustrated in Figure 4.

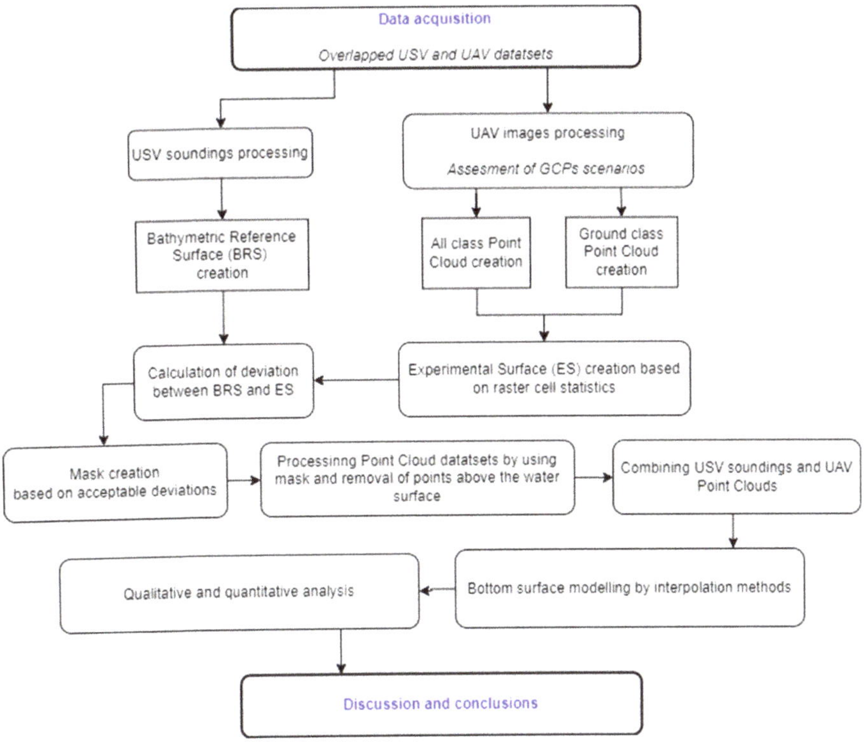

Figure 4. The research process adopted in this study.

2.5. USV Data Processing

Data acquisition was performed in the Hypack 2021 hydrographic software installed on the onboard computer. The operator, through a remote desktop and a Wi-Fi network, had access to all parameters and data recording status in near real time. Planning and viewing of platform telemetry parameters were performed through radio communications and Mission Planner software installed on the operator's computer. On a background chart, the operator determined the planned route by plotting runlines. In the next stage, data acquisition, the operator supervised the progress of the mission.

All necessary corrections were applied to the single-beam echosounder system to acquire reliable bathymetric data before data recording began. Vertical and horizontal offsets were entered relative to the platform's center reference point (CRP). The draft of the echosounder transducer was taken into account, and the average value of the speed of sound in water was included. The correctness of the entered parameters was confirmed by performing a calibration of the echosounder using the bar check method [63]. Positioning of the platform was performed with the RTK system. For data acquisition, depth data and raw echogram data were recorded. The recorded data set from the single-beam echosounder system was processed in the SBMAX64 module of the Hypack software. First, the data from the GNSS positioning system were analyzed after data import. The trajectory of the moving surveying platform in places of temporary loss of delivered RTK corrections was corrected.

In the second step of bathymetric data analysis, manual analysis of the echograms was performed against the indications of the bottom tracking algorithm [64] of the echosounder (Figure 5), which is based on the amplitude of the reflected signal. Due to the heavily vegetated bottom, especially in the nearshore zone, the study required manual digitization of the echogram. The processed data set was vertically referenced to the nearest water-level gauge, Most Długi Szczecin (Długi Bridge in Szczecin). An offset for depth was implemented using the water-level gauge zero as the chart datum.

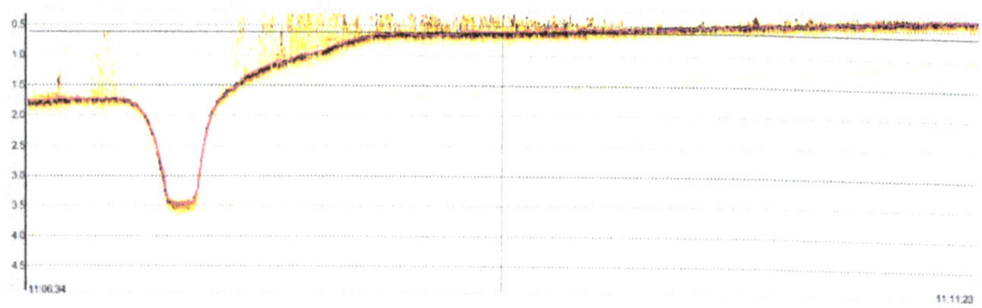

Figure 5. Example of an echogram for a single-beam echosounder with a manually digitized depth (purple line) for a single profile (own elaboration based on conducted measurements).

2.6. UAV Data Processing

The data acquired from the UAV platform were processed in PIX4D Mapper software (v. 4.4.12), designed for low-altitude data processing. The project was georeferenced by measuring the GCPs signalized in the field before the aerial flight and then measured on the images. A dense point cloud was created from the photographs, which was the input product for the development of the DTM. Point cloud classification was also performed in PIX4D Mapper software to separate ground points from the full point cloud. The classification is performed based on the algorithm proposed by [65], which uses geometric and color features for the classification process, allowing each point in the point cloud to be assigned to one of six predefined classes: unclassified, ground, road surface, high vegetation, buildings, and human made object. In the software, it is not possible to select the parameters for the point cloud classification; it is only possible to choose whether the classification process should be performed (this is recommended by the developer if one of the products is to be a DTM) or whether the cloud should be left unclassified in order to build a digital surface model (DSM).

When determining the bathymetry of a water body using UAVs, a land-based survey network is usually set up [66–68]. Other approaches are also reported, e.g., [69] land-based GCPs are used and two buoys are placed on water for the study. Another approach is to use both land and underwater GCPs [70]. The authors of [70], while conducting surveys in the coastal waters of one of the Maldives islands, pointed out that the best results can be obtained using an underwater network without refraction corrections. In the present study, we used this option, taking into account the underwater photogrammetric network. Our case can be regarded as an extension of this type of research in another type of water body. The study area included inland waters with lower transparency (darker color). Another method of assessing the accuracy of the elevation models was used, measuring the CPs' coordinates along the underwater profile.

To assess the applicability of a photogrammetric underwater network, the process of data alignment was carried out according to two independent measurement scenarios. The first measurement scenario assumed using only a land control network (6 land GCPs) for data alignment. The second scenario assumed using both land and underwater network points (10 GCPs) for data alignment. Table 2 summarizes the results for the data georeferencing process using the two measurement scenarios described. Obtained RMSE errors for X- and Y-coordinates have values below or around a centimeter for both scenarios. The Z-coordinate for the method using land GCPs gives a smaller error, about 3 cm, while the Z error (RMS) for the second scenario gives an error of about 6 cm. In addition, errors for underwater GCPs are included, which are similar for scenario 2.

Table 2. GCP errors for two scenarios of photogrammetric data acquisition.

GCP Type	RMS Errors [m]
Land GCP	X = 0.004, Y = 0.005, Z = 0.031
Land and underwater GCP	X = 0.007, Y = 0.012, Z = 0.059
Underwater GCP	X = 0.005, Y = 0.013, Z = 0.057

The final products obtained in the form of DTMs were checked for accuracy (height deviation) on the basis of the bottom points measured on the profile with an RTK receiver (Figure 2). Figure 6 summarizes the results showing the height differences on the profile obtained between these points and the height points measured on the DTM.

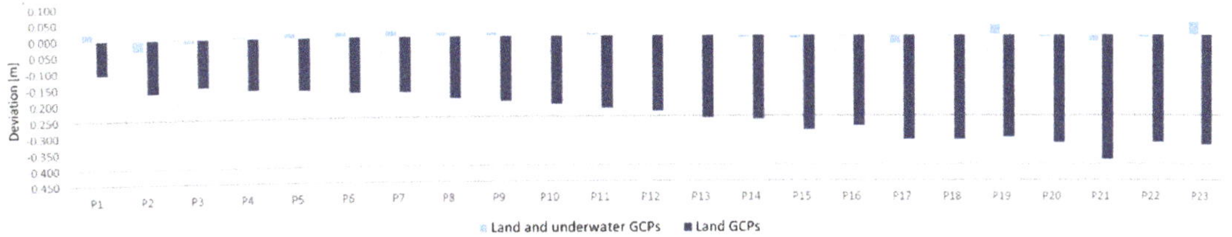

Figure 6. Height differences at the check points obtained from DTMs for land and underwater GCPs (light blue) and land GCPs (dark blue). P1–P23 means the points along the profile.

The minimum deviation, the maximum deviation, and the mean deviation for the survey with the land and underwater control networks were 0.00, 0.04, and 0.01 m, respectively. In a study for land GCPs only, the values were 0.11, 0.39, and 0.24 m, respectively. It should be noted that as the depth increases, the deviations for the two cases tend to increase. As an additional DTM control, an altitude check of the models on the land area was performed. The verification was performed on 29 independent control points. The minimum deviation was 0.00. The maximum deviation and the mean deviation for the study with underwater network were 0.45 and 0.05 m, respectively. For the survey without an underwater network, the values were 0.003, 0.23, and 0.04 m, respectively. Analysis of the height control points confirms the high accuracy of the developed products not only in the underwater area but also in the land area. On the basis of the above results, it can be stated that determining bathymetry by the photogrammetric method with an underwater control network without taking into account refraction correction gives good results.

2.7. Characteristics of Data Sets

In the study, it was assumed that the surface reconstruction would be carried out in the final stage using interpolation techniques. The analysis of spatial distribution and density of data is important in this case, as it may affect the accuracy of the modeled surfaces. It is especially important in the case of interpolation methods with parameters usually adjusted to the measurement set of similar or the same spatial distribution. In our case, we have a data set with varying spatial distribution and density. In the part of the body of water accessible to the USV platform, data were acquired on planned parallel profiles 10 m apart. On the profile, the data were spaced every 30–40 cm. Such a distribution can be considered as regular with a high density of equally spaced data. The situation was different for the UAV data. The data had a scattered spatial distribution with a high density per square meter, ranging from 1 to over 300 points. It should be noted here that this density was variable over the UAV data coverage area, which was related to the processing of the full point cloud. This resulted in two point clouds, the PCA cloud and the PCG cloud, with different densities. The second factor of density change was the depth, which was related to the possibility of data acquisition by the photogrammetric method. With its increase, the density of points decreased. The densities of the data are shown in Figure 7, while the spatial distributions are shown in Figure 8.

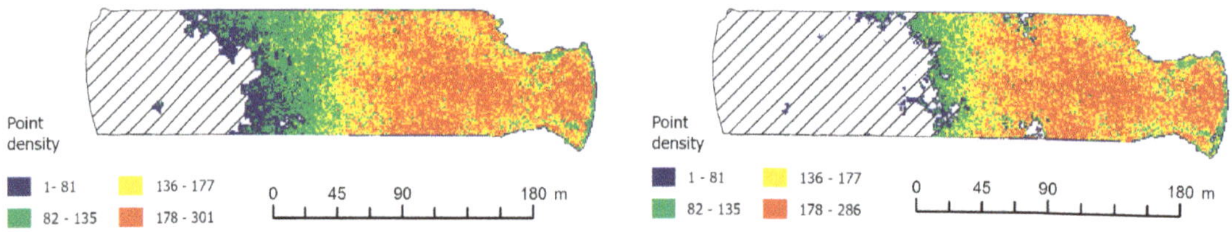

Figure 7. Density of point cloud per 1 m^2 for the PCA cloud (**left**) and the PCG cloud (**right**).

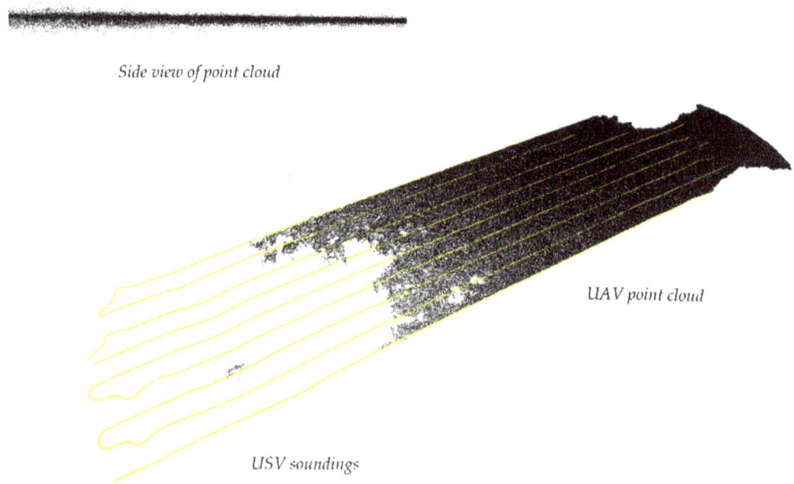

Figure 8. 3D view of a UAV point cloud (black) and USV soundings (yellow).

Analysis of the vertical characteristics of the data is also important, especially in terms of creating a bathymetric surface. Analyzing the data sets, it was found that USV data after processing do not have local extremes or major deviations. In the case of UAV data, the situation is different. These data are characterized by height variation, creating a layer of points. A comparison between USV and UAV data is illustrated in Figure 8. The high roughness of the point cloud may indicate the presence of submerged vegetation, bottom microtopography, or imprecise matching of the images. It is also important to consider the way the point cloud is created, which is based on the collinearity equation and epipolar geometry, where it is important to find homologous points in at least two images. When mapping shallow water areas, it is important to take measurements under favorable conditions, such as clear water, a calm water surface, and visible bottom texture [71], which should facilitate the acquisition of reliable height information from the point cloud. However, given the frequent variability of weather conditions during measurements and the diversity of water bodies, it is difficult to assess which parameters influence the height determination of the measurement points.

2.8. Creation of a Bathymetric Reference Surface

To carry out the surveys, a reference surface was created from the USV data, against which the deviations were calculated in subsequent steps, the motive being that bathymetric data generated from photogrammetric material have greater errors with increasing water depth, until they disappear. The second factor is that they are quite noisy. Therefore, in this case, it is difficult to determine to what depth they are generated correctly and what depth value should be taken as a reference. In the case of USV data, such a problem does not exist as the measurement is performed using hydroacoustic methods, which are recognized and proven measurement techniques. Additional recording of the echogram enables accurate processing of the depth data and removal of noise or measurement errors [72]. It also has the advantage of recording data on profiles, which facilitates the final data processing.

The BRS surface was created in the USV measurement point domain using the TIN method, in a GRID structure with a 0.5 m resolution. The Delaunay algorithm was used in the creation of the TIN mesh. Delaunay triangulation has the following properties: in the area of the circle described on any points of an elementary triangle, there is no other point from the whole set of measurements; it maximizes the smallest angles of elementary triangles, which makes their shape the best fit to the set of depth points [73]. The reference surface from the USV data and the surface from the UAV data are illustrated in Figure 9. As can be seen, the TIN surface is characterized by smoothness and continuity of the bathymetric data, while the surface created from the UAV cloud has a rough structure and generation is not possible at greater depths.

Figure 9. BRS surface developed by the TIN method from USV data in the sounding domain (**a**) and the corresponding surface developed from UAV data (**b**). View within the limits of the complete development of the digital bathymetric model.

A characteristic of the bathymetric surface developed with the UAV is also that in deeper waters, the amount of erroneous data increases significantly. In Figure 10, large deviations, both positive and negative, can be observed at these locations, ranging from −2.79 to 3.21 m.

Figure 10. Deviation of depth values between photogrammetric surface and BRS.

2.9. UAV Point Cloud Processing

As mentioned earlier, low-altitude photogrammetry methods are characterized by limited possibilities for obtaining bathymetric data. Analyzing our case, we can state that the range of larger discrepancies starts from a depth of 0.7 to 1 m and the approximate limit of depth generation is 1.3 m. The purpose of this stage of data processing is to remove points from the data set that do not fall within the assumed threshold value (vertical tolerance). In this study, the threshold value was assumed at the uncertainty level for the special category area according to the IHO regulations published in Publication No. 44 [74], which for depths of 4 m and less is 0.25 m. In the final stage of the study, eight data sets were prepared, which differed in the way that the points were removed using developed masks based on the assumed tolerance. The subsequent stages of point cloud processing are presented below.

Stage 1: Creation of experimental bathymetric surfaces (ES)

These surfaces were created in raster form with a resolution of 0.5 m. Rules for aggregation of points in the resulting raster cell were based successively on three statistical operations: arithmetic mean value (M), highest value (H), and lowest value (L). Surface generation was performed for both PCA and PCG clouds. Six bathymetric models were developed, denoted successively for the PCA cloud as ES(A) M, ES(A) H, and ES(A) L and for the PCG cloud as ES(G) M, ES(G) H, and ES(G) L. The data processing is illustrated in Figure 11.

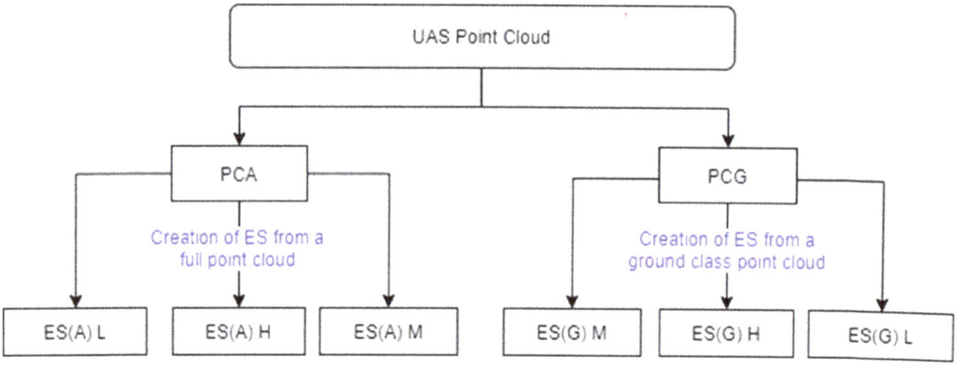

Figure 11. Creation of an ES surface based on the aggregation of values in the resulting raster cell based on the statistics M (mean value), H (highest value), and L (lowest value).

Stage 2: Calculation of deviations from the BRS

The operation was carried out using raster algebra tools, subtracting successive rasters of experimental surfaces with M, H, and L values from the cell values of the reference surface raster. The result of this operation was a raster with deviation values. The task was carried out using ArcGIS Pro software.

Stage 3: Creation of masks

Rasters with deviation values were the basis for determining the set of pixels with acceptable deviation values. These ranges were determined using the quantization method, setting two threshold levels for raster pixels: +0.25 and −0.25 m. The determined raster area based on pixels with acceptable deviations was used to create masks with M, H, and L values. One more mask type, HL, was included in the study, which was a common part of H and L masks. The motivation for its creation was that the resulting pixel in this case had to meet a threshold condition for both maximum and minimum values. In the final stage, four masks were created based on ES(A) surfaces, MASK(A) M, MASK(A) H, MASK(A) L, and MASK(A) HL, and four masks based on ES(G) surfaces, MASK(G) M, MASK(G) H, MASK(G) L, and MASK(G) HL.

Stage 4: Processing of point clouds using masks

The PCA and PCG clouds were processed using masks, which finally enabled the preparation of eight sets of UAV points: PCA M, PCA H, PCA L, PCA HL, PCG M, PCG H, PCG L, and PCG HL. The points contained by the corresponding mask were preserved in the geodata sets. This process is illustrated in Figure 12.

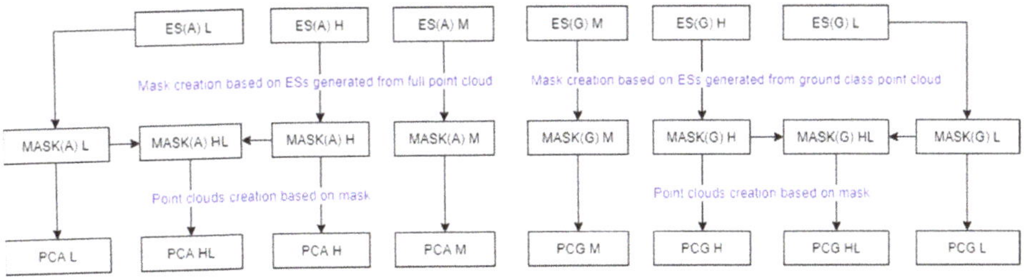

Figure 12. Scheme for creating masks from ES(A) and ES(G) surfaces and processing point clouds using them.

The masks and the processed sets of PCA and PCG points are illustrated in Figure 13. As observed, a smaller depth spread was obtained for the PCG cloud, which is related to the creation of ES from ground class points only.

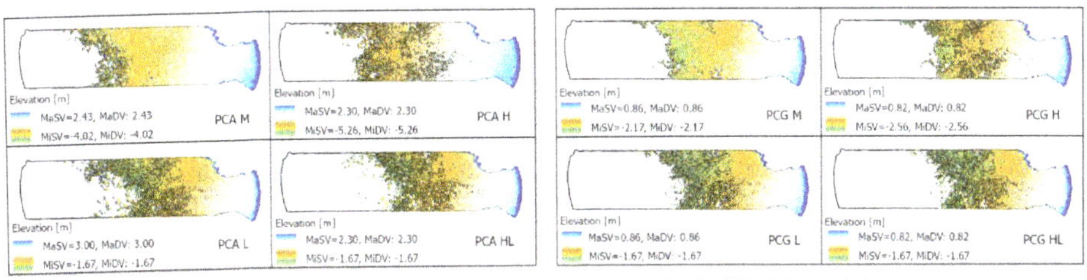

Note: MaSV- Maximum Scale Value, MiSV- Minimum Scale Value, MaDV - Maximum Data Value, MiDV - Minimum Data Value

Figure 13. Range of masks and processed point clouds (color scale represents individual 10 elevation classes based on Natural Breaks (Jenks) method).

Stage 5: Filtration of points

The next step was to analyze points above the water level. The height of the water level was determined by referencing to a water gauge (a hydrographic reference was used). For photogrammetric surveys, water surfaces generate noise in the form of points above their surface. These were reduced in the proposed data processing. The threshold filtering value for such points was set at a predetermined tolerance of 0.25 m, the aim being to preserve points that are part of the bathymetric surface but may be above the water surface due to a model elaboration error in the z-axis. This approach allows the continuity of the model to be maintained without removing points above the water surface in ultra-shallow parts of the water area. In addition, a higher number of such points were observed near the shoreline. The reason for this is, in most cases, the problem of spectral separation of the water boundary and the presence of organic objects (e.g., sediments of dead vegetation) that accumulate there due to the wave impact. A schematic of this data processing step is shown in Figure 14. The prefix F in front of the point cloud names denotes the data set with the points above the water surface removed according to the assumed criteria.

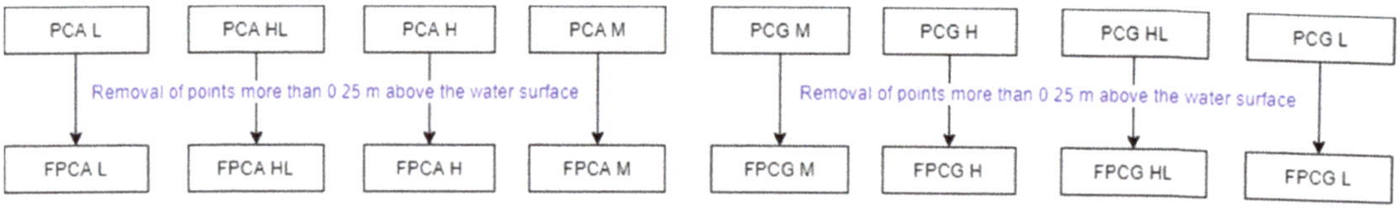

Figure 14. Schematic of the process of removing points more than 0.25 m above the water surface.

Figure 15 illustrates the spatial distribution of points above the water surface. As can be seen, the mask created from the mean values, MASK(A) M, leaves most such points. Significantly fewer points of this type remain for PCG clouds. An interesting observation is that for two types of point clouds, PCA and PCG, masks of H and HL type do not leave such points in the BRS surface domain. Quite a number of such points, however, can be observed in a narrow strip of the shoreline, although in most cases, their deviations from the water surface do not exceed 3 cm.

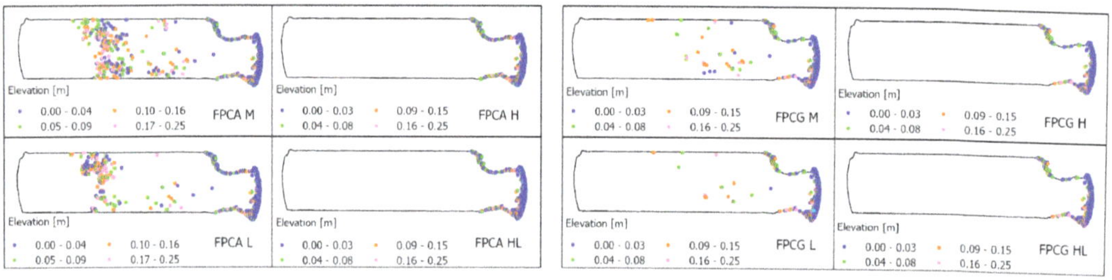

Figure 15. Distribution of points with height above zero for FPCA clouds (**left**) and FPCG clouds (**right**).

Stage 6: Merging data in a file

To create a single file integrating data from the two sensors, the USV data were converted to LAS format and merged with the UAV data set in Global Mapper software (v.21). The input photogrammetric data sets were FPCA and FPCG point clouds, into which the USV bathymetric data were imported. The final result was eight combined sets (CSPCA and CSPCG), which are the material for creating digital surface models by interpolation methods. The scheme for merging the data sets is illustrated in Figure 16, while the final data sets are shown in Figure 17.

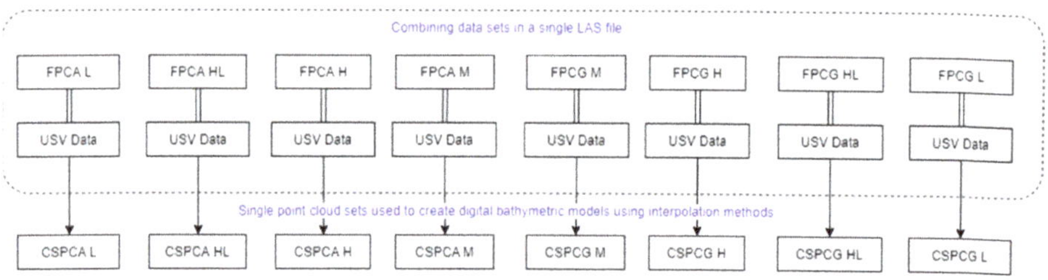

Figure 16. Scheme for the integration of data sets.

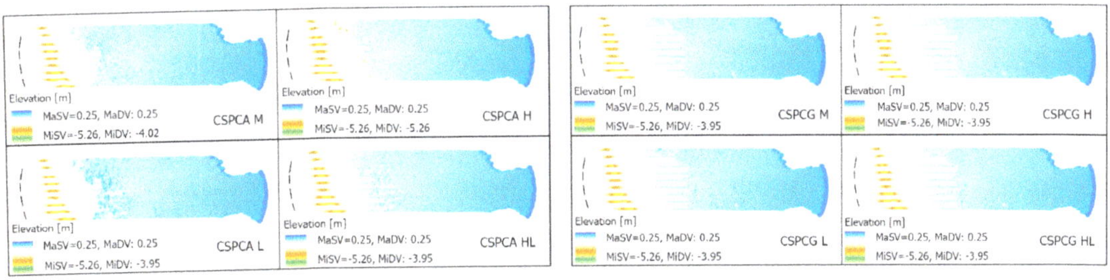

Note: MaSV- Maximum Scale Value, MiSV- Minimum Scale Value, MaDV - Maximum Data Value, MiDV - Minimum Data Value

Figure 17. Integrated USV and UAV data (for better profile visualization, larger point symbols were used).

2.10. Creation of Digital Bathymetric Models Using Interpolation Methods

The combined sets of USVs and UAVs data were used to model the surfaces using different interpolation methods. Five interpolation methods were tested: triangulation (TRI) [75], natural neighbor (NAN) [76], inverse distance to a power [77], kriging (KRI) [78], and radial basis function (RBF) method [79]. Considering the complex spatial structure of the geodata, consisting of regular survey profiles and a point cloud with a high density and scattered distribution, this case was also included in the research. The first two of the mentioned methods are parameter-free methods, while the others require them. From the operator's point of view, the parameter-free methods are certainly a better choice, as they do not require knowledge of the influence of parameters on the final shape of the modeled surface. This advantage can also be considered as a disadvantage because parameters usually allow for a better fit of the model to the real surface. In contrast, methods with parameters require the knowledge of parameters and an assessment of their influence on the final shape of the model. Due to the multivariate nature of the research carried out in this study, the influence of the values of various parameters of such methods, such as kriging, IDW, and RBF, was not analyzed, but the default values were used. The study of these parameters can be a further development of the research undertaken in this paper. The method parameters used were as follows: IDW power = 2; KRI, linear semi-variogram model, point kriging type, polynomial drift order = 0; and RBF, kernel type multiquadric, shape factor (R2) calculated according the formula (length of diagonal of the data extent)2/(25 × number of data points). IDW, KRI, and RBF methods used a four-sector search, 12 points per sector, resulting in interpolations from a total of 48 measurement points. The four-sector search divides the GRID mesh node space into four equal sectors with an opening angle of 90°, which makes it possible to select samples for interpolation evenly distributed around it. The usefulness of this way of searching samples was indicated in the work [45]. Modeling was carried out in Surfer 20 software. Figure 18 illustrates the basic steps involved in processing USV and UAV data.

Figure 18. Visualization of subsequent stages of USV and UAV data processing: (**a**) USV survey and UAV PCG point cloud, (**b**) combined USV and UAV data: CSPCG HL, and (**c**) digital model of the bathymetric surface created by the kriging method.

3. Results

3.1. Qualitative Analysis

To perform a qualitative analysis considering the influence of all completed operations on the final surface models, a scheme was developed and is presented in Steps 1–5.

1. Evaluate interpolation methods for their ability to filter out points above the water surface.
2. Select a data set (CSPCA or CSPCG) with better visual effect.
3. Select the best interpolation method(s) on the data set selected in Step 2.
4. Select a mask based on the conclusions of Steps 2 and 3.
5. Return to Step 1 and apply Steps 2, 3, and 4 to the second data set rejected in the first approach.

As interpolation methods are also considered as filtering methods, in a first step, the possibility of filtering points above the water surface was analyzed. Two cases were considered: the occurrence of positive depth values in the BRS surface domain and that in the shoreline area. In the case of the BRS surface domain, CSPCA points above the water were observed for mask M and L. After the modeling process, it was found that the natural neighborhood method and the IDW method coped with this problem for mask M. In the case of a cloud formed from a class of ground-type points (CSPCG), no points above the water surface were observed in the BRS domain area for all cases and methods except for the RBF method. In contrast, a few positive values were observed near the shoreline for both cases. The percentage of pixels with positive values is summarized in Table 3. As seen, with the exception of the RBF method, they represent a small percentage of the total area. For CSPCA sets, they range from 0.04% to 0.12%. For CSPCG, these values range from 0.04% to 0.07%. Considering this criterion, the use of the CSPCG cloud is preferable.

Table 3. Summary of the contribution of pixels with positive depth values in percentage (%).

Point Cloud Type	TIN				NAN				KRI				RBF				IDW			
	M	H	L	HL	M	H	L	HL	M	H	L	HL	M	H	L	HL	M	H	L	HL
CSPCA	0.08	0.06	0.11	0.06	0.06	0.06	0.08	0.06	0.09	0.07	0.12	0.07	0.88	0.75	1.06	0.91	0.04	0.04	0.05	0.04
CSPCG	0.06	0.06	0.06	0.06	0.05	0.05	0.05	0.07	0.07	0.07	0.07	0.07	0.62	0.71	0.79	0.86	0.04	0.04	0.04	0.04

For the CSPCA data set with regard to the choice of interpolation method, the surfaces obtained with the RBF method do not give satisfactory results, while the other interpolation methods tested (KRI, IDW, TRI, and NAN) allow one to obtain qualitatively similar models with correct plasticity, which we define as the visual reasonableness of the generated surface shape, without the occurrence of local distortions or other artefacts. The use of L and H masks results in rough surfaces, while the use of the HL mask results in the smoothest surfaces. Mask M provides slightly better results than masks L and H, but worse than mask HL. The use of this mask with the IDW method provides a model with reasonably good plasticity of the terrain with a small amount of roughness.

The CSPCG data set helps obtain surfaces with a largely uniform structure. Considering interpolation methods, the best surfaces are obtained for kriging, triangulation, and IDW methods. The models obtained with these methods are similar in terms of quality. Surfaces obtained using the NAN method produce correct models, but the terrain plasticity at the interface between USV and UAV data is characterized by low smoothness. The RBF method visually produces the roughest surfaces. Analyzing the use of masks, the best visual surfaces for each of the selected methods are given by the HL mask, although the M and L masks also help achieve a satisfactory final effect of the modeled surfaces.

When analyzing the correctness of the surface reconstruction in the area covered by USV measurements (profiles), the quality was comparable for all methods. In the case of the IDW method, a tendency toward a slight curvature of the surface along the profiles was observed. However, this method preserved the highest real depth, of 0.95 m, and the positive values of height had the lowest value (0.09 m).

In conclusion, the HL mask can be considered the most effective, as it allows the correct representation of surfaces derived from both CSPCA and CSPCG clouds. The visual evaluation also concluded that creating models using CSPCG point clouds reduces the extent of UAV points used while increasing the extent of USV points. Considering the criterion of minimizing the number of pixels above the water surface, the IDW method performs the best. Figures 19–23 show the final models obtained from the CSPCA and CSPCG data sets using the masks M, H, L, and HL. The minimum (min) and maximum (max) values were presented additionally. Figures 19–23, respectively, show the models for the methods triangulation, natural neighbor, inverse distances, kriging, and radial basis functions. In Figure 23, due to the large local extrema generated by the RBF method, the digital bathymetric model is in the brown color range.

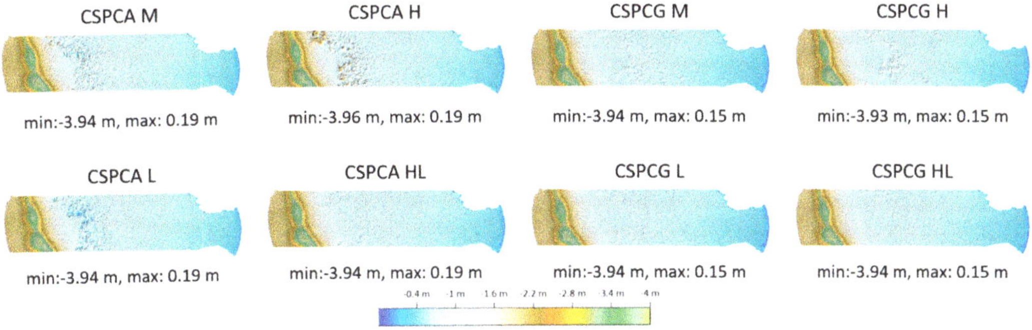

Figure 19. Models obtained by triangulation.

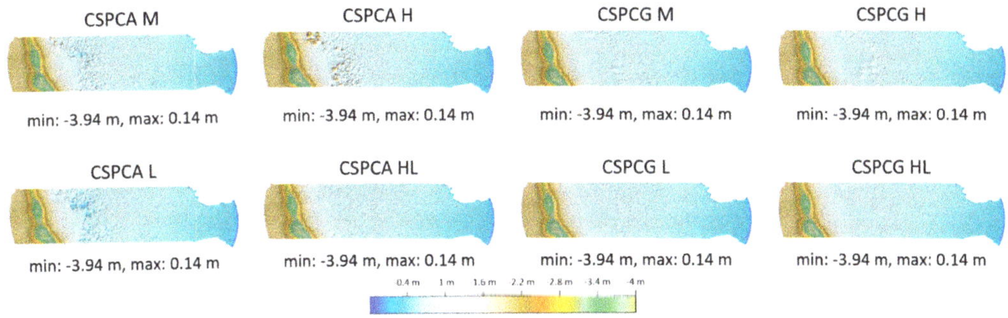

Figure 20. Models obtained using the natural neighbor method.

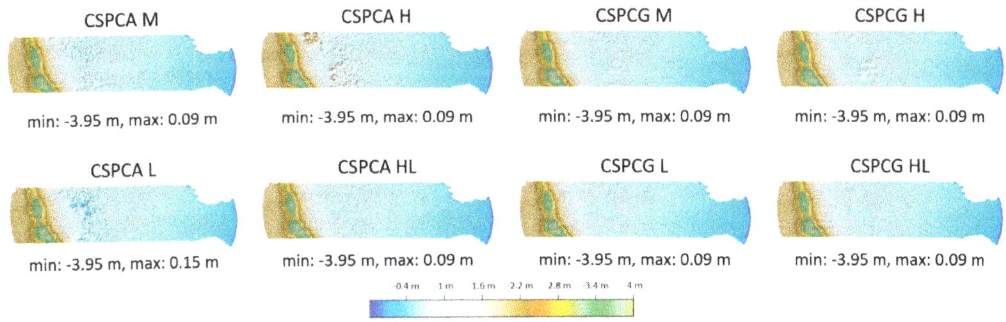

Figure 21. Models obtained by the inverse distance method.

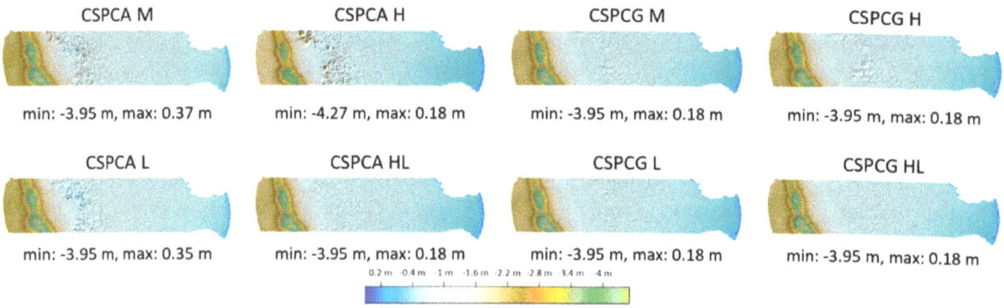

Figure 22. Models obtained by kriging.

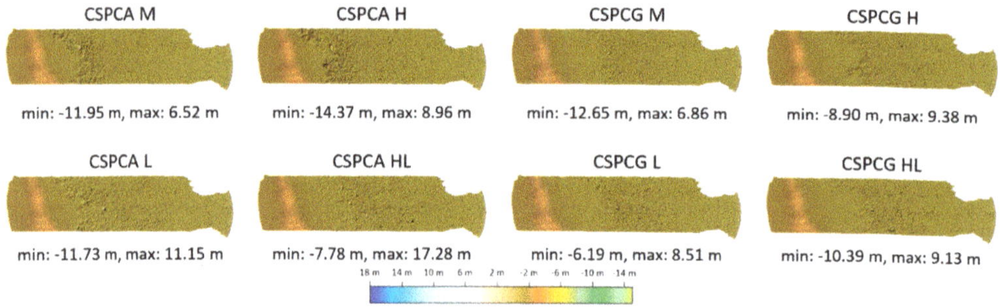

Figure 23. Models obtained by the RBF method.

3.2. Quantitative Analysis

Quantitative analysis was performed using points extracted from the USV survey data set, which included 56 points, distributed evenly across the test area. For the quantitative analysis, the methods chosen were IDW, KRI, NAN, and TRI, which allow correct surface reconstruction. For each generated surface from the test data set, the difference between the heights of the test points and the corresponding points located on the surfaces of the developed digital surface model was calculated. The maximum error (MaxE), the minimum error (MinE) and the mean error (ME) were calculated for each analyzed case according to Equation (1):

$$ME = \frac{\sum_{i=1}^{n}(Z_i - z_i)}{n} \quad (1)$$

where Z_i is the height measured at the CP_i point with coordinates (x_i, y_i), z_i is the height of the modeled surface at the point with coordinates (x_i, y_i), and n is the number of CP points.

To assess the precision of the tests carried out, the root mean square error (RMSE) was also calculated according to Equation (2):

$$RMSE = \sqrt{\frac{\sum_{i=1}^{n}(Z_i - z_i)^2}{n}} \quad (2)$$

The results are presented in Figure 24. Analyzing the MaxEs, the worst results are obtained for the model interpolated with the IDW method created on the CSPCA data set and the H mask (0.57 m). High MaxEs are also obtained for all interpolation methods for models created from CSPCA data set using H mask (0.48–0.57 m). Better results are obtained for models from this set for mask M, where the biggest error, for the IDW method, is 0.17 m and the lowest, for the NAN method, is 0.16 m. The use of the HL and L masks and the IDW, KRI, TRI, and NAN interpolation methods decreased the MaxEs, which are in the range of 0.09 to 0.12 m. The MaxEs for the CSPCG data set for KRI, IDW, TRI, and

NAN, regardless of the mask, in the range of 0.11 to 0.16 m. The smallest error for this data set is obtained for the KRI and TRI methods using the HL and L masks (0.11 m).

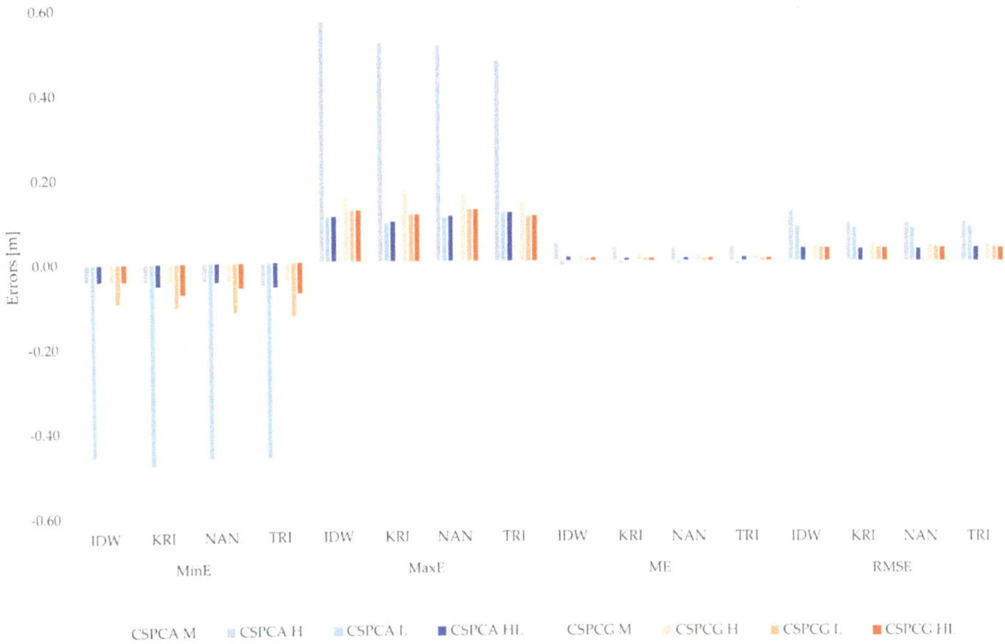

Figure 24. Summary of MinE, MaxE, ME, and RMSE for the analyzed tests.

Analyzing the MinEs, the worst results are obtained for the model interpolated with the KRI method created on the CSPCA data set and the L mask (−0.48 m). High MinEs are obtained for all interpolation methods for models created from CSPCA data set using L mask (−0.48–−0.46 m). Better results are obtained for models from this set for mask M, where the biggest error, for the KRI method, is −0.27 m and the lowest, for the IDW method, is −0.16 m. The use of the HL and H masks and the IDW, KRI, TRI, and NAN interpolation methods decreased the MinEs, which are in the range of −0.04 to −0.06 m. The MinEs obtained using the CSPCG data set for KRI, IDW, TRI, and NAN, regardless of the mask, are in the range of −0.04 to −0.13 m. The smallest error for this data set is obtained for the IDW, NAN, and TRI methods using the H mask and IDW using the HL masks (−0.04 m).

The MEs obtained for the CSPCA data set are comparable and are in the range from −0.01 to 0.04 m. For HL and L masks, their values are the smallest and for all methods and are in the range of 0.01 m to −0.01. The MEs for the CSPCG data set are smaller, ranging from 0.01 to 0.02 m, where the best masks for all methods are L and HL (0.01 m).

The value of the RMSE is the largest for the CSPCA data set (0.12 m) for the IDW method and the H mask. The smallest RMSE values (0.03 m) for this data set are obtained for all methods (IDW, KRI, NAN, and TRI) for the HL mask. For the CSPCG data set, the maximum value of the RMSE is 0.05 for the TRI interpolation method and M mask, while the smallest values (0.03 m) are obtained for HL mask and each interpolation method tested and L mask for the IDW and KRI method.

Analyzing the MinEs and MaxEs, it can be concluded that they are comparable. The largest values of these errors for CSPCA datasets are in the range of −0.48 and 0.57 m and for CSPCG datasets are in the range of −0.13 and 0.16 m. Values for these errors, which are within the accepted tolerance (+/−0.25 m), were obtained for the CSPCA HL dataset and all CSPCG datasets. For the above cases they are between −0.13 and 0.16 m. In the case of ME for all options, the values are also comparable, which are in the range of −0.01 to 0.04 m. The lowest values were achieved for all methods for CSPCA HL dataset (0.01 m), CSPCA L dataset (−0.01 m), and for all methods for CSPCG datasets (0.01 m). However, by analyzing the RMSE, it can be concluded that by far the best results can be achieved for the CSPCG dataset for HL mask for all methods (0.03 m), L mask for IDW and KRI (0.03 m), and the CSPCA for the HL mask (0.03 m). Based on the results obtained, it can be concluded that the most universal mask for all methods is the HL mask.

4. Discussion

More and more studies are being conducted in shallow water areas. This is linked to technological development and the creation of unmanned platforms that can perform measurements in areas previously inaccessible to larger hydrographic units. However, even a small submersion of these units does not eliminate the problems associated with the survey of shallow and ultra-shallow waters. Remote sensing techniques, such as bathymetric LIDAR, and passive optical imaging methods, can be used for the tasks of mapping bathymetry to the boundaries of the coastline. In the first case, the main limitation is the cost of obtaining data or the ability to measure larger depths in freshwater bodies, such as rivers [25] and lakes [33], while in the second case, often the optical complexity of the water bodies requires additional measurements in situ [80]. An additional constraint is the noticeable decrease in accuracy with increasing depth, water turbidity, and diverse bottom material [61]. The research undertaken presents a photogrammetric approach that confirms the possibility of obtaining data using an underwater photogrammetric control network [70] and fills the gap in coastal water mapping where the opportunities for obtaining bathymetric data with passive remote sensing sensors are significantly limited. The need to create hybrid methods using hydrographic and optical methods was pointed out in [13]. In the present case, the acquisition of bathymetry was limited to the depth of 1.3 m. We have, therefore, proposed an original method for combining hydrographic and low altitude remote sensing data. Thus, we point to the possibility of mapping bathymetry with high spatial resolution in shallow waters, which is a limitation in obtaining bathymetric data for platforms equipped with hydroacoustic sensors. The presented method is also another technique that complements the research on the fusion of various data types related to the creation of bathymetric models [44,81,82].

The method presented has been studied in terms of creating the final product, which is a digital bathymetric model. The results of the proposed method can be considered good if we take as a criterion the deviation threshold value based on the vertical bathymetric measurement uncertainty [74]. At the final stage of the quantitative evaluation, the error values for the selected methods and datasets were low (ME = 0.01 m, RMSE = 0.03 m). The studies were carried out on data with a complex spatial distribution. The hybrid data set consisted of USV data recorded along the profiles and scattered points from UAVs. The results of the studies of the interpolation methods were good for most cases, except for the RBF method. At the same time, this is an interesting case that should be taken into account in the creation of digital models. The data set itself is not sufficient material to create a continuous bathymetric model. The correct modeling process should take into account appropriate interpolation techniques that can be used to correctly reconstruct the bottom topography. The impact of data density is also an important factor to consider. In previous research, interpolation techniques have been studied for various data reduction methods [45]; the RBF method was one of those that produced good surface modeling results. This case study showed that the structure of high-density data adversely affects the interpolation capabilities of the RBF method, which complements the knowledge and possibilities of practical use of interpolation methods.

The method developed also has operational limitations. The study further confirmed the importance of the underwater photogrammetric control network included in the study [70]. However, in some cases, the assumption of such a photogrammetric control network may prove problematic. Already during the measurements, local bottom sediments were found to be an obstacle, which covered and partly covered the GCP when the water became cloudy. This required looking for a suitable GCP location. Another limitation related to the establishment of the underwater photogrammetric network in water bodies with a high degree of siltation or that are difficult to access. Setting it in water is also more time consuming than on land. Hence, this approach greatly increases the time of measurements. In addition, for the validation of the final product, check points should be measured. Practice also shows that a required element is accurate mapping of the location of GCPs on a mobile device (up to approximately 1 m of horizontal accuracy). Even in such shallow waters, with low water transparency, it may not be possible to locate the GCPs again. In summary, the limitations of bathymetry mapping using a photogrammetric control network are the type of bottom, the accessibility of the measuring site from land, and the amount of time associated with GCP and CP measurements. Another limitation of the method is its applicability to large areas due to the technical difficulties and the time involved in establishing an underwater photogrammetric network.

A disputed area of development is the coastline, covering both water and land. It usually contains a large variety of vegetation (e.g., submerged, partially submerged, and organic deposits) as well as large outflows. Given the measurement errors and the presence of these objects, some of the models developed may have positive depth values. Studies have shown that pixels containing positive values are small in number, indicating good results of the modeling process. However, shallow waters can make it difficult to map bathymetry to the coastline, which will require manual correction in most cases. Therefore, based on the data modeling process carried out, the possibility of full automation of data development can be excluded. Coastline variability will also be an additional factor in the ambiguity of the bathymetric model boundary region. The most common definition of coastline is that it is the boundary between land and water [83]. This definition is not precise because it refers only to its instantaneous state. The coastline is characterized by short-term and long-term variability. Coastline variability in marine areas is affected by wave movements, tides, winds, erosion, deposition and accumulation [84], and storm surges. To precisely define the concept of coastline, the authors [83] identified 45 indicators and divided them into three groups: visual (visible in the image to the operator), tidal (relating to the intersection of the data on the flows with the digital terrain model or the edge profile), and digital (identified by automatic algorithms). In [85], the shore is defined as a zone of water–land interaction, where on the water side, the factors affecting the land are hydrodynamic factors, while on the land side, they are geodynamic and morphodynamic factors. Attention should be paid to these factors when carrying out work on other water types or in other weather conditions.

One imperfection of the proposed method can be considered to be the non-uniformity of the surface structure of the obtained bathymetric models. This model was created from data recorded by different sensors, which translated into a differentiated spatial structure of the measurement sets. The surface obtained from USV data was smoother, while that from the UAV data was rougher. From the point of view of visualization, a different structure of the surface can influence the inference regarding, e.g., the shape of the bottom or its structure.

A key element of the proposed method is the bathymetric reference surface, developed from echosounder measurements with a USV. The TIN method was used to develop it. Taking into account that it is one of the interpolation methods used to create, among others, geographical surfaces, the question can be raised about whether there is a better method. The application of parametric methods is connected with an appropriate selection of their values, which influences the final shape of the surface. This aspect opens up further optimization criteria that, on the one hand, can indeed improve the final results, but on the other hand, increase the uncertainty of the created model. Considering the presented experiment with the TIN reference surface, it is also important to refer to the results obtained, which far exceed the requirements related to the uncertainty of the bathymetric measurements. Nevertheless, further research in this area could bring interesting conclusions, especially in cases of a different type of spatial distribution of soundings or bottom shape. Densification or a different shape and direction of the ASV sounding profiles could be considered.

In the field of interpolation methods, the impact of data reduction on the efficiency of surface reconstruction using interpolation methods can be studied. Given the spatial complexity of hybrid data sets (spatial distribution and data density), this will be an interesting topic.

We also recommend the continuation of research on the harmonization of the structure of the hybrid model, which should affect the clarity of the visual message and the final aesthetics of the model. Taking into account the possibility of using different geovisualization techniques of terrain surfaces [86], this is an important aspect of end-use data. Assuming that the ideal data development process usually involves full automation, further research should address the problem of bathymetric surface modeling in close proximity to the coastline. As indicated in the research, the obtained model has some imperfections, such as the occurrence of few pixels with positive depth values in close proximity to the coastline. As the main objective was the integration of USV and UAV data, the work did not include the final process of model development, which could be model cleaning using mathematical morphology methods or smoothing operations.

It is necessary to continue research related to the development of hybrid methods, using both photogrammetric techniques, based on an underwater network, as well as remote sensing, using correction for refraction of electromagnetic waves at the air–water junction. Taking into account the optical complexity of waters as well as their types (rivers, lakes, seas, and oceans), one should expect diverse results as well as validation of different research approaches.

5. Conclusions

According to the results of this study, data integration can be achieved using the presented geoprocessing process (Figure 25). Important elements of the proposed method are the acquisition of UAV data using underwater GCP points, the use of a ground class point cloud, the creation of a mask based on the bathymetric reference surface, and the filtering of points above the water surface. In the presented methodology, we propose the use of the HL mask as more universal. The interpolation methods KRI, IDW, NAN, and TRI were found to be comparable, any of which can be used. The final bathymetric surface modelling process should be assessed visually.

The advantage of the proposed solution is the development of a digital bathymetric model within the assumed depth deviations. The analysis of the final results shows that the values of the obtained average and maximum errors are relatively small. The whole process of geoprocessing can also be automated in GIS software, provided the presented functions are available. The developed model of the bottom may be used for creating navigation charts requiring planning of shallow water USV surveys, e.g., small areas of archaeological mapping (about 0.03 km^2) as well as for the analysis of the shape of the bottom for hydrotechnical purposes. The data obtained in this way could also be used to create bathymetric databases for depth mapping in shallow and ultra-shallow waters.

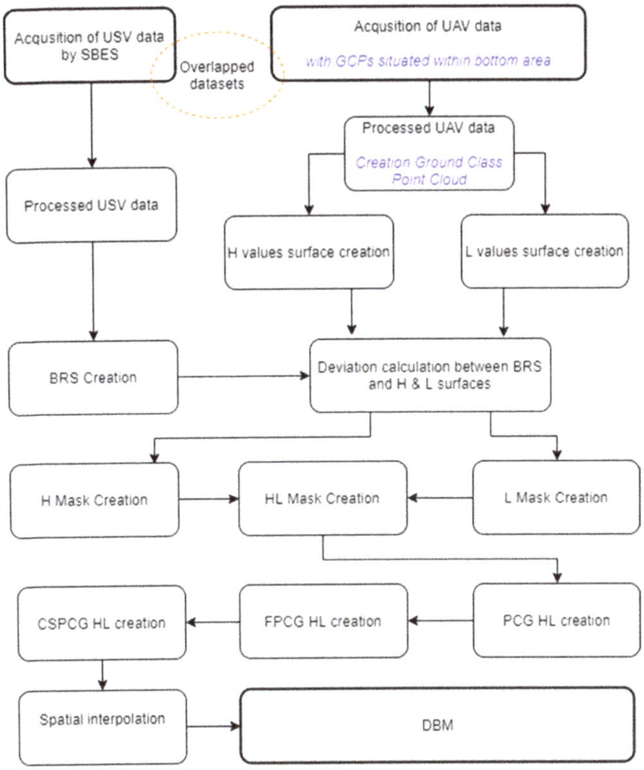

Figure 25. Proposed method for integrating USV and UAV data in terms of geoprocessing bathymetric and photogrammetric data.

Author Contributions: Conceptualization, J.L.; methodology J.L.; validation, J.L. and M.L.; formal analysis, J.L., W.K., M.L. and G.Z.; investigation, J.L., M.L., W.K. and G.Z.; resources, W.K., J.L., M.L. and G.Z.; data curation, J.L., G.Z. and M.L.; writing—original draft preparation, J.L., W.K., M.L. and G.Z.; writing—review and editing, J.L., W.K., M.L. and G.Z.; visualization, J.L. and G.Z.; supervision, J.L.; project administration, J.L. All authors have read and agreed to the published version of the manuscript.

Funding: The study presented was cofinanced by the European Union from the European Regional Development Fund under the 2014–2020 Operational Programme Smart Growth. The project titled "Development of technology for acquisition and exploration of gravimetric data of foreshore and seashore of Polish maritime areas" was implemented as part of the National Centre for Research and Development competition: 1/4.1.4/2018. This research was conducted under grant no. 2/S/KG/21, financed from a subsidy of the Ministry of Science and Higher Education for statutory activities in the Maritime University of Szczecin.

Data Availability Statement: Not applicable.

Conflicts of Interest: The authors declare no conflict of interest.

References

1. Yeu, Y.; Yee, J.-J.; Yun, H.S.; Kim, K.B. Evaluation of the Accuracy of Bathymetry on the Nearshore Coastlines of Western Korea from Satellite Altimetry, Multi-Beam, and Airborne Bathymetric LiDAR. *Sensors* **2018**, *18*, 2926. [CrossRef]
2. Genchi, S.A.; Vitale, A.J.; Perillo, G.M.E.; Seitz, C.; Delrieux, C.A. Mapping Topobathymetry in a Shallow Tidal Environment Using Low-Cost Technology. *Remote Sens.* **2020**, *12*, 1394. [CrossRef]

3. Guth, P.L.; Van Niekerk, A.; Grohmann, C.H.; Muller, J.-P.; Hawker, L.; Florinsky, I.V.; Gesch, D.; Reuter, H.I.; Herrera-Cruz, V.; Riazanoff, S.; et al. Digital Elevation Models: Terminology and Definitions. *Remote Sens.* **2021**, *13*, 3581. [CrossRef]
4. Zwolak, K.; Wigley, R.; Bohan, A.; Zarayskaya, Y.; Bazhenova, E.; Dorshow, W.; Sumiyoshi, M.; Sattiabaruth, S.; Roperez, J.; Proctor, A.; et al. The Autonomous Underwater Vehicle Integrated with the Unmanned Surface Vessel Mapping the Southern Ionian Sea. The Winning Technology Solution of the Shell Ocean Discovery XPRIZE. *Remote Sens.* **2020**, *12*, 1344. [CrossRef]
5. Janowski, L.; Wroblewski, R.; Dworniczak, J.; Kolakowski, M.; Rogowska, K.; Wojcik, M.; Gajewski, J. Offshore benthic habitat mapping based on object-based image analysis and geomorphometric approach. A case study from the Slupsk Bank, Southern Baltic Sea. *Sci. Total Environ.* **2021**, *801*, 149712. [CrossRef]
6. Kasvi, E.; Salmela, J.; Lotsari, E.; Kumpula, T.; Lane, S.N. Comparison of remote sensing based approaches for mapping bathymetry of shallow, clear water rivers. *Geomorphology* **2019**, *333*, 180–197. [CrossRef]
7. Bahaa, S.A.; Salem, S.S.; Hany, E.A. Accuracy Assessment of Global Bathymetry and Topography at 15 Arc Sec (SRTM15 + V2) Using Red Sea Floor Topography, Egypt. *IJERT* **2021**, *10*, 271–276.
8. Alvarez, L.V.; Moreno, H.A.; Segales, A.R.; Pham, T.G.; Pillar-Little, E.A.; Chilson, P.B. Merging Unmanned Aerial Systems (UAS) Imagery and Echo Soundings with an Adaptive Sampling Technique for Bathymetric Surveys. *Remote Sens.* **2018**, *10*, 1362. [CrossRef]
9. Legleiter, C.J.; Overstreet, B.T.; Kinzel, P.J. Sampling Strategies to Improve Passive Optical Remote Sensing of River Bathymetry. *Remote Sens.* **2018**, *10*, 935. [CrossRef]
10. Legleiter, C.J.; Roberts, D.A.; Lawrence, R.L. Spectrally based remote sensing of river bathymetry. *Earth Surf. Process. Landf.* **2009**, *34*, 1039–1059. [CrossRef]
11. Williams, R.D.; Brasington, J.; Vericat, D.; Hicks, D.M. Hyperscale terrain modelling of braided rivers: Fusing mobile terrestrial laser scanning and optical bathymetric mapping. *Earth Surf. Process. Landf.* **2014**, *39*, 167–183. [CrossRef]
12. Lyzenga, D.R. Passive remote sensing techniques for mapping water depth and bottom features. *Appl. Opt.* **1978**, *17*, 379–383. [CrossRef]
13. Legleiter, C.J.; Fosness, R.L. Defining the Limits of Spectrally Based Bathymetric Mapping on a Large River. *Remote Sens.* **2019**, *11*, 665. [CrossRef]
14. Slocum, R.K.; Wright, W.; Parrish, C.; Costa, B.; Sharr, M.; Battista, T.A. *Guidelines for Bathymetric Mapping and Orthoimage Generation Using sUAS and SfM, An Approach for Conducting Nearshore Coastal Mapping*; NOAA Technical Memorandum NOS NCCOS 265; NOAA: Silver Spring, MD, USA, 2019. [CrossRef]
15. Starek, M.J.; Giessel, J. Fusion of uas-based structure-from-motion and optical inversion for seamless topo-bathymetric mapping. In Proceedings of the 2017 IEEE International Geoscience and Remote Sensing Symposium (IGARSS), Fort Worth, TX, USA, 23–28 July 2017.
16. Kedzierski, M.; Wierzbicki, D.; Sekrecka, A.; Fryskowska, A.; Walczykowski, P.; Siewert, J. Influence of Lower Atmosphere on the Radiometric Quality of Unmanned Aerial Vehicle Imagery. *Remote Sens.* **2019**, *11*, 1214. [CrossRef]
17. Burdziakowski, P. Polymodal Method of Improving the Quality of Photogrammetric Images and Models. *Energies* **2021**, *14*, 3457. [CrossRef]
18. Partama, I.Y.; Kanno, A.; Ueda, M.; Akamatsu, Y.; Inui, R.; Sekine, M.; Yamamoto, K.; Imai, T.; Higuchi, T. Removal of water-surface reflection effects with a temporal minimum filter for UAV-based shallow-water photogrammetry. *Earth Surf. Process. Landf.* **2018**, *43*, 2673–2682. [CrossRef]
19. Doukari, M.; Katsanevakis, S.; Soulakellis, N.; Topouzelis, K. The Effect of Environmental Conditions on the Quality of UAS Orthophoto-Maps in the Coastal Environment. *ISPRS Int. J. Geo-Inf.* **2021**, *10*, 18. [CrossRef]
20. Woodget, A.S.; Dietrich, J.T.; Wilson, R.T. Quantifying Below-Water Fluvial Geomorphic Change: The Implications of Refraction Correction, Water Surface Elevations, and Spatially Variable Error. *Remote Sens.* **2019**, *11*, 2415. [CrossRef]
21. Woodget, A.S.; Carbonneau, P.E.; Visser, F.; Maddock, I.P. Quantifying submerged fluvial topography using hyperspatial resolution UAS imagery and structure from motion photogrammetry. *Earth Surf. Process. Landf.* **2015**, *40*, 47–64. [CrossRef]
22. Dietrich, J.T. Bathymetric Structure-from-Motion: Extracting shallow stream bathymetry from multi-view stereo photogram-metry. *Earth Surf. Process. Landf.* **2017**, *42*, 355–364. [CrossRef]
23. Shintani, C.; Fonstad, M.A. Comparing remote-sensing techniques collecting bathymetric data from a gravel-bed river. *Int. J. Remote Sens.* **2017**, *38*, 2883–2902. [CrossRef]
24. Marcus, W.A.; Fonstad, M.A. Remote sensing of rivers: The emergence of a subdiscipline in the river sciences. *Earth Surf. Process. Landf.* **2010**, *35*, 1867–1872. [CrossRef]
25. Legleiter, C.J.; Harrison, L.R. Remote Sensing of River Bathymetry: Evaluating a Range of Sensors, Platforms, and Algorithms on the Upper Sacramento River, California, USA. *Water Resour. Res.* **2019**, *55*, 2142–2169. [CrossRef]
26. Agrafiotis, P.; Skarlatos, D.; Georgopoulos, A.; Karantzalos, K. DepthLearn: Learning to Correct the Refraction on Point Clouds Derived from Aerial Imagery for Accurate Dense Shallow Water Bathymetry Based on SVMs-Fusion with LiDAR Point Clouds. *Remote Sens.* **2019**, *11*, 2225. [CrossRef]
27. Agrafiotis, P. Image-Based Bathymetry Mapping for Shallow Waters. Ph.D. Dissertation. 2020. Available online: https://dspace.lib.ntua.gr/xmlui/handle/123456789/50897 (accessed on 22 November 2021). [CrossRef]
28. Skarlatos, D.; Agrafiotis, P. A Novel Iterative Water Refraction Correction Algorithm for Use in Structure from Motion Photogrammetric Pipeline. *J. Mar. Sci. Eng.* **2018**, *6*, 77. [CrossRef]

29. Mandlburger, G.; Pfennigbauer, M.; Schwarz, R.; Flöry, S.; Nussbaumer, L. Concept and Performance Evaluation of a Novel UAV-Borne Topo-Bathymetric LiDAR Sensor. *Remote Sens.* **2020**, *12*, 986. [CrossRef]
30. Gesch, D.B.; Brock, J.C.; Parrish, C.E.; Rogers, J.N.; Wright, C.W. Introduction: Special Issue on Advances in Topobathymetric Mapping, Models, and Applications. *J. Coast. Res.* **2016**, *76*, 1–3. [CrossRef]
31. Lee, Z.; Shangguan, M.; Garcia, R.A.; Lai, W.; Lu, X.; Wang, J.; Yan, X. Confidence Measure of the Shallow-Water Bathymetry Map Obtained through the Fusion of Lidar and Multiband Image Data. *J. Remote Sens.* **2021**, *2021*, 1–16. [CrossRef]
32. Tysiac, P. Bringing Bathymetry LiDAR to Coastal Zone Assessment: A Case Study in the Southern Baltic. *Remote Sens.* **2020**, *12*, 3740. [CrossRef]
33. Chormański, J.; Nowicka, B.; Wieckowski, A.; Ciupak, M.; Jóźwiak, J.; Figura, T. Coupling of Dual Channel Waveform ALS and Sonar for Investigation of Lake Bottoms and Shore Zones. *Remote Sens.* **2021**, *13*, 1833. [CrossRef]
34. Quadros, N.; Collier, P.; Fraser, C. Integration of Bathymetric and Topographic Lidar: A Preliminary Investigation. *Int. Arch. Photogramm. Remote Sens. Spat. Inf. Sci.* **2008**, *37*, 1299–1304.
35. Madore, B.; Imahori, G.; Kum, J.; White, S.; Worthem, A. NOAA's use of remote sensing technology and the coastal mapping program. In Proceedings of the OCEANS 2018 MTS/IEEE Charleston, Charleston, SC, USA, 22–25 October 2018; pp. 1–7. [CrossRef]
36. Salameh, E.; Frappart, F.; Almar, R.; Baptista, P.; Heygster, G.; Lubac, B.; Raucoules, D.; Almeida, L.P.; Bergsma, E.W.J.; Capo, S.; et al. Monitoring Beach Topography and Nearshore Bathymetry Using Spaceborne Remote Sensing: A Review. *Remote Sens.* **2019**, *11*, 2212. [CrossRef]
37. Tozer, B.; Sandwell, D.T.; Smith, W.H.F.; Olson, C.; Beale, J.R.; Wessel, P. Global Bathymetry and Topography at 15 Arc Sec: SRTM15+. *Earth Space Sci.* **2019**, *6*, 1847–1864. [CrossRef]
38. Danielson, J.J.; Poppenga, S.K.; Brock, J.C.; Evans, G.A.; Tyler, D.J.; Gesch, D.B.; Thatcher, C.A.; Barras, J.A. Topobathymetric Elevation Model Development using a New Methodology: Coastal National Elevation Database. *J. Coast. Res.* **2016**, *76*, 75–89. [CrossRef]
39. Zhang, C. Applying data fusion techniques for benthic habitat mapping and monitoring in a coral reef ecosystem. *ISPRS J. Photogramm. Remote Sens.* **2015**, *104*, 213–223. [CrossRef]
40. Liu, Y.; Xu, S.; Zhu, T.; Wang, T. Application of Multi-Source Data Fusion Method in Updating Topography and Estimating Sedimentation of the Reservoir. *Water* **2020**, *12*, 3057. [CrossRef]
41. Hogrefe, K.R.; Wright, D.J.; Hochberg, E. Derivation and Integration of Shallow-Water Bathymetry: Implications for Coastal Terrain Modeling and Subsequent Analyses. *Mar. Geod.* **2008**, *31*, 299–317. [CrossRef]
42. Chowdhury, E.H.; Hassan, Q.K.; Achari, G.; Gupta, A. Use of Bathymetric and LiDAR Data in Generating Digital Elevation Model over the Lower Athabasca River Watershed in Alberta, Canada. *Water* **2017**, *9*, 19. [CrossRef]
43. Dost, R.J.J.; Mannaerts, C.M.M. Generation of lake bathymetry using SONAR, satellite imagery and GIS. In Proceedings of the 2008 ESRI International User Conference: GIS, Geography in Action, San Diego, FL, USA, 4–8 August 2008; pp. 1–5.
44. Włodarczyk-Sielicka, M.; Bodus-Olkowska, I.; Łącka, M. The process of modelling the elevation surface of a coastal area using the fusion of spatial data from different sensors. *Oceanologia* **2021**. [CrossRef]
45. Lubczonek, J.; Wlodarczyk-Sielicka, M.; Lacka, M.; Zaniewicz, G. Methodology for Developing a Combined Bathymetric and Topographic Surface Model Using Interpolation and Geodata Reduction Techniques. *Remote Sens.* **2021**, *13*, 4427. [CrossRef]
46. Bagheri, O.; Ghodsian, M.; Saadatseresht, M. Reach scale application of UAV+SfM methods in shallow rivers hyperspatial bathymetry. In Proceedings of the ISPRS—International Archives of the Photogrammetry, Remote Sensing and Spatial Information Sciences, Kish Island, Iran, 23–25 November 2015; Copernicus GmbH: Göttingen, Germany, 2015; Volume XL-15, pp. 77–81.
47. Visser, F.; Woodget, A.; Skellern, A.; Forsey, J.; Warburton, J.; Johnson, R. An evaluation of a low-cost pole aerial photography (PAP) and structure from motion (SfM) approach for topographic surveying of small rivers. *Int. J. Remote Sens.* **2019**, *40*, 9321–9351. [CrossRef]
48. Legleiter, C.; Pavelsky, T.; Durand, M.; Allen, G.; Tarpanelli, A.; Frasson, R.; Guneralp, I.; Woodget, A. Editorial for the Special Issue "Remote Sensing of Flow Velocity, Channel Bathymetry, and River Discharge". *Remote Sens.* **2020**, *12*, 2304. [CrossRef]
49. Bandini, F.; Olesen, D.; Jakobsen, J.; Kittel, C.M.M.; Wang, S.; Garcia, M.; Bauer-Gottwein, P. Technical note: Bathymetry observations of inland water bodies using a tethered single-beam sonar controlled by an unmanned aerial vehicle. *Hydrol. Earth Syst. Sci.* **2018**, *22*, 4165–4181. [CrossRef]
50. Stateczny, A.; Burdziakowski, P.; Najdecka, K.; Domagalska-Stateczna, B. Accuracy of Trajectory Tracking Based on Nonlinear Guidance Logic for Hydrographic Unmanned Surface Vessels. *Sensors* **2020**, *20*, 832. [CrossRef] [PubMed]
51. Specht, M.; Stateczny, A.; Specht, C.; Widźgowski, S.; Lewicka, O.; Wiśniewska, M. Concept of an Innovative Autonomous Unmanned System for Bathymetric Monitoring of Shallow Waterbodies (INNOBAT System). *Energies* **2021**, *14*, 5370. [CrossRef]
52. Shao, G.; Ma, Y.; Malekian, R.; Yan, X.; Li, Z. A Novel Cooperative Platform Design for Coupled USV–UAV Systems. *IEEE Trans. Ind. Inform.* **2019**, *15*, 4913–4922. [CrossRef]
53. Maleika, W. Inverse distance weighting method optimization in the process of digital terrain model creation based on data collected from a multibeam echosounder. *Appl. Geomat.* **2020**, *12*, 397–407. [CrossRef]
54. Wlodarczyk-Sielicka, M.; Blaszczak-Bak, W. Processing of Bathymetric Data: The Fusion of New Reduction Methods for Spatial Big Data. *Sensors* **2020**, *20*, 6207. [CrossRef] [PubMed]
55. Wolski, W.I. Changes in Dąbie Lake bathymetry in the period 1962–1996. *Limnol. Rev.* **2005**, *5*, 258.

56. Phantom 4 Pro—Product Information—DJI. 2021. Available online: https://www.dji.com/sg/phantom-4-pro/info#specs (accessed on 3 November 2021).
57. Specht, M.; Specht, C.; Wąż, M.; Naus, K.; Grządziel, A.; Iwen, D. Methodology for Performing Territorial Sea Baseline Measurements in Selected Waterbodies of Poland. *Appl. Sci.* **2019**, *9*, 3053. [CrossRef]
58. Stateczny, A.; Specht, C.; Specht, M.; Brčić, D.; Jugović, A.; Widźgowski, S.; Wiśniewska, M.; Lewicka, O. Study on the Positioning Accuracy of GNSS/INS Systems Supported by DGPS and RTK Receivers for Hydrographic Surveys. *Energies* **2021**, *14*, 7413. [CrossRef]
59. Giordano, F.; Mattei, G.; Parente, C.; Peluso, F.; Santamaria, R. Integrating Sensors into a Marine Drone for Bathymetric 3D Surveys in Shallow Waters. *Sensors* **2015**, *16*, 41. [CrossRef] [PubMed]
60. Philpot, W.D. Bathymetric mapping with passive multispectral imagery. *Appl. Opt.* **1989**, *28*, 1569–1578. [CrossRef] [PubMed]
61. Gao, J. Bathymetric mapping by means of remote sensing: Methods, accuracy and limitations. *Prog. Phys. Geogr. Earth Environ.* **2009**, *33*, 103–116. [CrossRef]
62. De Smith, M.J.; Goodchild, M.F.; Longley, P. *Geospatial Analysis: A Comprehensive Guide to Principles, Techniques and Software Tools*; Leicester: Matador, UK, 2009.
63. Lekkerkerk, H.-J.; Theijs, M. *Handbook of Offshore Surveying Volume III: Acquisition Sensors*; Skilltrade BV: Voorschoten, The Netherlands, 2011.
64. Lubis, M.Z.; Wulandari, P.D.; Mujahid, M.; Hargreaves, J.; Pant, V. Echo Processing and Identifying Surface and Bottom Layer with Simrad Ek/Ey 500. *J. Biosens. Bioelectron.* **2016**, *7*, 2–7. [CrossRef]
65. Becker, C.; Rosinskaya, E.; Häni, N.; D'Angelo, E.; Strecha, C. Classification of Aerial Photogrammetric 3D Point Clouds. *Photogramm. Eng. Remote Sens.* **2018**, *84*, 287–295. [CrossRef]
66. Tsukada, F.; Shimozono, T.; Matsuba, Y. UAV-based mapping of nearshore bathymetry over broad areas. *Coast. Eng. J.* **2020**, *62*, 285–298. [CrossRef]
67. Matsuba, Y.; Sato, S. Nearshore bathymetry estimation using UAV. *Coast. Eng. J.* **2018**, *60*, 51–59. [CrossRef]
68. Agrafiotis, P.; Karantzalos, K.; Georgopoulos, A.; Skarlatos, D. Correcting Image Refraction: Towards Accurate Aerial Image-Based Bathymetry Mapping in Shallow Waters. *Remote Sens.* **2020**, *12*, 322. [CrossRef]
69. Rossi, P.; Castagnetti, C.; Capra, A.; Brooks, A.J.; Mancini, F. Detecting change in coral reef 3D structure using underwater photogrammetry: Critical issues and performance metrics. *Appl. Geomat.* **2020**, *12*, 3–17. [CrossRef]
70. David, C.G.; Kohl, N.; Casella, E.; Rovere, A.; Ballesteros, P.; Schlurmann, T. Structure-from-Motion on shallow reefs and beaches: Potential and limitations of consumer-grade drones to reconstruct topography and bathymetry. *Coral Reefs* **2021**, *40*, 835–851. [CrossRef]
71. Mandlburger, G. Through-Water Dense Image Matching for Shallow Water Bathymetry. *Photogramm. Eng. Remote Sens.* **2019**, *85*, 445–455. [CrossRef]
72. Sandoval-Erazo, W.; Toulkeridis, T.; Morales-Sanchez, Á.; Mora, M.J.M. Sedimentological study of the reservoir of the Manduriacu hydroelectric project, northern Ecuador. *IOP Conf. Ser. Earth Environ. Sci.* **2018**, *191*, 012119. [CrossRef]
73. Stateczny, A.; Lubczonek, J. Digital Terrains Models. In *Methods of Comparative Navigation*; Stateczny, A., Ed.; Gdańsk Scientific Society: Gdynia, Poland, 2004.
74. IHO—International Hydrographic Organization. *S-44: IHO Standards for Hydrographic Surveys*, 6th ed.; IHO Publication: Monaco, Monaco, 2020.
75. Lee, D.T.; Schachter, B.J. Two algorithms for constructing a Delaunay triangulation. *Int. J. Parallel Program.* **1980**, *9*, 219–242. [CrossRef]
76. Sibson, R. A Brief Description of Natural Neighbor Interpolation. In *Interpreting Multivariate Data*; Barnett, V., Ed.; John Wiley and Sons: New York, NY, USA, 1981; pp. 21–36.
77. Franke, R. Scattered Data Interpolation: Test of Some Methods. *Math. Comput.* **1982**, *33*, 181–200.
78. Cressie, N. The origins of kriging. *Math Geol.* **1990**, *22*, 239–252. [CrossRef]
79. Powell, M.J.D. The Theory of Radial Basis Function Approximation in 1990. In *Advances in Numerical Analysis II*; Oxford University Press: Oxford, UK, 1992; Volume 2, pp. 105–210.
80. Kutser, T.; Paavel, B.; Verpoorter, C.; Ligi, M.; Soomets, T.; Toming, K.; Casal, G. Remote Sensing of Black Lakes and Using 810 nm Reflectance Peak for Retrieving Water Quality Parameters of Optically Complex Waters. *Remote Sens.* **2016**, *8*, 497. [CrossRef]
81. Leon, J.X.; Phinn, S.; Hamylton, S.; Saunders, M. Filling the 'white ribbon'—A multisource seamless digital elevation model for Lizard Island, northern Great Barrier Reef. *Int. J. Remote Sens.* **2013**, *34*, 6337–6354. [CrossRef]
82. Cheng, L.; Ma, L.; Cai, W.; Tong, L.; Li, M.; Du, P. Integration of Hyperspectral Imagery and Sparse Sonar Data for Shallow Water Bathymetry Mapping. *IEEE Trans. Geosci. Remote Sens.* **2014**, *53*, 3235–3249. [CrossRef]
83. Boak, E.H.; Turner, I. Shoreline Definition and Detection: A Review. *J. Coast. Res.* **2005**, *214*, 688–703. [CrossRef]
84. Alicandro, M.; Baiocchi, V.; Brigante, R.; Radicioni, F. Automatic Shoreline Detection from Eight-Band VHR Satellite Imagery. *J. Mar. Sci. Eng.* **2019**, *7*, 459. [CrossRef]
85. Subotowicz, W. *Beaches as the Element of Marine Environment (in Polish: Brzeg morski—Element środowiska morskiego)*; Inżynieria Morska i Geotechnika: Gdańsk, Poland, 2018; Volume 2, pp. 79–81.

86. Ruzinoor, C.M.; Shariff, A.R.M.; Mahmud, A.R.; Pradhan, B. 3D Terrain Visualisation for GIS: A Comparison of Different Techniques. In *True-3D Cartography*; Buchroithner, M., Ed.; Lecture Notes Geoinformation and Cartographty; Springer: Berlin/Heidelberg, Germany, 2012; pp. 43–65. [CrossRef]

remote sensing

MDPI

Article

Application of Unmanned Aerial Vehicles and Image Processing Techniques in Monitoring Underwater Coastal Protection Measures

Jakub Śledziowski [1,*], Paweł Terefenko [1], Andrzej Giza [1], Paweł Forczmański [2], Andrzej Łysko [2], Witold Maćków [2], Grzegorz Stępień [3], Arkadiusz Tomczak [3] and Apoloniusz Kurylczyk [4]

1 Institute of Marine and Environmental Sciences, University of Szczecin, 70-383 Szczecin, Poland; pawel.terefenko@usz.edu.pl (P.T.); andrzej.giza@usz.edu.pl (A.G.)
2 Faculty of Computer Science and Information Technology, West Pomeranian University of Technology, 71-210 Szczecin, Poland; pforczmanski@wi.zut.edu.pl (P.F.); andrzej.lysko@zut.edu.pl (A.Ł.); witold.mackow@zut.edu.pl (W.M.)
3 Faculty of Navigation, Maritime University of Szczecin, 70-500 Szczecin, Poland; g.stepien@am.szczecin.pl (G.S.); a.tomczak@am.szczecin.pl (A.T.)
4 Institute of Spatial Management and Socio-Economic Geography, University of Szczecin, 70-383 Szczecin, Poland; apoloniusz.kurylczyk@usz.edu.pl
* Correspondence: jakub.sledziowski@usz.edu.pl; Tel.: +48-91-444-23-54

Citation: Śledziowski, J.; Terefenko, P.; Giza, A.; Forczmański, P.; Łysko, A.; Maćków, W.; Stępień, G.; Tomczak, A.; Kurylczyk, A. Application of Unmanned Aerial Vehicles and Image Processing Techniques in Monitoring Underwater Coastal Protection Measures. *Remote Sens.* **2022**, *14*, 458. https://doi.org/10.3390/rs14030458

Academic Editor: Paolo Ciavola

Received: 30 November 2021
Accepted: 17 January 2022
Published: 19 January 2022

Abstract: A prerequisite for solving issues associated with surf zone variability, which affect human activity in coastal zones, is an accurate estimation of the effects of coastal protection methods. Therefore, performing frequent monitoring activities, especially when applying new nature-friendly coastal defense methods, is a major challenge. In this manuscript, we propose a pipeline for performing low-cost monitoring using RGB images, accessed by an unmanned aerial vehicle (UAV) and a four-level analysis architecture of an underwater object detection methodology. First, several color-based pre-processing activities were applied. Second, contrast-limited adaptive histogram equalization and the Hough transform methodology were used to automatically detect the underwater, circle-shaped elements of a hybrid coastal defense construction. An alternative pipeline was used to detect holes in the circle-shaped elements with an adaptive thresholding method; this pipeline was subsequently applied to the normalized images. Finally, the concatenation of the results from both the methods and the validation processes were performed. The results indicate that our automated monitoring tool works for RGB images captured by a low-cost consumer UAV. The experimental results showed that our pipeline achieved an average error of four pixels in the test set.

Keywords: unmanned aerial vehicle (UAV); coastal monitoring; image processing; object detection; underwater reef

1. Introduction

Nearshore morphology is continuously changing. Occasionally, this is due to drastic transformations during extreme storm events, and gradual change occurs because of day-to-day interactions with waves and currents [1–4]. Numerous coastal change models have been based on simplified global shoreline movement patterns [4–6] and a site-specific local and regional analysis that implements real coastal measurements—which are mostly surveyed as a cross-section of the coast [7–12]—to provide evidence for the vulnerability of sandy coasts to erosion. Therefore, the appropriate management of coastal zones is closely linked to the balance between natural processes and protection techniques [13]. For years, coastal protection measures have been dominated by hard structural solutions, such as seawalls, breakwaters, groins, and dikes [14]. However, traditional coastal defenses can no longer keep pace with climate change and continuous coastal population growth [15]. The support of new nature-friendly methods of coastal protection planning is becoming a major challenge for the research community, which is involved in knowledge-based shaping of environmental policies [6]. The use of hybrid solutions, such as artificial reefs or constructed dunes, combines the strength of hard infrastructures and the sustainability of natural approaches to minimize structural weaknesses [15]. Intensive monitoring and availability of data, showing how various constructions behave in different coastal landscapes, could benefit the wider implementation of these nature-based protection techniques. Thus, there is a growing need for reliable data in this rapidly changing dynamic coastal environment [16].

Therefore, frequent monitoring techniques are necessary to capture the variability of the surf zone and the impact of implementing protection solutions to enable in-depth and complete coastal analyses [13,17]. Topography acquisition is an extremely common approach that has been performed in a straightforward manner [18]. However, mapping changes during the construction of underwater protection projects in shallow coastal waters is generally expensive, and frequent surveying can pose a financial challenge for national coastal protection services. The conventional methods used in subtidal zones for underwater monitoring generally comprise either a vessel equipped with a hydroacoustic device [3,19] or video tools and other types of SCUBA-operated devices [20]. Here, we present new possibilities for cost-effective monitoring of underwater construction in the surf zone using unmanned aerial vehicles (UAVs).

UAVs are now used in many areas to monitor and protect seacoasts. Research areas include the analysis of subaerial morphology [21–23], progressive changes in the coastal zone [24], advantages of using UAVs for quick mapping of beach areas after the occurrence of storms and large waves [25], nearshore hydrodynamics [26], and estimation of volumetric changes [27]. UAV-based photogrammetry is becoming a fundamental monitoring and analytical tool, thanks to its low-cost and user-friendly techniques, which are comparable to traditional topographic surveys [28–30]. Unmanned aerial vehicle imagery, combined with the structure from motion (SfM) algorithm, which is based on the variability of the beach topographic profile caused by the transition from water to sand, is also used for shoreline measurements [31]. Furthermore, structure from motion [32] photogrammetry applied to data captured from unmanned aerial vehicles [33,34] is one of the most frequently used survey technologies for coastal areas [35].

The remarkable potential of UAV image-based analysis has been proven by the implementation of surf zone bathymetry estimation methods [3,36] and monitoring techniques for submerged objects [37]. The above-mentioned applications deserve special attention, and they not only present an analysis of water areas, but also provide descriptions of the various zones of the beach boundary structures. Following this trend, we present a method that can be employed for low-cost UAV monitoring of hybrid protection measures, such as underwater artificial reefs, using an artificial threshold in the Baltic Sea in Poland. Poland has no standards or guidelines for assisting with basic research or monitoring the requirements of new artificial reef modules. Thus, we aim to design the first basic tool for the low-cost monitoring of such structures, which could be employed to analyze the overall correctness of the designs or materials used to develop protective structures and evaluate the effects of ecosystem recovery and its impact on the modification of the coastal processes.

To achieve the main purpose, we tested fast and cheap geomatic techniques for monitoring underwater coastal protection measures. Three UAV photogrammetric surveys integrated with GNSS measurement of GCPs and a 2D transformation method were performed and analyzed. Although performing photogrammetric studies on underwater objects is extremely difficult (i.e., based only on aerial photographs), the implemented technique demonstrates the usefulness of UAVs.

However, a significant proportion of coastal areas worldwide, including Poland, are eroding [38], especially due to sea level rise [4,39], and no single research study has been performed concerning the underwater coastal protection measures in Poland, despite its regional significance. Furthermore, only a few studies have highlighted the usefulness of UAVs within integrated coastal zone management, especially for the documentation of post-storm coastal erosion, as well as monitoring damage to coastal infrastructures [33,40,41]. Thus, reliable information about the behavior of underwater structures benefits and enables numerous strategic decisions in coastal protection processes.

2. Study Area and Materials

Artificial reefs, constructed from concrete habitat modules, are new to the Baltic Sea, and no monitoring of such underwater structures has been conducted before. They were constructed under the project "Protection of the sea shore at Łeba, Rowy, and Ustka" in 2014–2016. During the construction phase, aerial images using UAV were taken of all underwater thresholds in Ustka, Rowy, and Łeba (Figure 1). The first orthophotos of all structures were created in 2016. Due to the apparent significant changes in the structure of the underwater thresholds between years and the high concentration of concrete elements in the structure, the thresholds in Rowy (Figure 2) were selected for further study.

Figure 1. Construction of underwater sill in Łeba (2016).

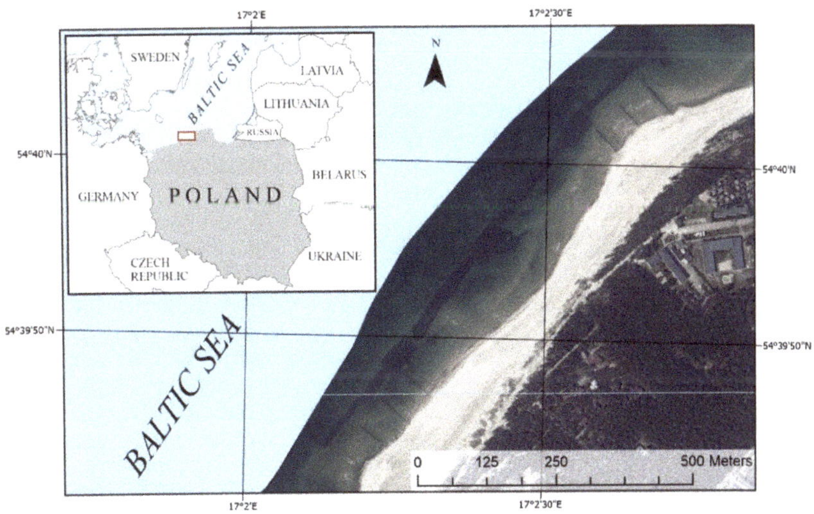

Figure 2. Location of the investigated area.

This section of the Polish coast exhibits diverse characteristics, but primarily assumes the form of eroded cliffs. The most dynamic coastal zone is the beach, which experiences cyclical changes (i.e., increasing from spring to summer and decreasing due to marine erosion in the autumn and winter months) [42].

The monitored underwater construction in Rowy is a 750 m long structure that is located approximately 170 m offshore at depths ranging from 2 to 4 m (Figure 3). It consists of 680 concrete coils, each 2.4 m in diameter, with a 0.8 m diameter void at the top. In 2016, the 1st module on the west side contained 121 elements, the 2nd and 3rd modules contained 143 elements each, the 4th module contained 150 elements, and the last module contained 123 elements. All groups were separated by four segments made of stone modules.

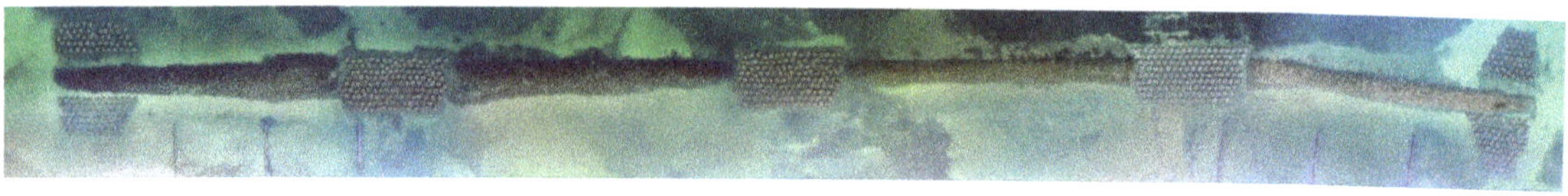

Figure 3. Monitored underwater construction in Rowy in 2016.

The data used in this study cover three separate UAV surveys conducted on 19 May 2016, 8 February 2018, and 10 September 2021. To collect aerial data from the first 2 flights, we used a low-budget DJI Inspire 1 unmanned aerial vehicle with a 12 MPix camera, 3.6 mm lens, F/2.8 aperture, ISO 200, and a shutter speed of 1/120 of a second. The last flight was made with a DJI Matrice 300 RTK drone, with a 12 MPix camera, 4.5 mm lens, wide-angle lens, F/2.8 aperture, ISO 110, shutter from 1/80 to 1/200, depending on the light. The time of data acquisition was determined by the angle of incidence of the solar rays to minimize the reflections on the water surface in relation to the position of the camera lens. Finally, the measurements were started at 8:30 a.m. in 2016, at 11:30 a.m. in 2018, and at 8:00 a.m. in 2021. The first and third survey were conducted without additional filters on the lens, and the second survey used a polarizing filter to eliminate light reflections on the water surface.

Flight parameters, such as flight altitude, spacing between the axes of the series, and distance between the centers of subsequent photos (i.e., the length of the shooting base, which determines where the shutter is turned on and off) were automatically calculated after entering the expected output pixel size. The values included the camera parameters, longitudinal and lateral coverage, and vehicle speed. The mission plan was formulated using the DJI software. Sidelap and overlap during the mission were set at 75%.

The experimental assumptions for using a low-cost UAV required the carefully planning of all surveys based on the weather; therefore, wind and wave conditions in the two surveys were similar, to allow a comparison of the acquired data. Constant observation of meteorological conditions was conducted at the Interdisciplinary Centre for Mathematical and Computational Modelling of the Warsaw University website; and the state of the sea was monitored using the high-resolution operational wave model (WAM), which was validated for the Baltic Sea in the framework of the Hindcast of Dynamic Processes of the Ocean and Coastal Areas of Europe (HIPOCAS) project [43]. Moreover, before the final decision to initiate the survey, images from the web cameras in the port of Rowy were analyzed.

The weather conditions during the 2016 survey were extremely suitable; that is, southern winds with a maximum speed of 2 m/s were observed, and no precipitation, cloud cover, or waves were identified. During the second survey in 2018, the weather conditions were extremely similar; that is, no waves, clouds, or precipitation, were spotted, and the maximum speed of the southerly wind was 2 m/s. The temperature difference was significant (12 °C and −12 °C in the first and second surveys, respectively). During the last survey in 2021, the weather condition was similar, with wind at a maximum speed of 3 m/s, and no precipitation, cloud cover, or waves were identified. The temperature was about 15 °C.

Obtaining aerial data required a perfect match to atmospheric conditions, while determining ground control points for the georeferencing process required careful consideration. To analyze and compare orthophotos from consecutive years, it was necessary to find common points on all developed mosaics. For this purpose, using a Trimble RTK (real time kinematic) receiver, fixed GCP points were determined along the wooden spurs (Figure 4).

Figure 4. Survey in Rowy, 2021.

In addition, vertical warning signs (yellow-painted metal pipes 20 cm in diameter with a cross on top), located in the axis of the entire structure and protruding about 1.5 m above the water surface, were used to align the orthophotos in the post-production process. The coordinates of these warning signs were measured with a Leica TS11 total station and the distance to the last wooden piles in the rows of piling spurs located along the entire structure was checked.

All orthophotos were compiled from the vertical images taken by UAV and using Agisoft's Metashape software. This software is used for automatic image processing and creates a dense point cloud using image matching and image georeferencing algorithms (Figure 5).

Figure 5. Point cloud in Agisoft Metashape software.

Alignment and matching of orthophotos from all measurements were performed using QGIS software. All orthophotos from 2016, 2018, and 2021 were cut into 5 sectors, presenting only the modules that contained concrete circles. The rest of the structure made of stones, located between these modules, was omitted from further analyses (Figure 6). A preliminary validation process was also performed in QGIS software, using manual determination of the centers of all circles visible underwater.

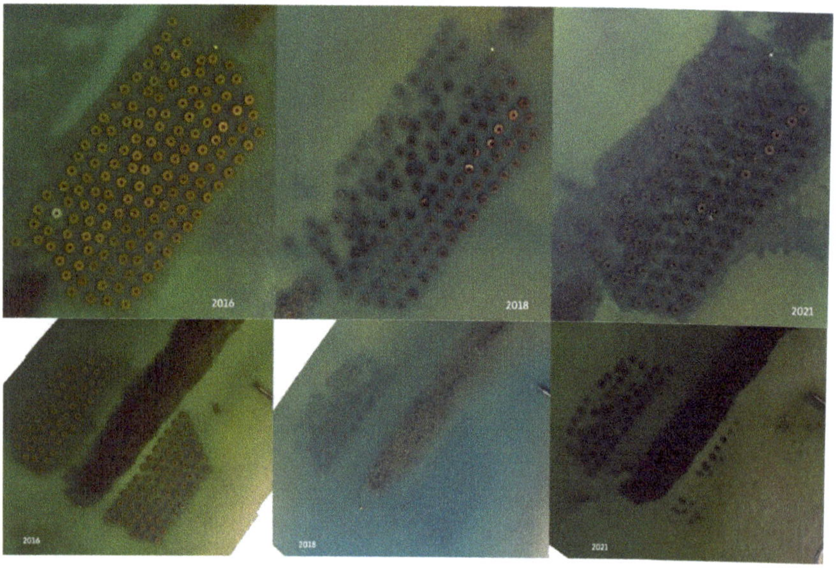

Figure 6. View of selected sample habitat modules placed in the central (**top**) and western (**bottom**) portions of the sill structure in Rowy over the years 2016–2021.

3. Methodology and Results

Owing to artificial reef construction, which includes a stone mound in addition to the habitat modules, further analysis included only the five groups of habitat modules, without considering the stone mound separating these modules in further work. The concrete blocks were arranged in fairly equal rows (the 1st module on the west side consists of 2 segments in which the blocks are arranged in 5 and 6 rows, the middle 3 modules have 9 rows each, and the eastern last module has 2 segments containing 8 and 9 rows). In several rows, concrete blocks were broken or overturned during the construction phase in 2016 and these elements were excluded from further analysis at the start. Only those elements where a circular shape could be recognized were included in the study. In 2018 and 2021, structural damage and displacement of the circles in each row are evident.

Each concrete block had a diameter of 2.4 m and different heights (i.e., 1.5, 2, and 2.5 m) depending on the depth of the bottom. The walls of the blocks were 15 cm thick and comprised 12 conical holes distributed evenly along the circumference and one conical hole on the upper surface. The characteristic shape and construction enabled the implementation of an automatic algorithm to detect submerged concrete circle-shaped elements in the UAV images acquired during the monitoring campaigns.

The proposed solution was based on image processing and analysis techniques. A four-step algorithm (Figure 7) was proposed, and an object detection methodology provided a list of circle-shaped origins stored in a floating point format (coordinates in the image plane) for each analyzed series of UAV images. The algorithm was implemented and validated in Python using the OpenCV, NumPy, Pillow, and Matplotlib standard libraries.

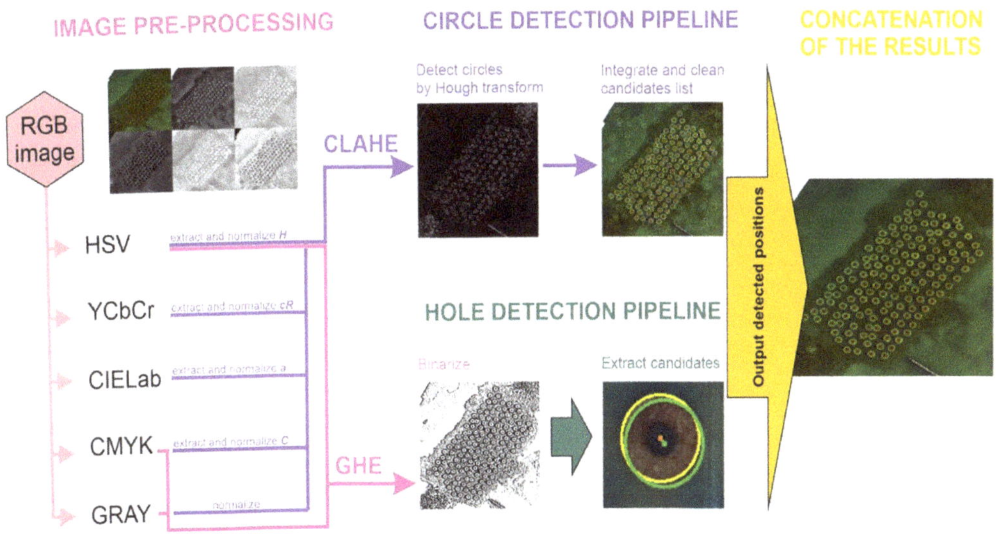

Figure 7. Monitored underwater construction in Rowy.

3.1. Image Processing

The original input images were delivered from UAV surveys in RGB color format with a possibly high spatial resolution. Because the color characteristics of individual images differed significantly (due to various environmental conditions, such as lighting, time of day, and season), several color-based pre-processing activities were required. Thus, each input image was converted into five dedicated representations: grayscale, CIE Lab, HSV, YCbCr, and CMYK. In this study, different channels from each of the five representations were tested and observed to provide a higher accuracy than that of a per-pixel-based approach [44]. Finally, the five most promising channels were chosen for further analysis: cyan from CMYK, Cr from YCbCr, hue from HSV, a channel from CIELab, and a grayscale image (Figure 8).

Figure 8. Example input image transformed into five dedicated representations.

3.2. Circle Detection Pipeline

The second step, which performs the detection of the circular elements of underwater construction, is the core of this approach. As the objects we are looking for are circles, we have proposed several methods to perform automatic circle recognition for each single-channel image [45]. First, we normalized all the extracted matrices to a 0–255 range and stored them in an unsigned int 8-bit format. Second, contrast-limited adaptive histogram equalization (CLAHE) [46] was applied to all the normalized images, with the following parameters: clip limit of 2.0 and tile grid size of 8 × 8 pixels. In each image, the detection of circular shapes with Hough transform (Figure 9) [47] was realized by assuming a very narrow (pre-defined) range of radii. For each individual list of detected candidates saved in the earlier step, we integrated circles that were closer than the distance equal to the radius by averaging their positions. Finally, we concatenated the five resulting lists (one for each pre-defined channel) and removed duplicates (Figure 9) (candidates that were close to each other) by using radius as a criterion.

Figure 9. Hough transformation (**left**) and elimination of duplicates (**right**).

3.3. Hole Detection Pipeline

In the next stage, we performed hole detection for the grayscale, cyan, and hue channels (Figure 10).

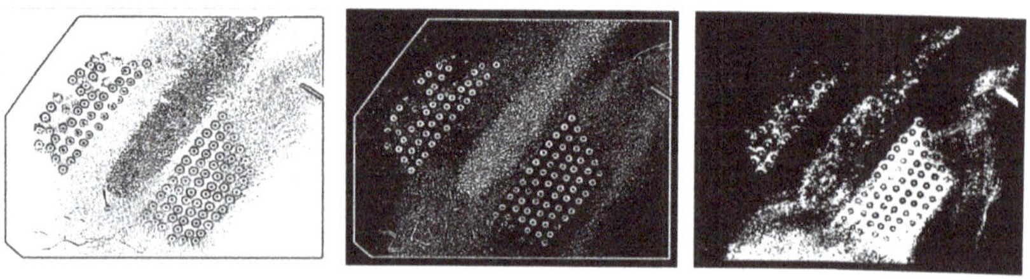

Figure 10. Preprocessing results for grayscale image (**left**), cyan channel (**middle**), and hue channel (**right**).

In this process, we detected the centers of circles that were significantly darker than their surroundings; the images were normalized to 0–255 and 8-bit representations and histogram equalization was performed according to global equalization criteria. The prepared images were segmented using adaptive thresholding. However, the images had to be blurred with a Gaussian mask of 15 × 15 pixels to smooth small objects that interrupted the threshold process. Prepared images were analyzed using a sequence of morphological opening and closing operations, with an ellipse of 5 × 5 pixels being used as a structuring element. This allowed us to find isolated contours (blobs) that satisfy the following conditions: a blob area with a larger range (600, 1400) and an aspect ratio of width to height in a specific range (0.8, 1.2). For each individual list of candidates, we integrated holes (candidates) that were closer than the distance equal to the radius by averaging their positions and removing duplicates (holes close to each other were removed by using radius as a criterion: Figure 11—left side).

Figure 11. Detection of circle centers (**left**) and detected objects of specific geometrical properties in respective color channels: green objects—gray channel; blue objects—cyan channel; red objects—hue channel.

3.4. Concatenation of the Results

The last analysis step included the concatenation of the results from the processing pipelines. First, the duplicates were removed by averaging the origins. Second, isolated candidates far from their neighbors were removed. We considered the distance between origins (based on the experiments, a value of 200 pixels was used). After removing all the duplicates and isolated candidates, we generated a list of circle origins for all test campaigns.

3.5. Validation Process

In the process of validating the algorithm, two qualified operators (specialists in the interpretation of remote sensing data of the coastal zone) independently identified the origin of the circles visible at the selected test site. In addition, a specially developed web-based program, based on the Leaflet Javascript library and accessible via a web browser, was used to benchmark the location recognition of individual infrastructure elements. This program works based on an algorithm and manual classification. A portal was created to allow multiple users (mostly senior students from various marine, geography, and computer science majors) to simultaneously enter identification points for the center of concrete circles. Each user, after registering, was given a random set of images with habitat modules from different years, and their task was to mark the centers of all circles visible in each image (Figure 12A). In this way, results were collected from approximately 40 additional operators who collectively marked over 25,000 points (Figure 12C). The resulting point coordinates were then stored in a PostgreSQL database (Figure 12B). The database allowed the rejection of significant outliers and averaged the data to centroids in the WGS84 system (EPSG:4326).

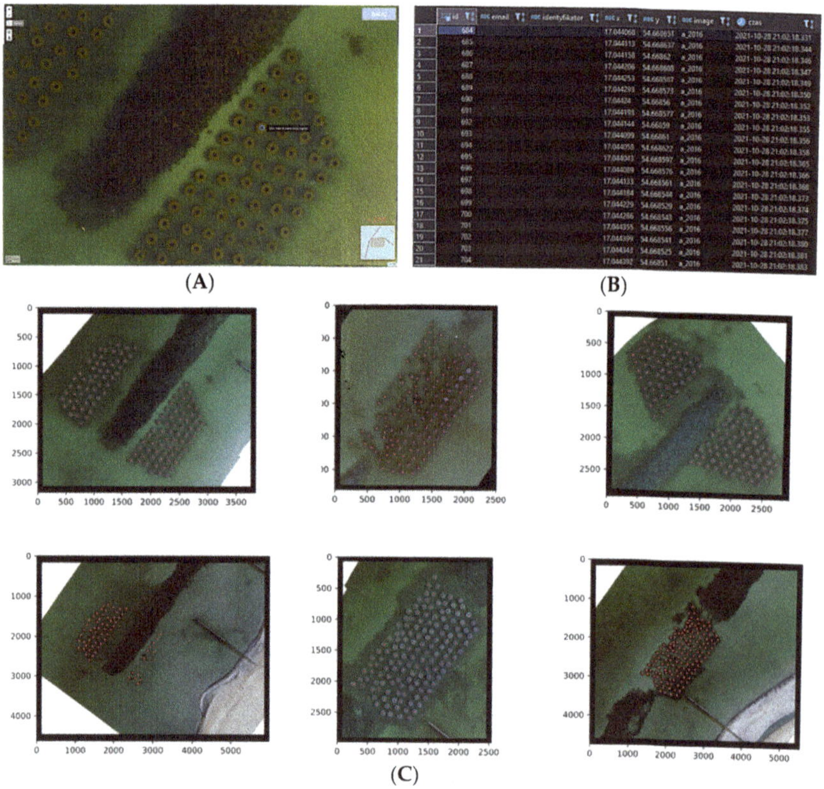

Figure 12. (**A**) Validation process in dedicated software accessible via a web browser. (**B**) Table view from PostgreSQL database. (**C**) Example validation results with points marked by test operators.

In the next step, the obtained data was transformed into a cartesian grid format corresponding to the pixel layout of the individual images in the X and Y axes system. The upper-left corner of the raster was taken as point 0,0. The program was written in the Python language. For analysis, the identified circle origins were averaged and taken as reference points, which were compared with the output of the origins obtained using the automatic method.

The origins of circles undetected by our algorithm were not considered during the accuracy estimation stage. The comparison utilized several stages. For each point in the reference list, the closest point in the list was searched for based on the coordinates in the image plane. The Euclidean distance between the points was calculated and stored as an error value. However, if several points were found within that distance, the closest one was chosen. Finally, if no points were found in the predefined tolerance area, the algorithm skipped to the next point in the reference list.

The study of variation in the position of points (geometric centers of circles) was based on the rejection of constancy of the position of any of the points. In this method, it was assumed that any number of test points could move in time in any direction. However, it was assumed that there would be a certain number of fixed points, and the constancy of these points could be confirmed by computational methods. At the same time, the variability of the position of the moving points will be determined in relation to the position of the points considered as a fixed. Therefore, research questions arose, as follows: How could we find fixed points? How could we determine the constancy of these points (reference datum)? How could we determine the displacements of the moving points relative to the fixed points? A method for answering these questions using coordinate transformation was proposed, as shown in Figure 13.

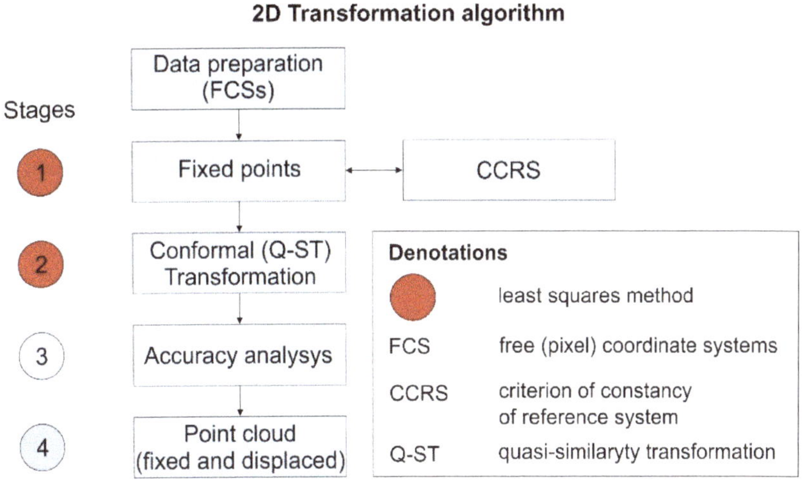

Figure 13. Algorithm: 2D transformation algorithm.

In the first stage of the 2D transformation (Figure 13), the criterion of points stability (reference datum) was determined. This criterion is defined as the uncertainty (mean error) of the point position after the transformation, not exceeding the double value of the standard deviation, which is described by Formula (1). This approach is applicable to the study of the deformation of engineering structures. The coordinates of the outlier points were not used to determine the coordinate transformation parameters, and these points were treated as moving and the displacement (deformation) values are determined for them [48,49].

$$V_i \leq 2\sigma_0 \tag{1}$$

Criterion (1) was applied iteratively, with the rejection of points that exceeded twice the standard uncertainty, until the set of points stabilized, i.e., the position uncertainties of all points met condition (1). These points were considered fixed, and the displacements of the moving points were determined relative to these points. For the transformation of coordinates, a conformal (Helemrt) transformation model with a center of gravity, in the form of a Q-ST (quasi similarity transformation), was adopted in Relation (2) [50], as follows:

$$\begin{aligned} X_{TRF} &= Q\cdot\left(\lambda\cdot X_{trf} - X_{ctrf}\right) + X_{CTRF} \\ &= Q\cdot\left[\left(D_{trf}{}^{T}\cdot D_{trf}\right)^{-1}\cdot D_{trf}{}^{T}\cdot D_{TRF}\right]\cdot X_{trf} - \frac{\sum_{i=1}^{N}(X_{ictrf})}{n} + X_{CTRF} \end{aligned} \tag{2}$$

where:

- X_{TRF}—coordinates in the target reference frame (TRF);
- Q—(quasi) rotation matrix;
- λ—scale factor;
- X_{trf}—coordinates in the transformed reference frame (trf);
- X_{ctrf}—coordinates of the center of gravity in the transformed reference frame (centroid of scaled transformed reference frame);
- X_{CTRF}—coordinates of the center of gravity in the target reference system (centroid of target reference frame);
- D_{TRF}—vector length in the target reference frame (TRF);

- D_{trf}—vector length in the transformed reference frame (trf).

The rotation matrix Q (quasi) in Relation (2), in 2D space, is given in Matrix (3), as follows:

$$Q = \begin{bmatrix} \lambda_X & 0 \\ 0 & \lambda_Y \end{bmatrix} \times \begin{bmatrix} \cos\varphi & -\sin\varphi \\ \sin\varphi & \cos\varphi \end{bmatrix} = \begin{bmatrix} q_{11} & q_{12} \\ q_{21} & q_{22} \end{bmatrix} \quad (3)$$

The coefficients of Matrix (3) were determined using the least squares method, according to Formula (4), as follows:

$$\begin{bmatrix} X_{C1} \\ Y_{C1} \\ X_{C2} \\ Y_{C2} \\ \cdots \\ X_{Cn} \\ Y_{Cn} \end{bmatrix} = \begin{bmatrix} x_{c1} & -y_{c1} & 0 & 0 \\ 0 & 0 & x_{c1} & -y_{c1} \\ x_{c2} & -y_{c2} & 0 & 0 \\ 0 & 0 & x_{c2} & -y_{c2} \\ & & \cdots & \\ x_{cn} & -y_{cn} & 0 & 0 \\ 0 & 0 & x_{cn} & -y_{cn} \end{bmatrix} \times \begin{bmatrix} q_{11} \\ q_{12} \\ q_{21} \\ q_{22} \end{bmatrix} \quad (4)$$

where

$$X = (A^T \cdot A)^{-1} \cdot A^T \cdot L$$

As a result of the iterative approach in the transformation in Relation (2), fixed points were selected to satisfy Condition (1). In the next step, the transformation in Relation (2) was performed for the moving points, that is, those that did not participate in the determination of the transformation coefficients. The mean errors after the Q-ST transformation in Relation (2), for the fixed points and the displacements for the moving points, are summarized in Table 1.

Table 1. Mean errors of transformations (standard deviation) and mean displacements for moving points.

Measurement Period	No. of Stable (Adjusted) Points	Mean Errors on Adjusted Points (pix)			Displacement (Mean Value) of Moving Points (pix)		
		m_X	m_Y m_Z	m_P	m_X	m_Y m_Z	m_P
2016	13	0.9	1.0	1.4	-	-	-
2018–2021	13	-	-	-	27.3	19.8	33.7

The data presented in Table 1 show that the average transformation error on points determined as fixed is within the values of determination of geometric centers of circles (Figure 14), which indicates high stability of the position of points—the constancy of the reference system. At the same time, it can be noted that the average displacement of points in the period 2018–2021 for moving points is 34 pix, which for a ground sampling distance (GSD) of 2 cm is about 0.7 m.

The overall detection accuracy (understood as a number of correctly detected objects in comparison to the actual ones) is equal to 0.987, varying for particular years (see Figure 13, left chart). According to the cumulative histogram of error distributions (Figure 13, right chart), the detection accuracy is quite high. The processing of all channels ("all") instead of individual ones ("c"—cyan; "gr"—grayscale; "h"—HSV hue; "cr"—YCbCr chroma; "a"—Lab) gives the highest detection accuracy. The errors distribution for the cr channel is slightly better than for all the concatenated channels. Detection accuracy for all the channels is higher, which makes such an approach more sensible. The majority of the errors did not exceed 4 pixels, with an average of 4.06 and a median of 4.12 pixels. Pixel value has been presented here as a field pixel representation—GSD (ground sampling distance)—a value of 0.02 m. This is primarily because the operators marked the circle origins by considering the shadow inside the central part of the objects, whereas the algorithm detected the center on the two-dimensional projection of the three-dimensional structure. It can be observed on the spatial distribution plot (Figure 15), where the majority of errors are within offset (2.2). The closer look unveils that the accuracy for images taken in 2016 and 2018 is quite high. Moreover, the errors are kept in significantly low range. On the other hand, images taken in 2021 characterize lower accuracy and larger spread of errors distribution, since in these images have lower contrast coming from the fact that the underwater structures have been covered by sand.

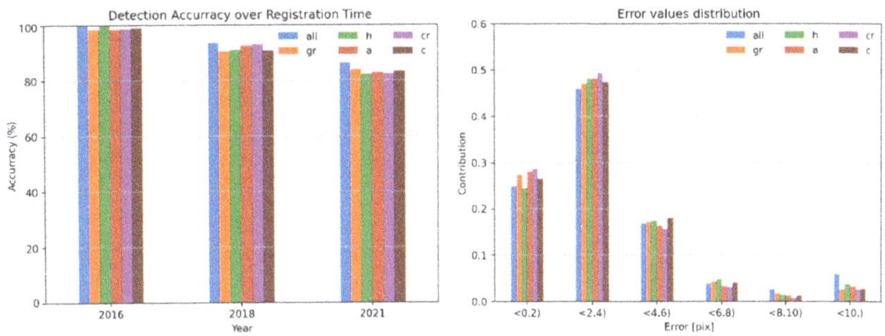

Figure 14. The accuracy (**left** chart) and the statistical distribution of errors (frequency histogram, **right** chart).

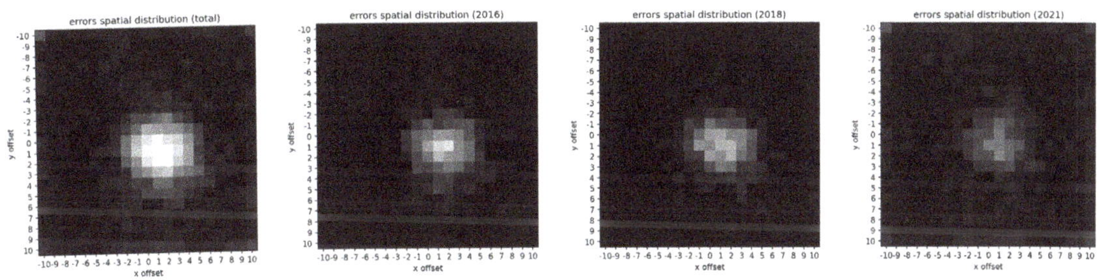

Figure 15. The spatial distribution of errors for specific registration years.

The additional verification of the employed approach involved the same set of experiments, yet with a focus on elementary algorithm stages (i.e., circle detection and hole detection only stages). We have calculated accuracy (as a percentage of detected objects in comparison to the actual ones) and error distribution as a distance between detected object position and the actual one. The detailed results are presented in Figures 16 and 17. It is worth noticing that the detecting holes in the hue channel for images taken in 2016 gave a very low accuracy, while for images taken in 2018 and 2021 it worked better. On the other hand, detecting holes in cyan channel for images taken in 2021 was also not efficient. As it can been seen, an approach founded on a single detection stage (no matter which one) cannot result in high accuracy and low detection errors. The simple solution, that makes the algorithm insensitive to color characteristics, is concatenation of the results from different channels.

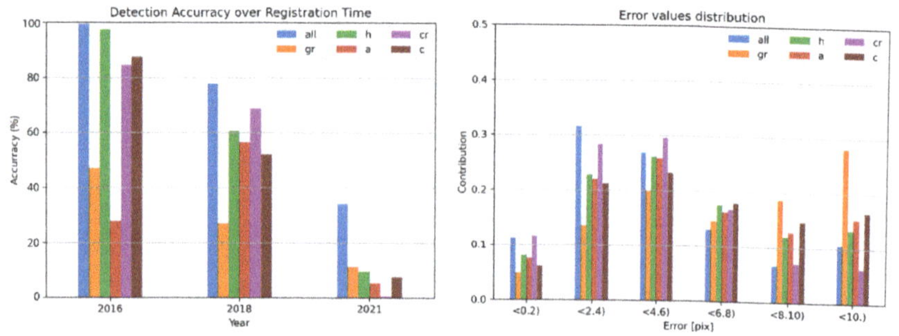

Figure 16. The accuracy of object detection and the distribution of errors (related to the elementary color channels)—employing only circle detection.

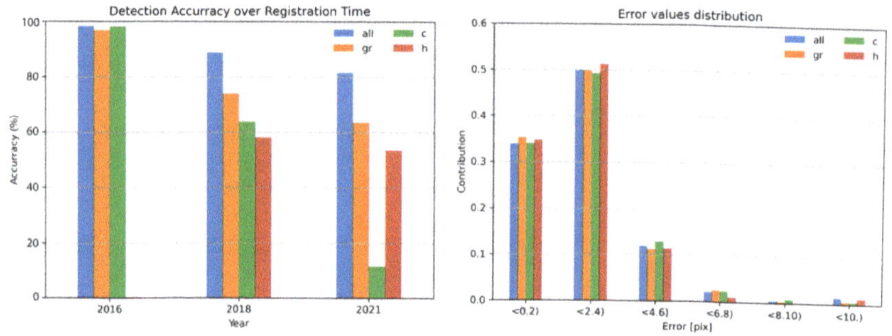

Figure 17. The accuracy of object detection and the distribution of errors (related to the elementary color channels)—employing only holes detection.

4. Discussion

In this paper, we present an innovative work, where an off-the-shelf UAV with a regular RGB camera was used to monitor shallow hybrid coastal protection measures. Moreover, the remarkable potential of UAV image-based analysis has been proven by the implementation of bathymetry estimation methods [3,36,51] and monitoring techniques for submerged objects [37,52,53]; however, none of the presented research have been used to identify underwater coastal protection measures. They are focused on military purposes [52] or water organisms and waste [52,53]. Furthermore, the employed image processing methods have been successfully applied to circular objects detection in other fields of science [54,55]; hence, the application of Hough transform and related image processing-based methods to analyze underwater objects is clearly a new approach.

The principal part of the analyzed underwater artificial reef in our experiment contained multiple regular circle textures. Based on these elements, the implemented algorithm was optimized and customized to provide good identification results with the proposed methodology. Inevitably, some elements require more attention; however, the sections identified according to the construction documentation are crucial for reef stability. Circular concrete blocks are the most exposed to hydrometeorological conditions; thus, monitoring these elements is obligatory for construction efficiency.

The proposed methodology was used for five different color separation stages in order to be invariant to as many factors as possible. Nevertheless, several thresholds to analyze the data have been implemented. The algorithm was tuned to specific characteristics of input data, while using thresholds and predefined values that can be easily adopted; this enables the proposed method to be used in different applications and in various environmental conditions. The pixel size of circles or holes depends on their physical dimensions and is related to the imaging altitude and spatial resolution of images. It primarily influences the averaging filter mask, morphological element size, and blob area. The thresholds used at the intensity segmentation stages are universal to some extent, since they have been calculated for several images acquired at different times and various seasons.

UAV, combined with GNSS measurements, has been shown in many works to be an efficient tool for high-resolution analysis in coastal zones [56–58]. It has been proved that UAV photogrammetry is useful when decimeter-level accuracy is required. Different surveys report photogrammetric model accuracy values on sub-centimeter values, respectively, for the horizontal component [59,60]. The accuracy we found in identification of underwater objects by UAV was very similar. The majority of errors did not exceed 4 pixels (8 cm) for the 3 surveys. Such a result is very similar to that found in [33,59,61,62], even though, in those surveys, UAV was used to identified elements on land instead of under water.

The realized experimentation demonstrates that the UAV is a suitable tool to monitor underwater structures withing accepted range of the error propagation. However, this is only valid if the accuracy is assessed and monitored throughout the entire process. Notably, both the circle and hole detection process documented here, and the considerations presented in this manuscript, were made possible because the weather conditions were thoroughly analyzed and tracked throughout the entire process. Although this research feature, in particular, made it possible to evaluate the usage of UAV monitoring of underwater structures, it is suggested that uncertainty is taken into account while processing UAV surveys for coastal monitoring of underwater objects.

The results reveal that the movement complexity of the identified objects is visible over time. This leads to a need for additional monitoring procedures and analyses of changes in reef construction.

5. Conclusions

This manuscript proposed a monitoring solution for underwater coastal protection measures by automated identification of crucial construction elements in RGB UAV photographs, which was performed using image processing algorithms. We applied two different pipelines and concatenated the outcomes. The experimental results showed that the accuracy of our algorithm meets the requirements of the monitoring process and can be used with readily available UAV equipment to save manpower and material costs. Furthermore, we will continue to conduct an analysis of various underwater protection measures to extend this research to include more application scenarios. Additionally, differences in the identified objects, which have been observed using UAV image time series, will be analyzed and compared against hydrometeorological conditions to determine the most significant parameters that influence the protection measures in coastal zones.

The UAV monitoring campaign turned out to be suitably accurate for studying coastal defense underwater structures, for two reasons especially: firstly, the practicality of these devices—this is due to their wide availability and operating velocity, especially in comparison with traditional expensive methods of mapping underwater objects with echosounders. Secondly, spectral RGB data represents a huge advantage when conducting 2D detection of underwater objects and their relationship to topography.

The realized UAV surveys in the underwater environment experience different types of limitations. They are represented by the complexity and, at a certain level, the impossibility of complete detection of single elements of analyzed construction, which are partly covered with sand. This factor strongly complicates the detection process, and final results can be assumed to be realistic when elements such as sand or vegetation have little or no influence.

Despite this, the fully automatic object detection algorithm presented here, which is able to work on RGB data, represents a rapid and useful way to monitor the underwater coastal structures. The high resolution of UAV data allowed rapid elaborations that, until today, were much more time consuming, and only in the best-case scenario (multibeam echosounders) did these attain a similar resolution. Here, we confirm the conclusions made by Fabbri et al., 2021, that the availability of small, high-resolution sensors that UAVs can transport, such as RGB cameras or multi-spectral cameras, is increasing, presenting a new approach for high- resolution spatial studies. Thus, the implementation of the presented methodology can deliver data on underwater coastal protection on demand, measuring comparably high spatial, and much higher temporal, resolutions than traditional measurements. These attributes are particularly useful for a range of coastal applications, such as monitoring and other activities within integrated coastal zone management.

Author Contributions: Conceptualization, J.Ś.; methodology, J.Ś., P.T., P.F. and A.G.; software, A.Ł. and W.M.; validation, J.Ś., P.T., G.S. and A.T.; formal analysis, J.Ś., P.T., A.G. and P.F.; data curation, J.Ś., A.Ł., A.K. and W.M.; writing—original draft preparation, J.Ś., P.T., A.G., P.F. and G.S.; writing—review and editing, J.Ś. and A.T.; visualization, J.Ś., A.Ł. and W.M.; project administration, P.T.; funding acquisition, P.T. and A.K. All authors have read and agreed to the published version of the manuscript.

Funding: This research was funded in part by the National Science Centre, Poland, through the project "Influence of surfzone and beach morphology on coastal cliff retreat—INSUMOR", project number 2020/39/B/ST10/01122.

Institutional Review Board Statement: Not applicable.

Informed Consent Statement: Not applicable.

Data Availability Statement: Data available on request.

Conflicts of Interest: The authors declare no conflict of interest.

References

1. Musielak, S.; Furmańczyk, K.; Bugajny, N. Factors and Processes Forming the Polish Southern Baltic Sea Coast on Various Temporal and Spatial Scales. In *Coastline Changes of the Baltic Sea from South to East*; Springer: Boca Raton, FL, USA, 2017; Volume 19, pp. 69–86.
2. Terefenko, P.; Zelaya Wziątek, D.; Dalyot, S.; Boski, T.; Pinheiro Lima-Filho, F. A High-Precision LiDAR-Based Method for Surveying and Classifying Coastal Notches. *ISPRS Int. J. Geo-Inf.* **2018**, *7*, 295. [CrossRef]
3. Tsukada, F.; Shimozono, T.; Matsuba, Y. UAV-based mapping of nearshore bathymetry over broad areas. *Coast. Eng. J.* **2020**, *62*, 285–298. [CrossRef]
4. Vousdoukas, M.I.; Ranasinghe, R.; Mentaschi, L.; Plomaritis, T.A.; Athanasiou, P.; Luijendijk, A.; Feyen, L. Sandy coastlines under threat of erosion. *Nat. Clim. Chang.* **2020**, *10*, 260–263. [CrossRef]
5. Ibaceta, R.; Splinter, K.D.; Harley, M.D.; Turner, I.L. Enhanced coastal shoreline modelling using an Ensemble Kalman Fil-ter to include non-stationarity in future wave climates. *Geophys. Res. Lett.* **2020**, *47*, 22–2020. [CrossRef]
6. Paprotny, D.; Terefenko, P.; Giza, A.; Czapliński, P.; Vousdoukas, M.I. Future losses of ecosystem services due to coastal ero-sion in Europe. *Sci. Total Environ.* **2021**, *760*, 144310. [CrossRef] [PubMed]
7. Terefenko, P.; Terefenko, O. Determining the role of exposure, wave force, and rock chemical resistance in marine notch de-velopment. *J. Coast. Res.* **2014**, *70*, 706–711. [CrossRef]
8. Uścinowicz, G.; Szarafin, T. Short-term prognosis of development of barrier-type coasts (Southern Baltic Sea). *Ocean Coast. Manag.* **2018**, *165*, 258–267. [CrossRef]
9. Terefenko, P.; Giza, A.; Paprotny, D.; Kubicki, A.; Winowski, M. Cliff Retreat Induced by Series of Storms at Międzyzdroje (Poland). *J. Coast. Res.* **2018**, *85*, 181–185. [CrossRef]
10. De Sanjosé Blasco, J.J.; Gómez-Lende, M.; Sánchez-Fernández, M.; Serrano-Cañadas, E. Monitoring Retreat of Coastal Sandy Systems Using Geomatics Techniques: Somo Beach (Cantabrian Coast, Spain, 1875–2017). *Remote Sens.* **2018**, *10*, 1500. [CrossRef]
11. Terefenko, P.; Giza, A.; Paprotny, D.; Walczakiewicz, S. Characteristic of Winter Storm Xavier and Its Impacts on Coastal Morphology: Results of a Case Study on the Polish Coast. *J. Coast. Res.* **2020**, *95*, 684–688. [CrossRef]
12. Bugajny, N.; Furmańczyk, K. Short-term Volumetric Changes of Berm and Beachface during Storm Calming. *J. Coast. Res.* **2020**, *95*, 398–402. [CrossRef]
13. Stronkhorst, J.; Huismana, B.; Giardinoa, A.; Santinellia, G.; Santosb, F.F. Sand nourishment strategies to mitigate coastal erosion and sea level rise at the coasts of Holland (The Netherlands) and Aveiro (Portugal) in the 21st century. *Ocean Coast. Manag.* **2018**, *156*, 266–276. [CrossRef]
14. van Rijn, L. Coastal erosion and control. *Ocean Coast. Manag.* **2011**, *54*, 867–887. [CrossRef]
15. Morris, R.L.; Boxshall, A.; Swearer, S. Climate-resilient coasts require diverse defence solutions. *Nat. Clim. Chang.* **2020**, *10*, 485–487. [CrossRef]
16. Splinter, K.D.; Harley, M.D.; Turner, I.L. Remote Sensing Is Changing Our View of the Coast: Insights from 40 Years of Monitoring at Narrabeen-Collaroy, Australia. *Remote Sens.* **2018**, *10*, 1744. [CrossRef]

17. Holman, R.; Plant, N.; Holland, T. cBathy: A robust algorithm for estimating nearshore bathymetry. *J. Geophys. Res. Oceans* **2013**, *118*, 2595–2609. [CrossRef]
18. Zelaya Wziątek, D.; Terefenko, P.; Kurylczyk, A. Multi-Temporal Cliff Erosion Analysis Using Airborne Laser Scanning Sur-veys. *Remote Sens.* **2019**, *11*, 2666. [CrossRef]
19. Capperucci, R.M.; Kubicki, A.; Holler, P.; Bartholomä, A. Sidescan sonar meets airborne and satellite remote sensing: Chal-lenges of a multi-device seafloor classification in extreme shallow water intertidal environments. *Geo-Mar. Lett.* **2020**, *40*, 117–133. [CrossRef]
20. Vacchi, M.; Rovere, A.; Schiaffino, C.F.; Ferrari, M. Monitoring the effectiveness of re-establishing beaches artificially: Meth-odological and practical insights into the use of video transects and SCUBA operated coring devices. *Underw. Technol.* **2012**, *30*, 201–206. [CrossRef]
21. Brouwer, R.L.; de Schipper, M.A.; Rynne, P.F.; Graham, F.J.; Reniers, A.J.H.M.; MacMahan, J.H. Surfzone Monitoring Using Rotary Wing Unmanned Aerial Vehicles. *J. Atmos. Ocean. Technol.* **2015**, *32*, 855–863. [CrossRef]
22. Talavera, L.; Del Río, L.; Benavente, J.; Barbero, L.; López-Ramírez, J.A. UAS as tools for rapid detection of storm-induced morphodynamic changes at Camposoto beach, SW Spain. *Int. J. Remote Sens.* **2018**, *39*, 5550–5567. [CrossRef]
23. Brodie, K.L.; Palmsten, M.L.; Hesser, T.J.; Dickhudt, P.J.; Raubenheimer, B.; Ladner, H.; Elgar, S. Evaluation of videobased linear depth inversion performance and applications using altimeters and hydrographic surveys in a wide range of environ-mental conditions. *Coast. Eng.* **2018**, *136*, 147–160. [CrossRef]
24. Turner, I.L.; Harley, M.; Drummond, C.D. UAVs for coastal surveying. *Coast. Eng.* **2016**, *114*, 19–24. [CrossRef]
25. Cheng-Hao, L. Applying UAV and photogrammetry to monitor the morphological changes along the beach in Penghu Islands. *Int. Arch. Photogramm. Remote Sens. Spat. Inf. Sci.* **2016**, *XLI-B8*, 1153–1156.
26. Derian, P.; Almar, R. Wavelet-Based Optical Flow Estimation of Instant Surface Currents From Shore-Based and UAV Videos. *IEEE Trans. Geosci. Remote Sens.* **2017**, *55*, 5790–5797. [CrossRef]
27. Gómez-Gutiérrez, Á.; Gonçalves, G.R. Surveying coastal cliffs using two UAV platforms (multirotor and fixed-wing) and three different approaches for the estimation of volumetric changes. *Int. J. Remote Sens.* **2020**, *41*, 8143–8175. [CrossRef]
28. Brunetta, R.; Duo, E.; Ciavola, P. Evaluating Short-Term Tidal Flat Evolution Through UAV Surveys: A Case Study in the Po Delta (Italy). *Remote Sens.* **2021**, *13*, 2322. [CrossRef]
29. Fabbri, S.; Grottoli, E.; Armaroli, C.; Ciavola, P. Using High-Spatial Resolution UAV-Derived Data to Evaluate Vegetation and Geomorphological Changes on a Dune Field Involved in a Restoration Endeavour. *Remote Sens.* **2021**, *13*, 1987. [CrossRef]
30. Terefenko, P.; Lubczonek, J.; Paprotny, D. Editorial on Special Issue "Remote Sensing Applications in Coastal Environment". *Remote Sens.* **2021**, *13*, 4734. [CrossRef]
31. Luppichini, M.; Bini, M.; Paterni, M.; Berton, A.; Merlino, S. A New Beach Topography-Based Method for Shoreline Identifi-cation. *Water* **2020**, *12*, 3110. [CrossRef]
32. Westoby, M.J.; Brasington, J.; Glasser, N.F.; Hambrey, M.J.; Reynolds, J.M. "Structure-from-Motion" photogrammetry: A low-cost, effective tool for geoscience applications. *Geomorphology* **2012**, *179*, 300–314. [CrossRef]
33. Laporte-Fauret, Q.; Marieu, V.; Castelle, B.; Michalet, R.; Bujan, S.; Rosebery, D. Low-Cost UAV for High-Resolution and Large-Scale Coastal Dune Change Monitoring Using Photogrammetry. *J. Mar. Sci. Eng.* **2019**, *7*, 63. [CrossRef]
34. Casella, E.; Drechsel, J.; Winter, C.; Benninghoff, M.; Rovere, A. Accuracy of sand beach topography surveying by drones and photogrammetry. *Geo-Mar. Lett.* **2020**, *40*, 255–268. [CrossRef]
35. Rossi, L.; Mammi, I.; Pelliccia, F. UAV-Derived Multispectral Bathymetry. *Remote Sens.* **2020**, *12*, 3897. [CrossRef]
36. Matsuba, Y.; Sato, S. Nearshore bathymetry estimation using UAV. *Coast. Eng. J.* **2018**, *60*, 51–59. [CrossRef]
37. Taddia, Y.; Russo, P.; Lovo, S.; Pellegrinelli, A. Multispectral UAV monitoring of submerged seaweed in shallow water. *Appl. Geomat.* **2019**, *12*, 19–34. [CrossRef]
38. Luijendijk, A. The state of the world's beaches. *Sci. Rep.* **2018**, *8*, 6641. [CrossRef]
39. Cooper, J.A.G.; Masselink, G.; Coco, G.; Short, A.D.; Castelle, B.; Rogers, K.; Anthony, E.; Green, A.N.; Kelley, J.T.; Pilkey, O.H.; et al. Sandy beaches can survive sea-level rise. *Nat. Clim. Chang.* **2020**, *10*, 993–995. [CrossRef]
40. Kandrot, S.; Hayes, S.; Holloway, P. Applications of Uncrewed Aerial Vehicles (UAV) Technology to Support Integrated Coastal Zone Management and the UN Sustainable Development Goals at the Coast. *Estuaries Coasts* **2021**. [CrossRef]
41. Scarelli, F.M.; Sistilli, F.; Fabbri, S.; Cantelli, L.; Barboza, E.; Gabbianelli, G. Seasonal dune and beach monitoring using photogrammetry from UAV surveys to apply in the ICZM on the Ravenna coast (Emilia-Romagna, Italy). *Remote Sens. Appl. Soc. Environ.* **2017**, *7*, 27–39. [CrossRef]
42. Florek, W.; Kaczmarzyk, J.; Majewski, M. Dynamics of the Polish coast east of Ustka. *Geogr. Pol.* **2011**, *83*, 51–60. [CrossRef]
43. Cieślikiewicz, W.; Paplińska-Swerpel, B. A 44-year hindcast of wind wave fields over the Baltic Sea. *Coast. Eng.* **2008**, *55*, 894–905. [CrossRef]
44. Forczmanski, P.; Markiewicz, A. Two-stage approach to extracting visual objects from paper documents. *Mach. Vis. Appl.* **2016**, *27*, 1243–1257. [CrossRef]
45. Forczmański, P.; Markiewicz, A. Stamps Detection and Classification Using Simple Features Ensemble. *Math. Probl. Eng.* **2015**, *2015*, 367879. [CrossRef]
46. Pizer, S.M.; Amburn, E.P.; Austin, J.D.; Cromartie, R.; Geselowitz, A.; Greer, T.; ter Haar Romeny, B.; Zimmerman, J.B.; Zuiderveld, K. Adaptive histogram equalization and its variations. *Comput. Vis. Graph. Image Process.* **1987**, *39*, 355–368. [CrossRef]

47. Duda, R.O.; Hart, P.E. Use of the Hough transformation to detect lines and curves in pictures. *Commun. ACM* **1972**, *15*, 11–15. [CrossRef]
48. Hooijberg, M. *Practical Geodesy*; Springer: Berlin/Heidelberg, Germany, 1997. [CrossRef]
49. Schofield, W.; Breach, M. Engineering Surveying, 6th ed.CRC Press: New York, NY, USA, 2007.
50. Stępień, G.; Tomczak, A.; Loosaar, M.; Ziębka, T. Dimensioning Method of Floating Offshore Objects by Means of Quasi-Similarity Transformation with Reduced Tolerance Errors. *Sensors* **2020**, *20*, 6497. [CrossRef]
51. He, J.; Lin, J.; Ma, M.; Liao, X. Mapping topo-bathymetry of transparent tufa lakes using UAV-based photogrammetry and RGB imagery. *Geomorphology* **2021**, *389*, 107832. [CrossRef]
52. Matulewski, K.V.; Mcbride, W. Day/night underwater object detection from an airborne sensor using NOVAS (Non-acoustical Optical Vulnerability Assessment Software). In Proceedings of the OCEANS 2005 MTS/IEEE, Washington, DC, USA, 17–23 September 2005; Volume 3, pp. 2274–2278.
53. Watanabe, J.-I.; Shao, Y.; Miura, N. Underwater and airborne monitoring of marine ecosystems and debris. *J. Appl. Remote Sens.* **2019**, *13*, 044509. [CrossRef]
54. Smereka, M.; Duleba, I. Circular Object Detection Using a Modified Hough Transform. *Int. J. Appl. Math. Comput. Sci.* **2008**, *18*, 85–91. [CrossRef]
55. Strokina, N.; Matas, J.; Eerola, T.; Lensu, L.; Kälviäinen, H. Detection of bubbles as concentric circular arrangements. *Mach. Vis. Appl.* **2016**, *27*, 387–396. [CrossRef]
56. Moloney, J.G.; Hilton, M.J.; Sirguey, P.; Simons-Smith, T. Coastal Dune Surveying Using a Low-Cost Remotely Piloted Aerial System (RPAS). *J. Coast. Res.* **2018**, *345*, 1244–1255. [CrossRef]
57. Casella, E.; Rovere, A.; Pedroncini, A.; Stark, C.P.; Casella, M.; Ferrari, M.; Firpo, M. Drones as tools for monitoring beach topography changes in the Ligurian Sea (NW Mediterranean). *Geo-Mar. Lett.* **2016**, *36*, 151–163. [CrossRef]
58. Solazzo, D.; Sankey, J.B.; Sankey, T.T.; Munson, S.M. Mapping and measuring aeolian sand dunes with photogrammetry and LiDAR from unmanned aerial vehicles (UAV) and multispectral satellite imagery on the Paria Plateau, AZ, USA. *Geomorphology* **2018**, *319*, 174–185. [CrossRef]
59. Vecchi, E.; Tavasci, L.; De Nigris, N.; Gandolfi, S. GNSS and Photogrammetric UAV Derived Data for Coastal Monitoring: A Case of Study in Emilia-Romagna, Italy. *J. Mar. Sci. Eng.* **2021**, *9*, 1194. [CrossRef]
60. Zanutta, A.; Lambertini, A.; Vittuari, L. UAV Photogrammetry and Ground Surveys as a Mapping Tool for Quickly Monitoring Shoreline and Beach Changes. *J. Mar. Sci. Eng.* **2020**, *8*, 52. [CrossRef]
61. Mancini, F.; Dubbini, M.; Gattelli, M.; Stecchi, F.; Fabbri, S.; Gabbianelli, G. Using Unmanned Aerial Vehicles (UAV) for High-Resolution Reconstruction of Topography: The Structure from Motion Approach on Coastal Environments. *Remote Sens.* **2013**, *5*, 6880–6898. [CrossRef]
62. Long, N.; Millescamps, B.; Pouget, F.; Dumon, A.; Lachaussée, N.; Bertin, X. Accuracuy assessment of coastal topography derived from UAV images. *Int. Arch. Photogramm. Remote Sens. Spat. Inf. Sci.* **2016**, *XLI-B1*, 1127–1134. [CrossRef]

remote sensing

MDPI

Article

Simplified Method of Determination of the Sound Speed in Water on the Basis of Temperature Measurements and Salinity Prediction for Shallow Water Bathymetry

Artur Makar

Department of Navigation and Hydrography, Polish Naval Academy, Śmidowicza 69, 81-127 Gdynia, Poland; Artur.Makar@amw.gdynia.pl

Abstract: The aim of this paper is to present a method of determining sound speed in water, based on temperature measurements executed by means of a laboratory low-cost thermometer with a probe provided with a long cable. It has been assumed that the salinity variation in respect to depth, found in a shallow water area, has insignificant impact on the sound velocity distribution determined by the temperature changes. The salinity data were obtained via the Internet service from the closest measuring station that registers surface water parameters. The sound speed in water was determined based on the formulas widely adopted in hydroacoustics and compared with the results obtained from the measurements executed by means of a Conductivity/Salinity Temperature Depth (CTD/STD) probe. The impact of inaccuracy in determining the sound speed in respect to the SingleBeam EchoSounder (SBES) immersion depth, i.e., a method commonly used by unmanned surface vessels in seaport measurements, was estimated. The measurements were taken in water areas of the Baltic Sea of low salinity and then verified with measurements in the Mediterranean Sea representing quite high salinity. The method is an alternative for calibrating the SBES the bar check way and has the capacity to meet the requirements in respect to its application in hydrographic surveys.

Keywords: sound speed in water; sound velocity profiler; hydrographic surveys; reliability of digital sea bottom model

Citation: Makar, A. Simplified Method of Determination of the Sound Speed in Water on the Basis of Temperature Measurements and Salinity Prediction for Shallow Water Bathymetry. *Remote Sens.* **2022**, *14*, 636. https://doi.org/10.3390/rs14030636

Academic Editors: Jacek Lubczonek, Paweł Terefenko, Katarzyna Bradtke and Marta Wlodarczyk-Sielicka

Received: 7 December 2021
Accepted: 25 January 2022
Published: 28 January 2022

1. Introduction

Determination of the sound speed in water is a basic subject matter in hydrographic measurements. Operation of acoustic devices serving depth measurements is based, very simply stated, on a path equation in a straight-lined linear movement. The distance of the acoustic wave impulse path in water is a route between an electroacoustic transducer and the bottom or another obstacle with reflective qualities. When the sound velocity in water is known, the distance to the bottom is determined based on the measured travelling time along this route. For many years, it was sufficient to know the average speed of sound in the bathymetric measurements executed by means of SBES (SingleBeam EchoSounder). In the depth measurements requiring lower accuracy than in hydrography, e.g., for navigational needs, graphics being lines of constant sound velocity in respect to the temperature used to be applied for the given values of salinity.

It has become necessary to use vertical distribution of the sound speed in water, which is the dependence of the sound velocity on the depth, since MBES (MultiBeam EchoSounder) and underwater navigation systems used to be applied in the bathymetric measurements [1,2]. These devices, contrary to SBES, employ bias propagation of the acoustic wave. As a result of the change of the sound speed in water, its trajectory deflects towards the surface of lower velocity [3–11]. Nonlinear propagation results in more complicated determination of the depth and of the coordinates of the acoustic wave reflection from the bottom. The sound velocity in water profiler becomes an indispensable element of the measuring system.

In general, two types of devices—SVP (Sound Velocity Profiler) and CTD/STD (Conductivity/Salinity Temperature Depth)—are used to measure the sound velocity in water. The first one is an ultrasound device, and its operation principle is similar to the echosounder's one. The velocity is determined on a constant distance between the electroacoustic transducer and a reflective plate, based on the measured time elapsed between the impulse generation and the receipt of its echo after reflection from the plate. The second type—the CTD/STD sounders—measure original water parameters based on which the sound speed is determined. In the 1960s, many teams of hydroacoustic professionals used respective research—the most known relations describing the sound speed in water include Medwin [12,13], Wilson [14], Kinsler and Frey [15], DelGrosso [16], Chen and Millero [17–20], Mackenzie [21,22] and others [23–35].

Application of the sound velocity profiler in MBES systems and in underwater navigation is indisputable. These days, singlebeam echosounders are still in use, in spite of commercial availability of hydrographical multibeam echosounders. Due to emission of one vertical beam, labour intensity in the bathymetric measurements is higher in order to obtain high-density data in the water area under the research. However, they are still in use, especially on USVs (Unmanned Survey Vehicles), due to their small dimensions and uncomplicated calibration process. It is reasonable to look for solutions lowering costs of the sound velocity measurements in order to ensure the highest accuracy of the depth measurement by means of SBES. It may be achieved by determining the sound speed based on the temperature measurement and estimation of the salinity mean value based on the information available from the nearest measuring station [36–44].

Impact of change of the sound speed in water has more and more significance as the depth increases. Inaccuracies between the mean value or estimated distribution and the real distribution of the sound velocity have insignificant impact in shallow water and in restricted water areas of low depths. However, it is obligatory to calibrate SBES in the hydrographic operations and to prove it in the sound velocity measurement sheets. Measurement of the sound speed is the most accurate method of determining its distribution. A bar check calibration is another method. The bar check involves a metal cone or plate device lowered and recording the true depth versus the measured depth and compiling a depth correction table that will be used later to correct the measured depths [45–49].

Generally, the described method is dedicated to hydrographers utilising USV equipped with SBES in shallow waters. The surveys are realized in ports, especially small ones (marinas) with the depth no more 10 m and coastal areas. Deploying a USV in a marina enables manoeuvring between mooring places and yachts. The usage of SBES is sufficient under such circumstances. Similarly, in coastal area, it is easier to survey and manoeuvre a USV closer to the coastline as compared to open water.

Measurements of the sound velocity in water were executed in two water areas: on the Rivers Motława and Martwa Wisła in Gdańsk (A in Figure 1) and in Gdynia Marina (B in Figure 1). The measurements in Gdańsk were executed at the depths of 5 m and 10 m. The first water area serves to station large cruise ships and represents heavy sea traffic in the tourist season. The other one is located in the area of a publicly available slipway, which allowed for the analysis of the accuracy of the depth measurements to be performed in a large range of depths. The measurement station monitoring water parameters is located in the Northern Port in Gdańsk, far from the place the sound velocity measurements were executed. Moreover, it is located on the Gdańsk Bay side.

For the Gdynia area, the sound speed measurements were executed in the marina where the measurement station is located. The bathymetric measurements were carried out at a public beach.

The Baltic Sea belongs to the seas of the lowest salinity—it is often referred to as a subtly salty or brackish sea, not salty. Average salinity of the Baltic Sea is approx. 7 psu—this value usually varies from 2 psu to 12 psu. Salinity values of selected water areas of the Baltic Sea and other seas, including the Mediterranean Sea and salt lakes, are presented in Table 1 [50].

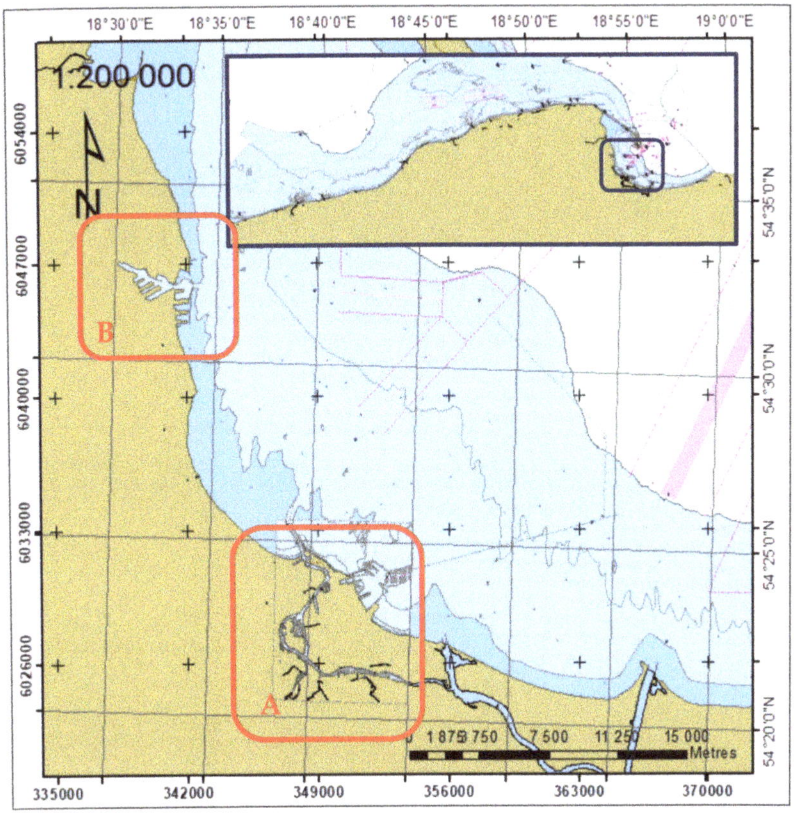

Figure 1. Location of sound speed in water and depth measurements: in Gdańsk (A) and Gdynia (B).

Table 1. Salinity of seas and lakes.

	Area	Salinity [psu]
Baltic Sea	Kattegat and Skagerrak	20
	Bay of Kiel	15 17
	Polish coast	7
	Bay of Puck	6.2
	Vistula Lagoon	1–3
	Gulf of Finland and Bothnia	2
Other seas	Black Sea	13–23
	Red Sea	40–80
	Mediterranean Sea	37–39
	North Sea	33–35
	World Ocean	34–36
Salt lakes	Lake Albert (Great Basin, OR, USA)	120
	Dead Sea (Israel, Jordan, West Bank)	337
	Gaet'ale Pond (Ethiopia)	433

2. Materials and Methods

2.1. Determination of the Sound Speed in Water on the Basis of Its Basic Parameters

In physics, the following formula is a basic equation describing velocity of propagation of elastic waves to which sound waves in the medium belong:

$$c = \frac{1}{\sqrt{k_{p,Q}\rho_0}} \tag{1}$$

where: c—sound speed (m/s), $k_{p,Q}$—the coefficient of compressibility, ρ_0—density and, in practice, it is not applicable in calculating the sound speed in hydrography.

Water density as well as its compressibility are compound functions of the salinity, temperature and pressure [51]. The sound velocity in water increases along with a growth of the temperature, salinity and static pressure. The temperature change has the greatest impact on the sound speed variations. The water compressibility module goes up, and the density of water decreases along with its growth. Change of the sea water temperature by 1 °C results in the sound speed change by the values ranging from 4.7 m/s at the temperature of 0 °C to 2.2 m/s at the temperature of 30 °C. The extent of the sound speed change Δct also depends on the initial temperature of the water in respect to the water in which its change takes place. The water salinity change by 1 psu has impact on the sound speed variation in the range of 1.0–1.4 m/s and the hydrostatic pressure change by 1 atm. (approximately 105 Pa), which means a depth change of 10 m results in the sound velocity change by 0.175 m/s.

Description of the formula c = f(T,S,D) is executed with a use of equations or tables. Medwin elaborated on one of the first and simplest formulas [12–14,52]:

$$\begin{aligned} c(S,T,D) = 1449 + 4.6T - 0.055TT^2 + 0.0003T^3 \\ +(1.39 - 0.012T)(S-35) + 0.017D \end{aligned} \tag{2}$$

where: T—temperature (°C), S—salinity (‰), D—depth (m). The other relationships commonly used are the following:

Wilson [15]

$$c(S,T,P) = 1449.14 + Dc_T + Dc_S + Dc_P + Dc_{STP}, \tag{3}$$

where: c_T, c_S, c_P, c_{STP}—coefficients described in Appendix A.

Del Grosso [17]

$$c(S,T,P) = 1402.392 + \Delta c_T + \Delta c_S + \Delta c_P + \Delta c_{STP}. \tag{4}$$

Mackenzie [21,22]

$$\begin{aligned} c(S,T,D) = 1448.96 + 4.591T - 5.304{\cdot}10^{-4}T^3 + 1.340(S-35) + 1.630{\cdot}10^{-2}D \\ +1.675{\cdot}10^{-7}D^2 - 1.025{\cdot}10^{-2}T(S-35) + 7.139{\cdot}10^{-13}TD^3. \end{aligned} \tag{5}$$

Coppens [53]

$$\begin{aligned} c(S,t,D) &= c(S,t,0) + (16.23 + 0.253t)D + (0.213 - 0.1t)D^2 \\ &\quad +[0.016 + 0.0002(S-35)](S-35)tD \\ c(S,t,0) &= 1449.05 + 45.7t - 5.21t^2 + 0.23t^3 + \left(1.333 - 0.126t + 0.009t^2\right)(S-35) \end{aligned} \tag{6}$$

where: t = T/10.

The international standard algorithm, often known as the UNESCO algorithm, is at-tributed to Chen and Millero [19,20] and has a more complicated form than the simple equations above but uses pressure as a variable rather than depth. For the original UNESCO paper, see Fofonoff and Millard [54]. Wong and Zhu [55] recalculated the coefficients in this algorithm following the adoption of the International Temperature Scale of 1990, and their form of the UNESCO equation is:

$$c(S,T,P) = C_W(T,P) + A(T,P)S + B(T,P)S^{\frac{3}{2}} + D(T,P)S^2. \tag{7}$$

The formula with coefficients is described in Appendix A.

The presented relations serving determination of the sound speed distribution are applicable in specific oceanographic conditions, i.e., in the given ranges of the temperature and salinity. The limits of their usage are presented in Table 2.

Table 2. Limits of formulas for determination the sound speed in water.

Depth	Temperature [°C]	Salinity [psu]
Medwin	0–35	0–45
Wilson	0–30	0–37
Maccenzie	2–30	25–40
Coppens	0–35	0–45
Del Grosso	0–30	30–40
Chen and Millero	0–40	0–40

2.2. Observations of Water Parameters

Monitoring and forecasting of environmental changes aim to extend data resources due to the sustainable development needs and in order to prevent risks. Contrary to traditional methods, a digital model controls the main characteristics of an ecosystem constantly, in terms of time and location. Such approach also allows the obtaining of a detailed quantification of variability of physical, dynamic and biochemical water parameters for the marine environment.

The water level is one of the observed parameters. Its significance is important in executing the bathymetric measurements to compare the depth measurement results with a chart datum. At the time of the measurements, the current water level may be obtained from a measurement point, tide gauge, hydrometric station or from GNSS measurements. Permanent monitoring in the observation point is available on the Internet with various intervals (usually 1 h) and age of the data (e.g., 3 days). They constitute national networks or form parts of larger-sized systems [56–58].

The other water parameters, being data sources for engineers in hydrology, oceanography and oceanology, are physical–chemical qualities used, for instance, in eco–hydrodynamic models. It consists of two modules. M3D_UG hydrodynamic one [59] and eco-system ProDeMo module [60,61]. The model works in a preoperational mode. Forty-eight-hour prognoses include fields of surface currents, temperature and salinity of sea water. Moreover, it predicts fields of biogenic salts: nitrates, ammonia, phosphates, silicates, nitrogen and total phosphorus and concentration of oxygen in sea water and phytoplankton biomass. The data from monitoring of the salinity in a surface layer were used to determine the vertical distribution of the sound velocity in water.

2.3. Determination of Sound Speed in Water Based on Temperature Measurement and Estimated Salinity

The presented mathematical relations serving to determine the sound speed in water employ the measured values of the temperature and salinity as a function to the depth, which is determined on the basis of hydrostatic pressure. The temperature measurement is easy and inexpensive. For this reason, one may use a laboratory thermometer of limited measuring range, with a probe on a long cable (Figure 2). The measurement is performed by reading the temperature at the depth of the probe immersion, and the depth is recorded by marking the cable with tracers.

Figure 2. Measurement of the water temperature.

The salinity is estimated on the basis of information available in the nearest measurement station or stations registering and providing various parameters of water. The Institute of Oceanology of the University of Gdańsk maintains multiyear registers of: the water temperature, salinity, level, primary production, cyanobacteria, phytoplankton and dissolved oxygen. It also maintains a 48 h prognosis of hydrological and hydrodynamic conditions for the water areas of: the Gdańsk Bay together with the Vistula Lagoon, the Bay of Pomerania and the Szczecin Lagoon as well as the South Baltic.

2.4. *Estimation of the Depth Measurement Accuracy, Utilising the Simplified Method of Sound Velocity in Water Measurement, in the Light of Hydrographical Organisations' Requirements*

Hydrographic organisations have defined minimal requirements for the hydrographic surveys, including an uncertainty of determining the position's coordinates (THU—total horizontal uncertainty) and of the depth measurement (TVU—total vertical uncertainty).

The TPU of a point is a measure of the accuracy to be expected for such a point when all relevant error/uncertainty sources are taken into account. Instead of "TPU", the term "error budget" is also used. Uncertainty sources that are to be considered are (GPS) position, draft, squat, load, tide (including spatial/temporal prediction to the depth measurement position), geoid model, bathy depth, node offsets, timing offsets, SOG (speed over ground) and CMG (course made good), gyro heading, pitch, roll and heave, mounting offsets, beam range, beam angle, beam width, beam steering, sound velocity at transducer head and sound velocity profile.

The accuracy requirements defined in [43,44] are given for five categories, depending on the water area designation. The categories of the highest requirements are as the following:

- Special Order is intended for those areas where underkeel clearances are critical. Therefore, 100% feature search and 100% bathymetric coverage are required. Examples of areas: berthing areas, harbours and critical areas of fairways and shipping channels.
- Exclusive Order hydrographic surveys are an extension of IHO Special Order with more stringent uncertainty and data coverage requirements. Their use is intended to be restricted to shallow water areas (harbours, berthing areas and critical areas of fairways and channels) where there is an exceptional and optimal use of the water column and where specific critical areas with minimum underkeel clearance and bottom characteristics are potentially hazardous to vessels. For this order, a 200% feature search and a 200% bathymetric coverage are required. The size of features to be detected is deliberately more demanding than for Special Order.

The formula below is used to compute the maximum allowable vertical measurement uncertainty [43,44], recognising that there are both depth-dependent and depth-independent error sources that affect the measurements of depths:

$$TVU_{max} = \sqrt{a^2 + (bh)^2} \tag{8}$$

for: Exclusive Order a = 0.15 b = 0.0040,
Special Order a = 0.25 b = 0.0075.

TVU for Exclusive Order and Special Order in the range of 1–10 m is presented in Table 3.

Table 3. TVU for Exclusive Order and Special Order in the range of 1–10 m.

Depth (m)	1	2	3	4	5	6	7	8	9	10
Exclusive Order	0.17	0.19	0.21	0.23	0.24	0.26	0.27	0.29	0.30	0.31
Special Order	0.26	0.28	0.29	0.30	0.32	0.33	0.34	0.35	0.36	0.37

2.5. STD/CTD Profiler for Sound Speed in Water Measurements

The water parameters were measured and recorded using SAIV A/S probe model SD204 (Figure 3). The STD/CTD model SD204 is a self-contained instrument that measures, calculates and records sea water conductivity, salinity, temperature, pressure and sound velocity in situ. The data recorded in the instrument are captured in physical units and can be copied to a PC and presented immediately after the measurements have been completed (or at any time later).

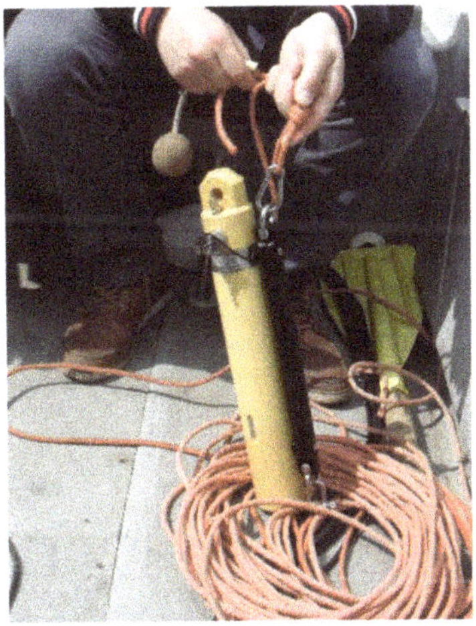

Figure 3. STD/CTD probe model SD204.

This makes the performance of the SD204 comparable to much more voluminous and expensive cable-based STD systems. The data from the instrument can also be transferred to a PC by cable or remotely via a modem and a telephone or satellite terminal. The specifications for STD/CTD model SD204 are presented in Table 4.

Table 4. Specifications for STD/CTD model SD204.

	Range	Resolution	Accuracy
Conductivity	0 to 70 mS/cm	0.01 mS/cm	±0.02 mS/cm
Salinity	0 to 40 ppt	0.01 ppt	0.02 ppt
Temperature	−2 to +40 °C	0.001 °C	±0.01 °C
Pressure	500, 1000, 2000, … 6000 m	0.01 mbar (m)	±0.02% of range
Sound velocity	1300 to 1700 m/s	5 cm/s	±10 cm/s

3. Results

3.1. Gdańsk—Motława River

Motława is a river running through the heart of Gdańsk's old town. Although there are insignificant currents in its flow, low depth and serious traffic of vessels in the tourist season have impact on its bed shape, hence the periodic bathymetric control.

Due to a great distance to the measurement station, determination of the sound velocity in water without application of SVP probe is difficult, especially as the station is located in the open water area—in the Gdańsk Bay, in the Northern Port (Figure 4). That is why, in order to determine the distribution of the sound speed in water, its parameters' values—the temperature and salinity—were analysed in distant places, thus taking advantage of the possibility of measuring the salinity in the surface layer.

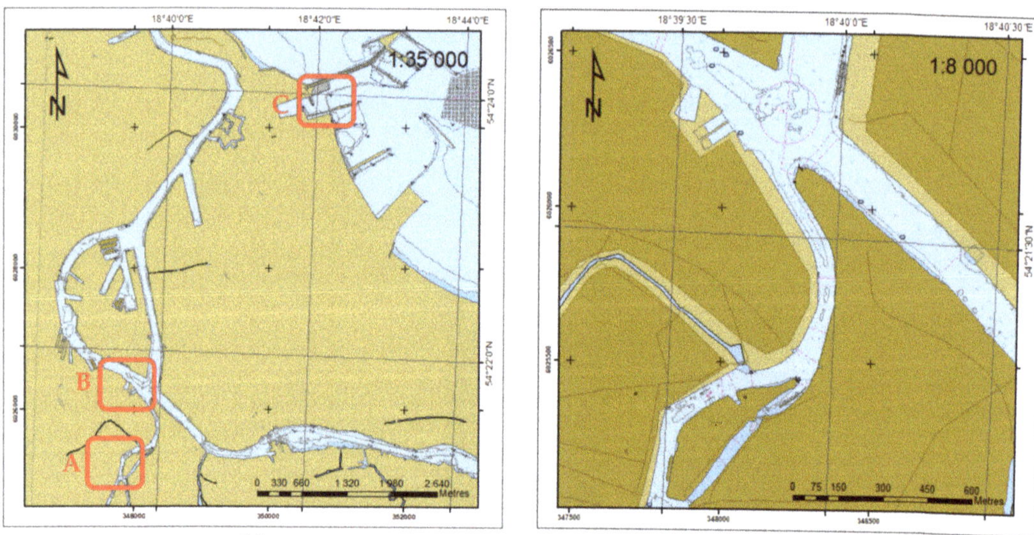

(**a**) (**b**)

Figure 4. Location of measurements of the sound speed in water in Gdańsk (**a**): Motława River (A), Martwa Wisła River (B) and water station in Gdańsk Northern Port (C); Motława River (**b**).

Time courses of water basic parameters: temperature and salinity are presented in Figure 5. A three-day range of changes of these parameters is not high, and it varies in the range of $\Delta t = 1.5$ °C and $\Delta S = 0.5$ psu. A value of $S = 6$ psu has been set to determine the sound velocity in water.

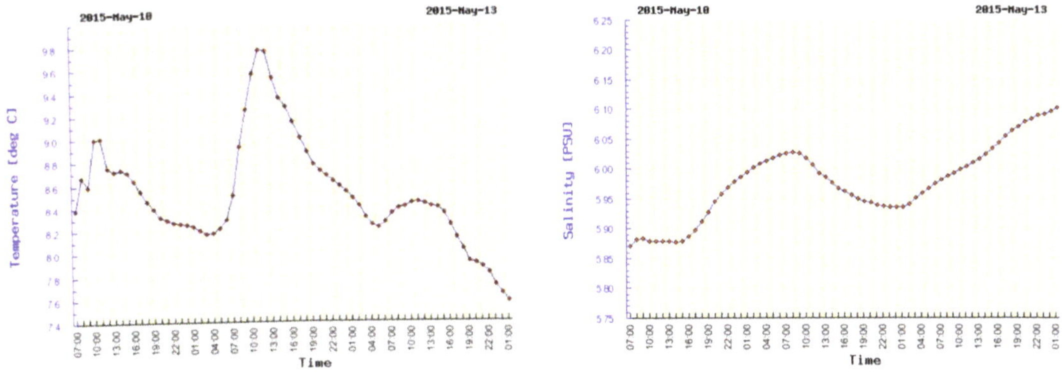

(a) (b)

Figure 5. Temperature (**a**) and salinity (**b**) recorded in the station in Gdańsk Northern Port in the Motława River for measurements in the Motława River.

The depth reaches 7m along the axis in this part of the Motława River. Shallowing to 5 m occurs in the place of measurement at the water tram pier. Therefore, the sound speed measurements were executed for such a value as a maximum, with an interval of 1m. The sound velocities were determined at those depths for the given, commonly used formulas (Table 5). Then, the sound velocities for the measured temperature values were determined, assuming that the water salinity was S = 6 psu.

Table 5. Sound speed in water determined using selected formulas.

Depth (m)	0	1	2	3	4	5
Temperature (°C)	14.6	13.8	13.1	12.5	11.6	10.8
Salinity (psu)	1	2.1	3.4	4.5	5.8	6.5
Medwin	1464.9	1463.4	1462.5	1461.7	1459.9	1458.0
Wilson	1465.9	1464.3	1463.2	1462.2	1460.4	1458.1
Maccenzie	1464.9	1463.4	1462.4	1461.6	1459.8	1457.7
Coppens	1465.6	1464.1	1463.1	1462.3	1460.4	1458.3
Del Grosso	1465.7	1464.1	1463.1	1462.2	1460.4	1458.2
Chen and Millero	1465.7	1464.2	1463.2	1462.3	1460.6	1458.4
MTPS [1] (S = 6 psu)	1471.6	1468.8	1466.3	1464.2	1460.8	1457.8

[1] MTPS—Measured Temperature-Predicted Salinity.

In graphic form, temperature, salinity and sound speed in water profiles in the Motława River are shown in Figure 6.

Based on the registered water parameters, one may observe a change of both temperature and salinity. Along with the depth growth, the temperature decreases within the range of 14.6–10.8 °C and the salinity within the range of 1–6.5 psu. It results in the sound speed change within the range of 1465–1458 m/s. With the assumption that the salinity is 6 psu, the difference between the determined sound velocity in water and its value obtained from the measurements of the temperature and salinity is the highest near the surface, when the salinity variation is the biggest.

In the Motława River, the difference between the determined (MTPS) and measured (on the basis of the temperature and salinity) sound velocity in the water is the highest near the surface. Table 6 shows the depth error as a difference between the depth determined using vertical distribution of the sound speed and on the basis of a simplified method with constant salinity of S = 6 psu. Although the difference in the sound speed is the highest, it has insignificant impact on the depth measurement error.

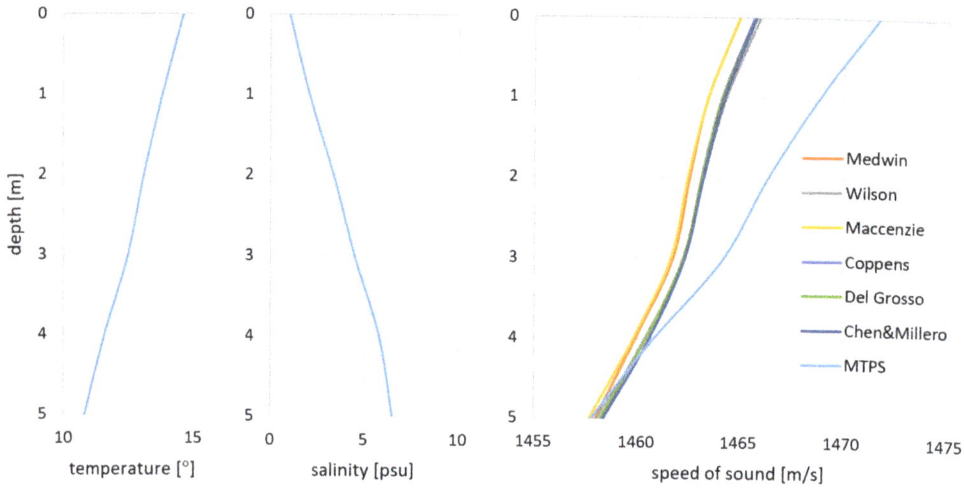

Figure 6. Temperature, salinity and sound speed in water profiles in the Motława River.

Table 6. Depth measurement error (cm) for selected formulas according to MTPS (S = 6 psu) in the Motława River.

Depth (m)	1	2	3	4	5
Medwin	0.4	0.9	1.5	2.4	3.7
Wilson	0.3	0.8	1.4	2.3	3.7
Maccenzie	0.4	0.9	1.5	2.5	3.8
Coppens	0.3	0.8	1.3	2.3	3.6
Del Grosso	0.3	0.8	1.4	2.3	3.6
Chen and Millero	0.3	0.8	1.3	2.2	3.6

3.2. Gdańsk—Martwa Wisła River

Bathymetric measurements of the slipway were executed in the area of confluence of the Rivers Motława and Martwa Wisła. It is located near the historic Imperial Shipyard. The registered SBES depth has reached as much as 7 m. The sound speed in water measurement was executed down to the depth of 10 m.

The time courses of the water basic parameters, temperature and salinity, registered by the Gdańsk Northern Port measurement station, are presented in Figure 7.

The extent of changes of these parameters is smaller than for the Motława River. The errors of the depth measurement with the simplified method applied, in respect to the described methods using the salinity measurement, are presented in the Table 7.

Table 7. Sound speed in water determined using selected formulas.

Depth (m)	0	1	2	3	4	5	6	7	8	9	10
Temperature (°C)	9.6	7.7	8.3	8.5	8.5	8.4	8.4	8.4	8.5	8.5	8.5
Salinity (psu)	2.1	4.4	5.4	5.7	5.9	6.0	6.2	6.3	6.4	6.4	6.4
Medwin	1447.6	1442.8	1446.6	1447.8	1448.0	1447.8	1448.0	1448.2	1448.7	1448.7	1448.7
Wilson	1448.1	1443.0	1446.8	1448.0	1448.3	1448.0	1448.2	1448.4	1449.0	1448.9	1449.0
Maccenzie	1447.5	1442.7	1446.4	1447.6	1447.9	1447.6	1447.9	1448.0	1448.6	1448.6	1448.6
Coppens	1448.2	1443.7	1447.0	1448.2	1448.5	1448.2	1448.5	1448.6	1449.2	1449.2	1449.2
Del Grosso	1448.2	1443.3	1447.0	1448.2	1448.4	1448.1	1448.4	1448.5	1449.0	1449.0	1449.0
Chen and Millero	1448.3	1443.4	1447.2	1448.3	1448.6	1448.3	1448.6	1448.7	1449.2	1449.2	1449.2
MTPS (S = 6 psu)	1453.1	1445.4	1447.9	1448.7	1448.7	1448.3	1448.3	1448.3	1448.7	1448.7	1448.7

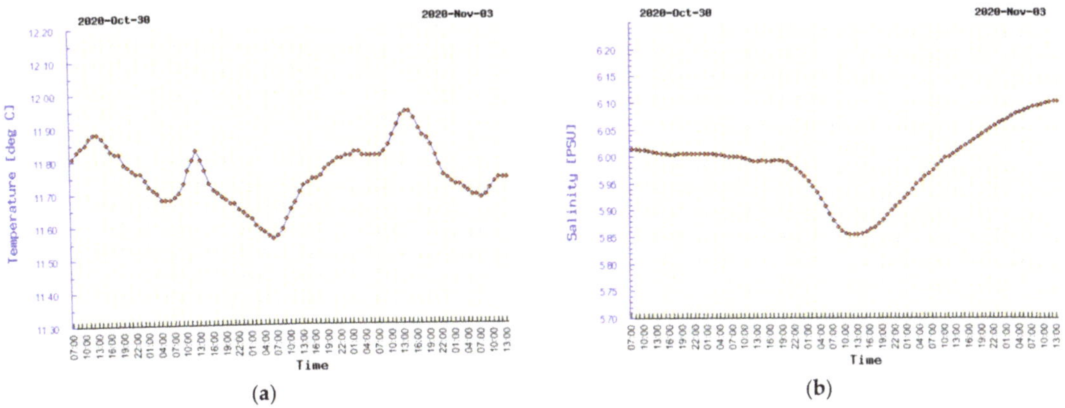

Figure 7. Temperature (**a**) and salinity (**b**) recorded in the station in Gdańsk Northern Port for measurements in the Martwa Wisła River.

In graphic form, temperature, salinity and sound speed in water profiles in the Martwa Wisła River are shown in Figure 8.

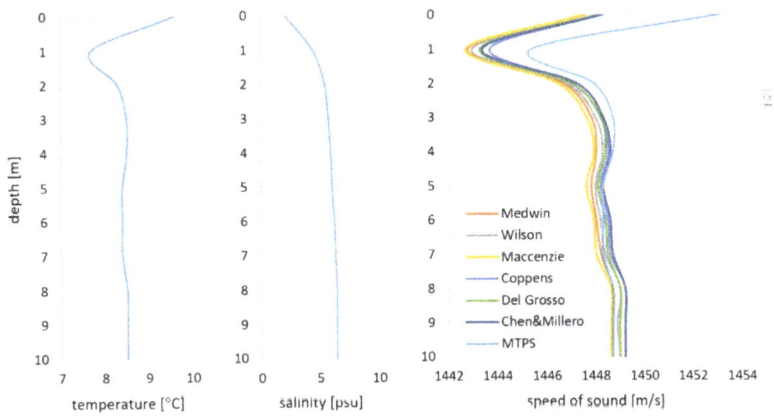

Figure 8. Temperature, salinity and sound speed in water profiles in the Martwa Wisła River.

Table 8 shows the depth error as a difference between the depth determined using vertical distribution of the sound speed and on the basis of the simplified method with constant salinity S = 6 psu in the Martwa Wisła River. In the depth range of 0–10 m, the difference is no more than 2.5 cm.

Table 8. Depth measurement error (cm) for selected formulas according to MTPS (S = 6 psu) in the Martwa Wisła River.

Depth (m)	1	2	3	4	5	6	7	8	9	10
Medwin	0.2	0.2	0.5	0.7	0.8	1.1	1.4	1.8	2.1	2.3
Wilson	0.2	0.2	0.5	0.8	0.9	1.2	1.4	2.0	2.2	2.5
Maccenzie	0.2	0.1	0.5	0.7	0.8	1.0	1.3	1.8	2.0	2.2
Coppens	0.1	0.2	0.6	0.9	1.0	1.3	1.5	2.4	2.4	2.6
Del Grosso	0.1	0.2	0.6	0.8	0.9	1.2	1.5	2.2	2.2	2.5
Chen and Millero	0.1	0.2	0.6	0.9	1.0	1.3	1.6	2.4	2.4	2.6

3.3. Gdynia—The Marina and the Public Beach

The measurement station in Gdynia is situated in the marina located in the southern part of the basins and piers of the Gdynia seaport (Figure 9). It has direct contact with the Gdańsk Bay and is shielded with two seawalls: one internal and one external.

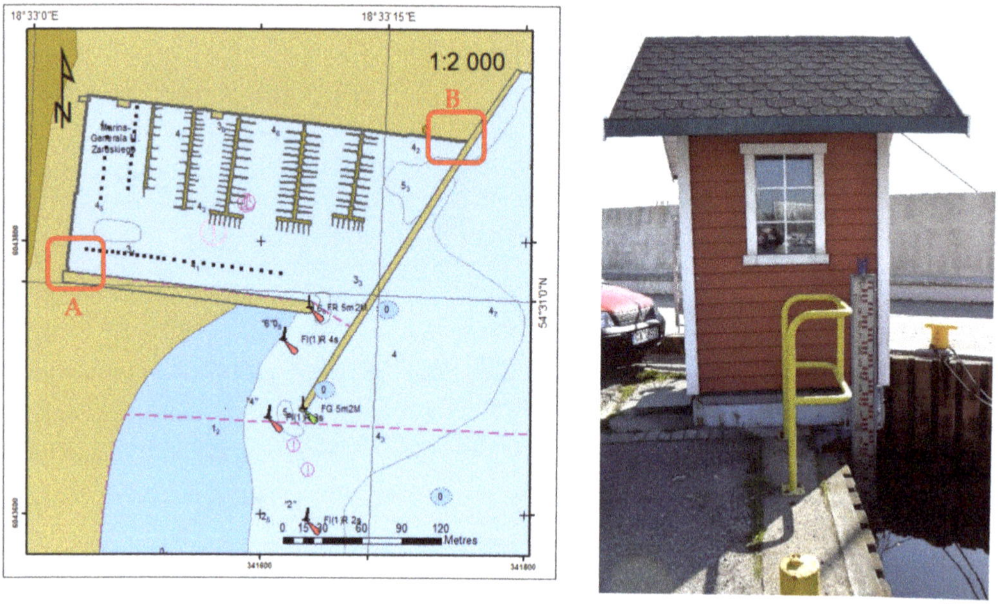

(a) (b)

Figure 9. Location of measurements of the sound speed in water in Gdynia (**a**), A and water station (**b**), B.

Measurements of vertical distribution of the sound speed in water were executed in that place and the bathymetric measurements at the area of the public beach. Similar to the marina, where the measurement of the sound velocity in water were executed, the depth reaches a value of 5 m in that place.

The temperature and salinity recorded in the station in Gdynia for measurements in marina and the public beach are shown in Figure 10.

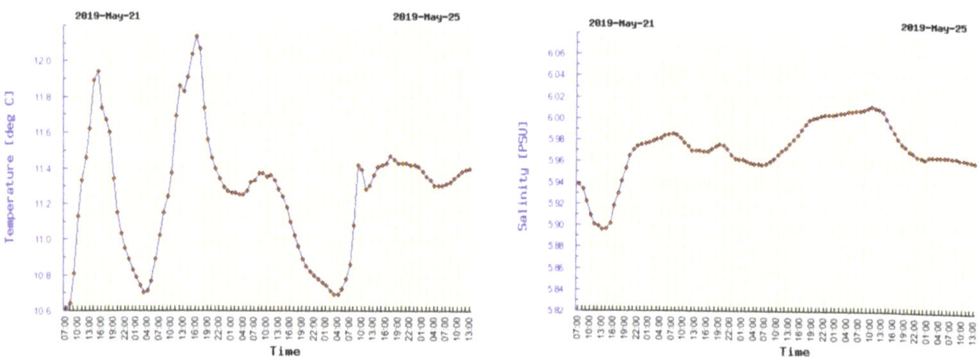

(a) (b)

Figure 10. Temperature (**a**) and salinity (**b**) recorded in the station in Gdynia for measurements in the marina and public beach.

The range of the temperature changes is insignificant, and the salinity is almost constant (Table 9). Thus, the estimated value of the salinity does not result in errors in the sound speed determination.

Table 9. Sound speed in water determined using selected formulas.

Depth (m)	0	1	2	3	4	5
Temperature (°C)	15.8	15.2	14.6	14.2	13.8	13.1
Salinity (psu)	6.9	6.9	7.0	7.0	7.1	7.1
Medwin	1476.1	1474.1	1472.2	1470.7	1469.5	1467.0
Wilson	1476.7	1474.7	1472.7	1471.3	1470.0	1467.5
Maccenzie	1476.1	1474.1	1472.1	1470.7	1469.5	1467.0
Coppens	1476.6	1474.6	1472.7	1471.3	1470.0	1467.5
Del Grosso	1476.7	1474.6	1472.7	1471.3	1470.0	1467.6
Chen and Millero	1476.8	1474.8	1472.8	1471.5	1470.2	1467.7
MTPS (S = 7 psu)	1476.9	1474.9	1472.8	1471.5	1470.1	1467.6

In graphic form, temperature, salinity and sound speed in water profiles in Gdynia are shown in Figure 11.

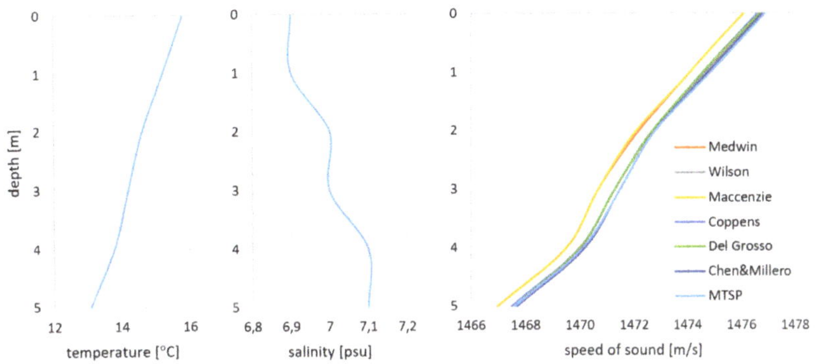

Figure 11. Temperature, salinity and sound speed in water profiles in Gdynia.

The depth errors for measurements in a public beach in Gdynia in the range of 0–5 m are presented in Table 10. For constant salinity S = 7 psu, the difference between the depth determined using vertical distribution of the sound speed and on the basis of the simplified method is not more than 3 cm.

Table 10. Depth measurement error (cm) for selected formulas according to MTPS (S = 7 psu) in Gdynia.

Depth (m)	1	2	3	4	5
Medwin	0.1	0.4	0.9	1.5	2.7
Wilson	0	0.3	0.7	1.3	2.5
Maccenzie	0.1	0.4	0.9	1.5	2.7
Coppens	0	0.3	0.7	1.3	2.5
Del Grosso	0	0.3	0.7	1.3	2.5
Chen and Millero	0	0.3	0.7	1.3	2.5

Depth error for measurements in Gdańsk and Gdynia are shown in Figure 12.

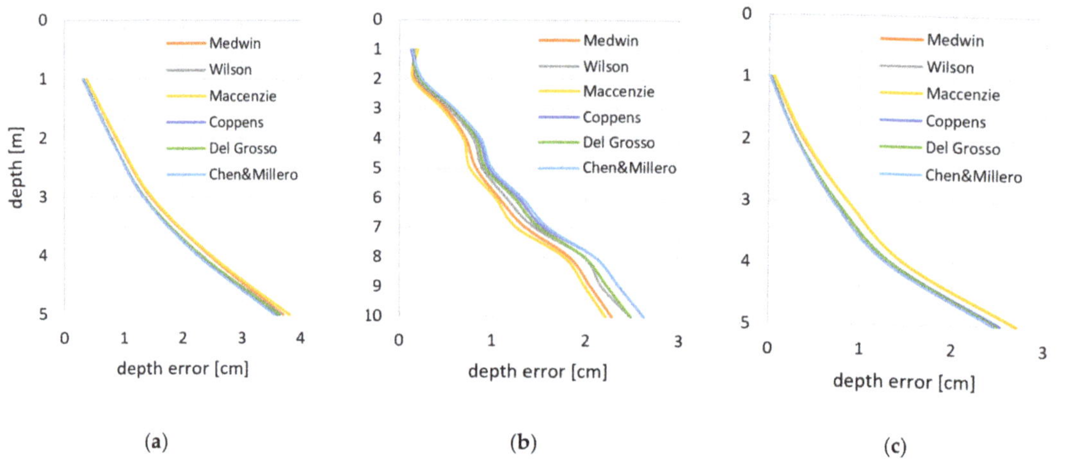

Figure 12. Depth error for measurements in the Motława (**a**) and Martwa Wisła (**b**) Rivers and Gdynia (**c**).

3.4. Mediterranean Sea—La Ciotat (France) and Barcelona (Spain)

The results given in Section 3.1–3.3 refer to the survey executed in waters of Gdansk Bay, the low-salinity part of the Baltic Sea (Table 1). The results of measurements of water parameters performed in the Mediterranean Sea [62–69] of much higher salinity were used to verify the method for determining the sound velocity distribution.

For validation of the simplified method, the measurements were executed by means of Valeport MIDAS ECM probe in three water areas: La Ciotat (France) 38 psu and Barcelona (Spain) 32 psu and 37 psu. Within the range of 0–15 m of the measured depths, the temperature and the salinity are constant. In such case, the sound velocity may be calculated for any depth, and it is constant in respect to the depth (Figure 13).

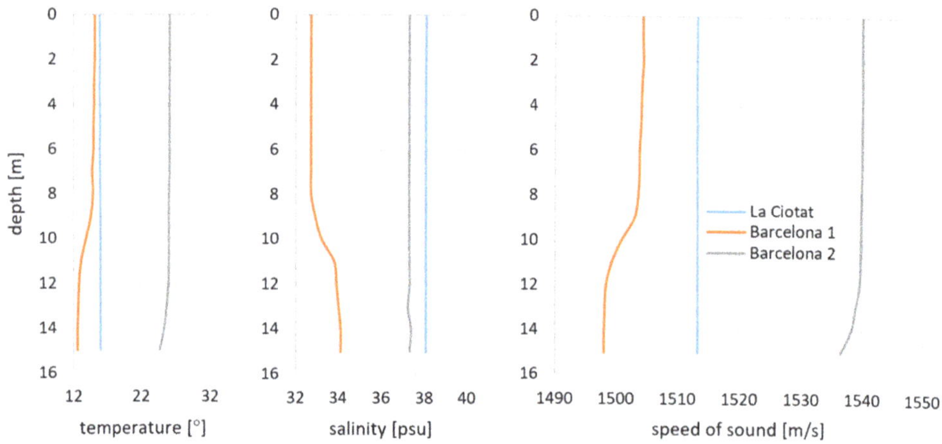

Figure 13. Temperature, salinity and sound speed in water profiles in the Mediterranean Sea.

Table 11 presents the results of the measured sound speed in the Mediterranean Sea using Midas SVP probe and determined on the basis of the simplified MTPS method. The difference between the two should be zero (it is 0.1 m/s due to the minimal difference of the salinity assumed for the calculations).

Table 11. Sound speed in water determined using selected formulas.

	Depth (m)	0	2	4	6	8	10	12	14	16
La Ciotat	Temperature (°C)	15.8	15.8	15.8	15.8	15.8	15.8	15.8	15.8	15.8
	Salinity (psu)	38.0	38.0	38.0	38.0	38.0	38.0	38.0	38.0	38.0
	Midas SVP (m/s)	1512.9	1512.9	1512.9	1512.9	1513.0	1513.1	1513.1	1513.1	1513.1
	MTPS (S = 32 psu) (m/s)	1512.8	1512.8	1512.8	1512.8	1512.9	1513.0	1513.0	1513.0	1513.0
Barcelona 1	Temperature (°C)	15.1	15.1	15.0	14.9	14.8	13.8	12.7	12,6	12.5
	Salinity (psu)	32.7	32.7	32.7	32.7	32.7	33.2	33.9	34.0	34.1
	Midas SVP (m/s)	1504.3	1504.3	1504.0	1503.7	153.5	1500.8	1498.3	1498.0	1498.0
	MTPS (S = 33 psu) (m/s)	1504.3	1504.3	1504.0	1503.6	1503.4	1500.7	1498.2	1497.9	1497.9
Barcelona 2	Temperature (°C)	26.2	26.2	26.2	26.2	26.1	26.1	26.0	25.4	24.7
	Salinity (psu)	37.3	37.3	37.3	37.3	37.3	37.3	67.3	37.4	37.3
	Midas SVP (m/s)	1539.8	1539.8	1539.8	1539.8	1539.8	1539.7	1537.5	1538.1	1536.5
	MTPS (S = 37 psu) (m/s)	1539.7	1539.7	1539.7	1539.7	1539.6	1539.6	1537.4	1538.0	1536.4

4. Discussion

The proposed method of determining the sound speed in water on the basis of the temperature measurement and salinity prognosis is dedicated to the measurements in shallow water. In such an area, the bathymetric measurements are executed, more and more often by USV equipped with SBES. Calibration of the echosounder is required, and one the available methods to do that is the bar check. The bar check involves lowering a flat plate below the echo sounder transducer to several known depths below the surface and comparing the actual versus measured depth. As the bar is moved down, the sound velocity in the echo sounder is adjusted until the measured depth matches the actual depth. At the end of the test, the echosounder is fixed with the average sound velocity over the water column. As the sound velocity error magnitude increases proportionally with depth, surveys in shallow water suffer from a smaller potential absolute error. This method is difficult to apply during the measurement executed by the SBES on board of the USV, although it is used in bathymetry, the results of which are presented in literature. The mean value of the sound speed in water is also used, and it causes lower accuracy (higher uncertainty) of the depth measurement. For bathymetric documentation, the sound speed in water or SBES calibration is mandatory. As far as the calibration being awkward for USV's surveys, the simplified MTPS method seems to be a low-cost, easy and effective solution.

The method presented in the article is dedicated to measurements in shallow water because of the length of the probe's cable, usually 1 or 5 m, seldom 10 m long. USVs are usually used in coastal surveys in shallow water, so this low-cost method can be used for supporting hydrographic measurement. During the surveys in open area on board a hydrographic motorboat or vessel, especially in bad weather conditions, measuring the temperature using a light thermometer could be difficult or impossible.

The environment can vary in time and space. In time variation with daily sound speed (temperature) fluctuations, sound speed has to be measured (determined) often. In space variation, e.g., flowing river to the sea, it is necessary to measure the sound of speed in numerous places. Thus, it is possible to model spatial–temporal sound speed in water distribution. In the future, local interpolation for determination local sound speed in water may be realized.

5. Conclusions

During hydrographic surveys, the sound speed in water is measured regularly. In daily surveys, it is measured at the beginning and at the end of the survey. In the summer, when the sun heats the water, the measurement period is shorter. When the temperature is measured more often, salinity is rather constant—actual sound speed is determined. Additionally, environmental conditions change in the same way for CTD/STD and simplified MTPS methods.

Singlebeam echosounders determine the depth on the basis of sound speed profile or its mean value. Mean value can be calculated on the basis of sound speed profile or estimated on the basis of constant temperature and salinity. It can also be determined on the basis of constant salinity and actual temperature measured in the transducer or in draught of the transducer. The best solution is to use an SVP probe, but it can be more expensive than the echosounder. The article presents solution on how to obtain accurate sound speed in the water (profile) using low-cost device. The solution is not equally as effective as the SVP probe, but it is better than using the mean value of sound speed and an alternative to bar check calibration.

The presented results of the research have been executed in the limited water area of various seabed shapes. It is important when sound velocity changes along with the depth growth. On one hand, the water salinity and temperature change along with the depth growth resulting in the change of the sound speed. Such an impact may be observed especially in the summertime and to a lesser extent in spring and autumn when the water temperature variations are smaller. On the other hand, in small depths, errors of their measurements are the slightest due to the insignificant change in the sound speed.

Two water areas of various distances to the measurement station were chosen for the research. As far as the first one is concerned (Gdańsk), the sound speed measurement was conducted in water of a river very distant from the station. This may have impact on the reliability of the parameter estimation, i.e., the difference between the values in the measurement place and in the measurement station. The value similar to the real one may be observed during the measurements carried out in Gdynia where the measurement station and the place of the measurements' execution are located in the marina basin.

The application of the proposed method for the hydrological conditions present in the area of South Baltic, of small differences in the salinity in respect to the depth, allows obtaining the accuracy of within several centimetres (but no more than 5cm) as compared to the sound velocity meter.

Positive results can be expected in other areas, where it is possible to estimate the water's salinity. The article presents results obtained in Gdynia, close to the water station, where the salinity spectrum is negligible. The salinity in the surveyed area and at the water station are equal. This is unlike the Motława River, where we can observe variable salinity profile dependent on the measurement's depth.

Funding: This research was supported by the Minister of National Defence of Poland as part of the program called Research Grant: "Backscattering of acoustic waves in the aquatic environmental".

Institutional Review Board Statement: Not applicable.

Informed Consent Statement: Not applicable.

Data Availability Statement: Not applicable.

Conflicts of Interest: The author declares no conflict of interest.

Appendix A. Selected Sound Speed in Water Formulas with Coefficients

Wilson (Equation (3)) [13]

$$c(S,T,P) = 1449.14 + Dc_T + Dc_S + Dc_P + Dc_{STP} \tag{A1}$$

$$Dc_T = 4.5721T - 4.4532\cdot10^{-2}T^2 - 2.6045\cdot10^{-4}T^3 + 7.9851\cdot10^{-6}T^4$$

$$Dc_S = 1.39799(S-35) - 1.69202\cdot10^{-3}(S-35)^2$$

$$Dc_P = 1.63432P - 1.06768\cdot10^{-3}P^2 + 3.73403\cdot10^{-6}P^3 - 3.6332\cdot10^{-8}P^4$$

$$\begin{aligned} Dc_{STP} = \quad & (S-35)(-1.1244\cdot10^{-2}T + 7.7711\cdot10^{-7}T^2 + 7.85344\cdot10^{-4}P \\ & -1.3458\cdot10^{-5}P^2 + 3.2203\cdot10^{-7}PT + 1.3101\cdot10^{-8}T^2P) \\ & +P(-1.8974\cdot10^{-3}T + 7.6287\cdot10^{-5}T^2 + 4.6176\cdot10^{-7}T^3 \\ & +P^2(-2.6301\cdot10^{-5}T + 1.9302\cdot10^{-7}T^2) + P^3(-2.0831\cdot10^{-7}T) \end{aligned}$$

Del Grosso (Equation (4)) [15]

$$c(S,T,P) = 1402.392 + \Delta c_T + \Delta c_S + \Delta c_P + \Delta c_{STP} \tag{A2}$$

$$\Delta c_T = C_{T1}T + C_{T2}T^2 + C_{T3}T^3$$

$$\Delta c_S = C_{S1}S + C_{S2}S^2$$

$$\Delta c_P = C_{P1}P + C_{P2}P^2 + C_{P3}P^3$$

$$\begin{aligned}\Delta c_{STP} = {} & C_{TP}TP + C_{T3P}T^3P + C_{TP2}TP^2 + C_{T2P2}T^2P^2 + C_{TP3}TP^3 + C_{ST}ST \\ & + C_{ST2}ST^2 + C_{STP}STP + C_{S2TP}S^2TP + C_{S2P2}S^2P^2\end{aligned}$$

Coefficients	Numerical Values
C_{T1}	5.012285
C_{T2}	$-0.551184 \cdot 10^{-1}$
C_{T3}	$0.221649 \cdot 10^{-3}$
C_{S1}	1.329530
C_{S2}	$0.1288598 \cdot 10^{-3}$
C_{P1}	0.1560592
C_{P2}	$0.2449993 \cdot 10^{-4}$
C_{P3}	$-0.8833959 \cdot 10^{-8}$
C_{ST}	$-0.1275936 \cdot 10^{-1}$
C_{TP}	$0.6353509 \cdot 10^{-2}$
C_{T2P2}	$0.2656174 \cdot 10^{-7}$
C_{TP2}	$-0.1593895 \cdot 10^{-5}$
C_{TP3}	$0.5222483 \cdot 10^{-9}$
C_{T3P}	$-0.4383615 \cdot 10^{-6}$
C_{S2P2}	$-0.1616745 \cdot 10^{-8}$
C_{ST2}	$0.9688441 \cdot 10^{-4}$
C_{S2TP}	$0.4857614 \cdot 10^{-5}$
C_{STP}	$-0.3406824 \cdot 10^{-3}$

Chen and Millero (Equation (7)) [19,20]

$$c(S,T,P) = C_W(T,P) + A(T,P)S + B(T,P)S^{\frac{3}{2}} + D(T,P)S^2 \tag{A3}$$

$$\begin{aligned}C_w(T,P) = {} & \left(C_{00} + C_{01}T + C_{02}T^2 + C_{03}T^3 + C_{04}T^4 + C_{05}T^5\right) \\ & + \left(C_{10} + C_{11}T + C_{12}T^2 + C_{13}T^3 + C_{14}T^4\right)P \\ & + \left(C_{20} + C_{21}T + C_{22}T^2 + C_{23}T^3 + C_{24}T^4\right)P^2 \\ & + \left(C_{30} + C_{31}T + C_{32}T^2\right)P^3\end{aligned}$$

$$\begin{aligned}A(T,P) = {} & \left(A_{00} + A_{01}T + A_{02}T^2 + A_{03}T^3 + A_{04}T^4\right) \\ & + \left(A_{10} + A_{11}T + A_{12}T^2 + A_{13}T^3 + A_{14}T^4\right)P \\ & + \left(A_{20} + A_{21}T + A_{22}T^2 + A_{23}T^3\right)P^2 \\ & + \left(A_{30} + A_{31}T + A_{32}T^2\right)P^3\end{aligned}$$

$$B(T,P) = B_{00} + B_{01}T + (B_{10} + B_{11}T)P$$

$$D(T,P) = D_{00} + D_{10}P$$

Coefficients	Numerical Values	Coefficients	Numerical Values
C_{00}	1402.388	A_{02}	$7.166{\cdot}10^{-5}$
C_{01}	5.03830	A_{03}	$2.008{\cdot}10^{-6}$
C_{02}	$-5.81090\ 10^{-2}$	A_{04}	$-3.21{\cdot}10^{-8}$
C_{03}	$3.3432{\cdot}10^{-4}$	A_{10}	$9.4742{\cdot}10^{-5}$
C_{04}	$-1.47797{\cdot}10^{-6}$	A_{11}	$-1.2583{\cdot}10^{-5}$
C_{05}	$3.1419{\cdot}10^{-9}$	A_{12}	$-6.4928{\cdot}10^{-8}$
C_{10}	0.153563	A_{13}	$1.0515{\cdot}10^{-8}$
C_{11}	$6.8999{\cdot}10^{-4}$	A_{14}	$-2.0142{\cdot}10^{-10}$
C_{12}	$-8.1829{\cdot}10^{-6}$	A_{20}	$-3.9064{\cdot}10^{-7}$
C_{13}	$1.3632{\cdot}10^{-7}$	A_{21}	$9.1061{\cdot}10^{-9}$
C_{14}	$-6.1260{\cdot}10^{-10}$	A_{22}	$-1.6009{\cdot}10^{-10}$
C_{20}	$3.1260{\cdot}10^{-5}$	A_{23}	$7.994{\cdot}10^{-12}$
C_{21}	$-1.7111{\cdot}10^{-6}$	A_{30}	$1.100{\cdot}10^{-10}$
C_{22}	$2.5986{\cdot}10^{-8}$	A_{31}	$6.651{\cdot}10^{-12}$
C_{23}	$-2.5353{\cdot}10^{-10}$	A_{32}	$-3.391{\cdot}10^{-13}$
C_{24}	$1.0415{\cdot}10^{-12}$	B_{00}	$-1.922{\cdot}10^{-2}$
C_{30}	$-9.7729{\cdot}10^{-9}$	B_{01}	$-4.42{\cdot}10^{-5}$
C_{31}	$3.8513{\cdot}10^{-10}$	B_{10}	$7.3637{\cdot}10^{-5}$
C_{32}	$-2.3654{\cdot}10^{-12}$	B_{11}	$1.7950{\cdot}10^{-7}$
A_{00}	1.389	D_{00}	$1.727{\cdot}10^{-3}$
A_{01}	$-1.262{\cdot}10^{-2}$	D_{10}	$-7.9836{\cdot}10^{-6}$

References

1. Makar, A. Influence of the Vertical Distribution of the Sound Speed on the Accuracy of Depth Measurement. *Rep. Geod.* **2001**, *5*, 31–34.
2. Makar, A. Vertical Distribution of the Sound Speed and its Mean Value in Depth Measurements Using a Singlebeam Echosounder. *Rep. Geod.* **2002**, *2*, 79–85.
3. Brekhovskikh, L.M.; Lysanov, Y.P. *Fundamentals of Ocean Acoustics*; Springer: New York, NY, USA; Berlin/Heidelberg, Germany, 2014.
4. Katsnelson, B.G.; Petnikov, V.G. *Shallow-Water Acoustics*; Springer & Praxis: London, UK, 2002.
5. Tolstoy, I.; Clay, C.S. *Ocean Acoustics. Theory and Experiment in Underwater Acoustics*; McGraw-Hill Book Company: New York, NY, USA, 1966.
6. Klusek, Z. *Propagation Conditions of the Sound Speed in the Southern Baltic*; Polish Academy of Science: Warsaw, Poland, 1990.
7. Costa, E.; Medeiros, E. Numerical Modeling and Simulation of Acoustic Propagation in Shallow Water. *Mecánica Comput.* **2010**, *XXIX*, 2215–2228.
8. Makar, A. Method of Determination of Acoustic Wave Reflection Points in Geodesic Bathymetric Surveys. *Annu. Navig.* **2008**, *14*, 1–89.
9. Makar, A. Refraction Correction of the Acoustic Wave in the Multibeam Systems. In Proceedings of the 5th World Multidisciplinary Earth Sciences Symposium WMESS 2019, Prague, Czech Republic, 9–13 September 2019.
10. Mohammadloo, T.; Snellen, M.; Renoud, W.; Beaudoin, J.; Simons, D. Correcting Multibeam Echosounder Bathymetric Measurements for Errors Induced by Inaccurate Water Column Sound Speeds. *IEEE Access* **2019**, *7*, 122052–122068. [CrossRef]
11. Plaa, J.; Snellen, M.; Simons, D. Correcting bathymetry measurements for water sound speed effects using inversion theory. *J. Acoust. Soc. Am.* **2008**, *123*, 3352. [CrossRef]
12. Medwin, H. Speed of Sound in Water: A Simple Equation for Realistic Parameters. *J. Acoust. Soc. Am.* **1975**, *58*, 1318–1319. [CrossRef]
13. Medwin, H. *Sounds in the Sea: From Ocean Acoustics to Acoustical Oceanography*; Cambridge University Press: Cambridge, UK, 2005.
14. Clay, C.S.; Medwin, H. *Acoustical Oceanography*; John Wiley & Sons: New York, NY, USA, 1977.
15. Wilson, W.D. Equation for the speed of sound in sea water. *J. Acoust. Soc. Am.* **1960**, *32*, 1357. [CrossRef]
16. Kinsler, L.; Frey, A.; Coppens, A.; Sanders, J. *Fundamentals of Acoustics*, 4th ed.; J. Wiley: New York, NY, USA, 1999.
17. Del Grosso, V.A. New Equation for the Speed of Sound in Natural Waters (with Comparisons to Other Equations). *J. Acoust. Soc. Am.* **1974**, *56*, 1084–1091. [CrossRef]
18. Millero, F.J. History of the equation of state of seawater. *Oceanography* **2010**, *23*, 18–33. [CrossRef]
19. Chen, C.T.; Millero, F.J. Speed of sound in seawater at high pressures. *J. Acoust. Soc. Am.* **1977**, *62*, 1129–1135. [CrossRef]
20. Chen, C.T.; Millero, F.J. Reevaluation of Wilson's sound-speed measurements for pure water. *J. Acoust. Soc. Am.* **1976**, *60*, 1270–1273. [CrossRef]
21. Mackenzie, K.V. Nine-term Equation for Sound Speed in the Oceans. *J. Acoust. Soc. Am.* **1981**, *70*, 807–812. [CrossRef]

22. Mackenzie, K.V. Nine Equation for Sound Speed in the Oceans. *J. Acoust. Soc. Am.* **1981**, *33*, 1498–1504. [CrossRef]
23. Jamshidi, S.; Abu Bakar, M.N.B. An Analysis on Sound Speed in Seawater using CTD Data. *J. Appl. Sci.* **2010**, *10*, 132–138. [CrossRef]
24. Spiesberger, J.L.; Metzger, K. New estimates of sound speed in water. *J. Acoust. Soc. Am.* **1991**, *89*, 1697. [CrossRef]
25. Calder, M. Calibration of echo sounders for offshore sounding using temperature and depth. *Aust. Surv.* **1974**, *26*, 108–112. [CrossRef]
26. Calder, M. Calibration of Echo Sounders for Offshore Sounding Using Temperature and Depth. *Int. Hydrogr. Rev.* **2015**, *52*, 23746.
27. Reha, M.; Alkan, R.; Alkan, Y.; Aykut, N. Sound Velocity Determination with Empirical Formulas & Bar Check. In Proceedings of the Shaping the Change XXIII FIG Congress, Munich, Germany, 8–13 October 2006.
28. Kozaczka, E.; Grelowska, G. Theoretical Model of Acoustic Wave Propagation in Shallow Water. *Pol. Mar. Res.* **2017**, *24*, 48–55. [CrossRef]
29. Leroy, C.; Robinson, S.; Goldsmith, M. A new equation for the accurate calculation of sound speed in all oceans. *J. Acoust. Soc. Am.* **2008**, *124*, 2774–2782. [CrossRef]
30. Yang, G.; Zheng, Q.; Hu, X.; Ma, D.-J.; Chen, Z.; Long, Y.; Ju, X.; Zhuang, Z.; Hui, Z.; Guo, Y.-L.; et al. Synoptic time-scale variability of sediment temperature profile and sound speed in shallow seas derived from in situ observations. Estuarine. *Coast. Shelf Sci.* **2020**, *244*, 106932. [CrossRef]
31. Brčić, J.; Grilli, F.; Sala, A. Influence of sea-bottom temperature and depth on swept area estimation from trawl surveys. *Sci. Mar.* **2014**, *78*, 135–142. [CrossRef]
32. Ali, M.M.; Jain, S.; Ramachandran, R. Effect of Temperature and Salinity on Sound Speed in the Central Arabian Sea. *Open Ocean. Eng. J.* **2011**, *4*, 71–76.
33. Kozaczka, E.; Grelowska, G.; Bittner, P.; Grelowski, A. The influence of Vistula water on the thermodynamic and acoustic parameters of the Gulf of Gdansk. *Oceanologia* **1995**, *37*, 227–248.
34. Grelowska, G. Study of Seasonal Acoustic Properties of Sea Water in Selected Waters of the Southern Baltic. *Pol. Mar. Res.* **2016**, *23*, 25–30. [CrossRef]
35. Allen, J.T.; Keen, P.W.; Gardiner, J.; Quartley, M.; Quartley, C. A new salinity equation for sound speed instruments. *Limnol. Oceanogr. Methods* **2017**, *15*, 810–820. [CrossRef]
36. Makar, A.; Specht, C.; Specht, M.; Dąbrowski, P.; Szafran, M. Integrated Geodetic and Hydrographic Measurements of the Yacht Port for Nautical Charts and Dynamic Spatial Presentation. *Geosciences* **2020**, *10*, 203. [CrossRef]
37. Specht, C.; Specht, M.; Cywiński, P.; Skóra, M.; Marchel, Ł.; Szychowski, P. A new method for determining the territorial sea baseline using an unmanned, hydrographic surface vessel. *J. Coast. Res.* **2019**, *35*, 925–936. [CrossRef]
38. Specht, M.; Specht, C.; Mindykowski, J.; Dąbrowski, P.; Masnicki, R.; Makar, A. Geospatial Modeling of the Tombolo Phenomenon in Sopot using Integrated Geodetic and Hydrographic Measurement Methods. *Remote Sens.* **2020**, *12*, 737. [CrossRef]
39. Specht, C.; Świtalski, E.; Specht, M. Application of an Autonomous/Unmanned Survey Vessel (ASV/USV) in Bathymetric Measurements. *Pol. Mar. Res.* **2017**, *24*, 36–44. [CrossRef]
40. Specht, M.; Specht, C.; Szafran, M.; Makar, A.; Dąbrowski, P.; Lasota, H.; Cywiński, P. The Use of USV to Develop Navigational and Bathymetric Charts of Yacht Ports on the Example of National Sailing Centre in Gdańsk. *Remote Sens.* **2020**, *12*, 2585. [CrossRef]
41. Acharya, B.; Bhandari, M.; Bandini, F.; Pizarro, A.; Perks, M.; Joshi, D.; Wang, S.; Dogwiler, T.; Ray, R.; Kharel, G.; et al. Unmanned Aerial Vehicles in Hydrology and Water Management: Applications, Challenges, and Perspectives. *Water Res. Res.* **2021**, *57*, e2021WR029925. [CrossRef]
42. Popielarczyk, D.; Templin, T.; Ciecko, A.; Grunwald, G. Application of GNSS and SBES techniques to investigate the Lake Suskie bottom shape. In Proceedings of the SGEM2016 Conference, Albena, Bulgaria, 28 June–6 July 2016; pp. 109–116. [CrossRef]
43. Popielarczyk, D.; Templin, T.; Lopata, M. Using the geodetic and hydroacoustic measurements to investigate the bathymetric and morphometric parameters of Lake Hańcza (Poland). *Open Geosci.* **2015**, *7*, 1–16. [CrossRef]
44. Popielarczyk, D.; Templin, T. Application of Integrated GNSS/Hydroacoustic Measurements and GIS Geodatabase Models for Bottom Analysis of Lake Hancza. *Pure Appl. Geophys.* **2014**, *171*, 997–1011. [CrossRef]
45. International Hydrographic Organization. *Standards for Hydrographic Surveys*, 6th ed.; Special Publication No. 44; International Hydrographic Bureau: Monaco, 2020.
46. Canadian Hydrographic Service. *Standards for Hydrographic Surveys*, 2nd ed.; Canadian Hydrographic Service: Ottawa, ON, Canada, 2013.
47. International Federation of Surveyors (FIG). *Guidelines for the Planning, Execution and Management of Hydrographic Surveys in Ports and Harbours*; FIG Commission 4 Working Group Hydrographic Surveying in Practice; The International Federation of Surveyors (FIG): Frederiksberg, Denmark, 2010.
48. International Hydrographic Organization. *Manual on Hydrography*, 1st ed.; Publication C-13, Corrections to 2011; International Hydrographic Bureau: Monaco, 2005.
49. Hydrographic Office of the Polish Navy. *Maritime Hydrography—Rules of Data Collecting and Results Presentation*; Hydrographic Office of the Polish Navy: Gdynia, Poland, 2009.
50. List of Bodies of Water by Salinity. Available online: https://en.wikipedia.org/wiki/List_of_bodies_of_water_by_salinity (accessed on 15 July 2021).

51. Dera, J. *Sea Physics*; Scientific Publication of PWN: Warsaw, Poland, 2003.
52. Talib, K.; Yusof, O.; Sulaiman, S.; Azizan, A. Determination of speed of sound using empirical equations and SVP. In Proceedings of the IEEE 7th International Colloquium on Signal Processing and Its Applications, CSPA 2011, Penang, Malaysia, 4–6 March 2011. [CrossRef]
53. Coppens, A.B. Simple equations for the speed of sound in Neptunian waters. *J. Acoust. Soc. Am.* **1981**, *69*, 862–863. [CrossRef]
54. Fofonoff, N.P.; Millard, R.C., Jr. *Algorithms for Computation of Fundamental Properties of Seawater, UNESCO Technical Papers in Marine Science*; No. 44; Division of Marine Sciences: Paris, France, 1983.
55. Wong, G.S.K.; Zhu, S. Speed of sound in seawater as a function of salinity, temperature and pressure. *J. Acoust. Soc. Am.* **1995**, *97*, 1732–1736. [CrossRef]
56. Merrifield, M.; Aarup, T.; Allen, A.; Aman, A.; Caldwell, P.; Bradshaw, E.; Fernandes, R.M.S.; Hayashibara, H.; Hernandez, F.; Kilonsky, B.; et al. The Global Sea Level Observing System (GLOSS). In Proceedings of the OceanObs'09: Sustained Ocean Observations and Information for Society, Venice, Italy, 21–25 September 2009; Volume 2.
57. Bradshaw, E.; Rickards, L.; Aarup, T. Sea level data archaeology and the Global Sea Level Observing System (GLOSS). *GeoResJ* **2015**, *6*, 9–16. [CrossRef]
58. Boski, T.; Wilamowski, A. Sea level changes–causes, time scales and the history of their recognition. *Prz. Geol.* **2020**, *68*, 820–823.
59. Ołdakowski, B.; Kowalewski, M.; Jędrasik, J.; Szymelfenig, M. Ecohydrodynamic Model of the Baltic Sea, Part I: Description of the ProDeMo model. *Oceanologia* **2005**, *47*, 477–516.
60. Kowalewska-Kalkowska, H.; Kowalewski, M.; Wiśniewski, B. The forecasting of hydrodynamic conditions in the Odra River mouth using hydrodynamic model of the Baltic Sea (M3D_UG). *Pol. J. Environ. Stud.* **2007**, *16*, 26–31.
61. Kowalewska-Kalkowska, H.; Kowalewski, M. Observations and modelling of extreme sea level events in the coastal waters of the Pomeranian Bay. In Proceedings of the Storm Surges Congress Conference, Hamburg, Germany, 13–17 September 2010.
62. Vargas-Yáñez, M.; García-Martínez, M.C.; Moya, F.; Balbín, R.; López-Jurado, J.L.; Serra, M.; Zunino, P.; Pascual, J.; Salat, J. Updating temperature and salinity mean values and trends in the Western Mediterranean: The RADMED project. *Prog. Oceanogr.* **2017**, *157*, 27–46. [CrossRef]
63. Naranjo, C.; Lafuente, J.; Sammartino, S.; Sanchez, J.; Sánchez, R.F.; Bellanco, M. Recent changes (2004–2016) of temperature and salinity in the Mediterranean outflow. *Geophys. Res. Lett.* **2017**, *44*, 5665–5672. [CrossRef]
64. Vasiliev, I.; Karakitsios, V.; Bouloubassi, I.; Agiadi, K.; Kontakiotis, G.; Antonarakou, A.; Triantaphyllou, M.; Gogou, A.; Kafousia, N.; de Rafélis, M.; et al. Large Sea Surface Temperature, Salinity, and Productivity-Preservation Changes Preceding the Onset of the Messinian Salinity Crisis in the Eastern Mediterranean Sea. *Paleoceanogr. Paleoclimatol.* **2019**, *34*, 182–202. [CrossRef]
65. Skliris, N.; Zika, J.D.; Herold, L.; Josey, S.A.; Marsh, R. Mediterranean sea water budget long-term trend inferred from salinity observations. *Clim. Dyn.* **2018**, *51*, 2857–2876. [CrossRef]
66. Gacic, M.; Schroeder, K.; Civitarese, G.; Cosoli, S.; Vetrano, A.; Eusebi Borzelli, G.L. Salinity in the Sicily Channel corroborates the role of the Adriatic-Ionian Bimodal Oscillating System (BiOS) in shaping the decadal variability of the Mediterranean overturning circulation. *Ocean Sci.* **2013**, *9*, 83–90. [CrossRef]
67. Salon, S.; Crise, A.; Picco, P.; De Marinis, E.; Gasparini, O. Sound speed in the Mediterranean Sea: An analysis from a climatological data set. *Ann. Geophys.* **2003**, *21*, 833–846. [CrossRef]
68. Salat, J.; Pascual, J.; Flexas, M.; Chin, T.M.; Vazquez-Cuervo, J. Forty-five years of oceanographic and meteorological observations at a coastal station in the NW Mediterranean: A ground truth for satellite observations. *Ocean Dyn.* **2019**, *69*, 1067–1084. [CrossRef]
69. Hussein, M.; Said, M.; Radwan, A. Estimation of Salinity Profiles in the Southeastern Mediterranean off the Egyptian Coast. *JKAU Mar. Sci.* **2011**, *22*, 79–95. [CrossRef]

remote sensing

Article

Multi-Temporal Analysis of Changes of the Southern Part of the Baltic Sea Coast Using Aerial Remote Sensing Data

Krystyna Michałowska [1,*] and Ewa Głowienka [2]

1 Department of Geodesy, Faculty of Civil and Environmental Engineering, Gdańsk University of Technology, 80-233 Gdansk, Poland
2 Faculty of Mining Surveying and Environmental Engineering, Department of Photogrammetry Remote Sensing of Environment and Spatial Engineering, AGH University of Science and Technology, al. Mickiewicza 30, 30-059 Krakow, Poland; eglo@agh.edu.pl
* Correspondence: krystyna.michalowska@pg.edu.pl

Abstract: Understanding processes that affect changes in the coastal zone and the ability to predict these processes in the future depends on the period for which detailed monitoring is carried out and on the type of coast. This paper analyzes a southern fragment of the Baltic coast (30 km), where there has been no anthropogenic impact (Slowinski National Park). The study was carried out covering a time interval of 65 years. Historic and current aerial photographs (orthophotomaps) from the following years: 1951, 1964, 1975, 1984, 1995, 2004, and 2016 were used. Changes in the position of the foredune toe line (FTL) in each years' images were used. For each time interval (1951–1964, 1964–1975, 1975–1984, 1984–1995, 1995–2004, 2004–2016), the coastal area over which morphodynamic processes (erosion and accumulation) took place was calculated. The calculated RL (reference line)—FTL positions allowed us to determine differences in the shoreline course in subsequent years and to determine the extent of shifts/changes of the coastline in each time period. The study results showed an equilibrium between the processes of accumulation and erosion, proving that the development of the studied natural coastline is balanced. There was only a change in the trend of the characteristics of changes from erosive into accumulative ones and vice versa. Moreover, along the studied coast section, a certain periodicity in the coastline changes can be observed. The intervals where predominant erosion occurs alternate with those when accumulation prevails, and then the cycle repeats. The analysis of historic/current aerial images and orthophotomaps from 1951–2016 indicated that strong storms have a significant impact on the magnitude of change (accumulation/erosion) and the formation of the studied coastline.

Keywords: coastal zone; shoreline changes; erosion/accretion; multi-temporal orthophotomaps; remote sensing

Citation: Michałowska, K.; Głowienka, E. Multi-Temporal Analysis of Changes of the Southern Part of the Baltic Sea Coast Using Aerial Remote Sensing Data. *Remote Sens.* **2022**, *14*, 1212. https://doi.org/10.3390/rs14051212

Academic Editors: Jacek Lubczonek, Paweł Terefenko, Katarzyna Bradtke and Marta Wlodarczyk-Sielicka

Received: 25 January 2022
Accepted: 26 February 2022
Published: 1 March 2022

1. Introduction

The coastal zone forms a specific system on the border between the terrestrial and marine environments, which is subjected to very dynamic natural and anthropogenic processes [1]. Studies on changes caused by those processes are extremely important for the maintenance of the sustainable functioning of coastal zones [2]. The problem of having proper management in the coastal zone occurs in many countries on various continents [3–13]. The literature shows numerous examples of coasts where the lack of care about the appropriate functioning of the coastal zone has resulted in serious social, economic, and financial consequences [14–18]. The methods applied in studies on changes in the shoreline are strictly connected with the characteristics of a given type of water body and the dynamics of phenomena and processes such as tides, storms, currents, and waves [11]. The monitoring of morphodynamic processes is carried out with statistical methods [12,19] based on direct measurement [20,21], RTK-GPS measurement [22], and methods based on optical images obtained from unmanned aerial vehicles (UAV) [23–29], airborne images [30–34], and satellite images [32,33,35–42]. In recent years, it has become popular to apply SAR data and laser scanning data (terrestrial—TLS and airborne—ALS) [5,43–49]. The use of remote sensing data makes it possible to monitor remotely with a good level of reliability of the obtained measurements for large coastal areas [50]. Pardo-Pascual [35] and his team presented a method for determining the shoreline with subpixel accuracy (<10 m) using multi-temporal satellite images from Landsat 7, Landsat 8, and Sentinel-2. In the case of multispectral, high-resolution WorldView-2 data, scientists can obtain an accuracy level of 1–5 m [36,37]. However, monitoring based on photogrammetric methods can obtain accuracy below 0.5 m [51]. The application of aerial image data raises the possibility of remote monitoring with the obtained measurements having good reliability for studying the dynamics of shoreline changes. The situations recorded in the aerial photos illustrate the total effect of the processes taking place on the shores of the non-tidal sea. The measurement carried out on this basis indirectly considers all factors influencing the changes in the position of the front dune base line in the analyzed period, e.g., changes in sea level, damage caused by storm surges, and others [52]. The foredune toe is one of the best indicators interpreted in aerial photographs to determine the state and long-term dynamics of the seashore [53]. The baseline of the foredune overlaps the vegetation line, an indicator of the limits of regular flooding by high water, and therefore, it represents an important indicator of shoreline movement [54]. The vegetation line is a natural line formed by the plants on the beach. It is easily identifiable, even in historical photographs [28,53–55].

The research was carried out for the southern shore of the Baltic Sea, which is a non-tidal sea with periodic sea level fluctuations. Climate change and the associated rising level of the Baltic Sea are causing an increase in the frequency of storm surges. This phenomenon may be the main natural threat to the southern coast in the coming decades. Monitoring these changes is extremely important. The changes of the Baltic sea coast (Figure 1) have been studied by many research teams [53,56,57] and included both long-term and short-term [58–60] changes mainly caused by meteorologically-enforced storm waves [21]. Understanding processes that affect changes in the coastal zone and the ability to predict those processes in the future depends on the period for which detailed monitoring is carried out. The fact whether the studies are carried out over short and medium periods (days, months, years, decades) [11,61] or long periods (several decades) is very important [6,62]. The research presented in the literature shows the results obtained for short sections of the Baltic coast using rare time-sampled measurements (up to 30–40 years) and measured points in small numbers (TLS, surveying every 500 or 1000 m).

A new aspect of the present study is that it explores, for the first time, changes of the natural shoreline of the southern Baltic from 1951–2016, long enough (65 years) to allow for capturing the trend of morphodynamic changes in the coastal zone. The analysis included the longest remote sensing data series obtainable.

Another novel aspect of the present work is using the short transects (every 10 m), which made it possible to obtain more precise measurement results (quasi-continuous) compared to measurements using 100–500 m profiles as applied in the state monitoring of the Baltic shore. The research presented herein focuses on a new quantitative approach using analysis of morphodynamic changes (erosion/accumulation) based on multi-temporal remote sensing data to determine trends in natural dune coast development (i.e., devoid of an anthropogenic influence). The work aimed to conduct a remote detection of changes in the coastal zone of the southern Baltic Sea at 6–13 year intervals. Detailed goals included the following:

- to calculate the area of coastal zone changes (erosion/accumulation) in each decade.
- to determine the spatial and temporal distribution of changes.
- to attempt to identify the main factors causing those changes.

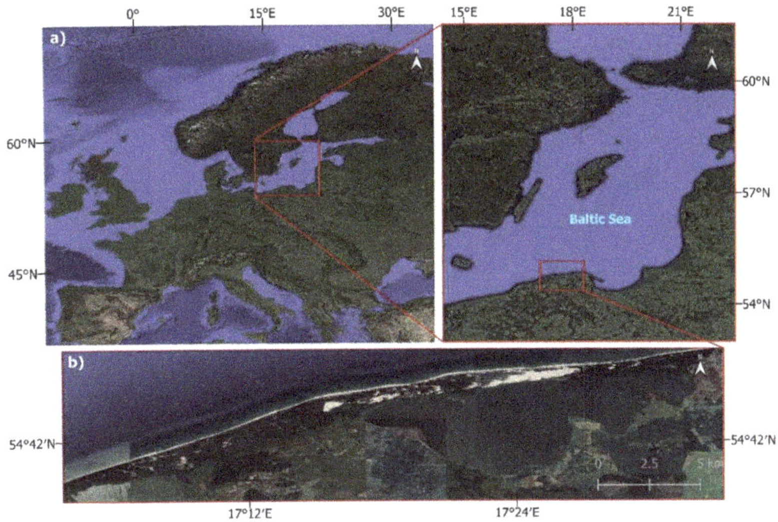

Figure 1. Study area: (**a**) Baltic Sea; (**b**) Location of examined part of the Polish Baltic Sea coast. The base map—Google Satellite.

2. Materials and Methods

2.1. Study Area

The research was conducted taking into account the southern coast of the non-tidal Baltic Sea, located within the strict protection limits of the Slowinski National Park. The examined coast section, 30 km long, constitutes a sandy sea barrier called the Gardno-Lebsko Spit (Figure 1). The spit, 0.6 to 2 km wide, separates two nearshore lakes, Gardno and Lebsko, from the Baltic Sea. The lakes are linked to the sea by means of short estuary sections of two very small rivers, the Lupawa and Leba. These rivers have an insignificant impact on changes occurring in the coastal zone. In its western portion, the spit is overgrown with forest and has gray dune fields. In the central and eastern parts of the spit, concerning an area of over 500 ha, there are moving dunes. The dunes travel with a speed of 3–10 m/year from west to east, in accordance with the direction of prevailing winds [63], and range in height from several meters up to 56 m above sea level [63,64].

The geological composition of surface formations, which affect the development of morphological processes, is little diversified. Four types of sedimentary rock mainly exist, which differ both in terms of their age and origin. The most common materials are Pleistocene sands of marine accumulation. Under the layer of sands, there are, in places, post-glacial biogenic formations (turfs, gyttjas) [63].

Due to the high variability of wind directions and speed occurring at the Polish coast, the wind wave is the most dynamic factor conveying the largest amounts of energy towards the bottom, controlling the intensity of deposit movements [65]. Winds blowing from the west to the east transport sand, from which foredunes and white linear dunes are formed, thus contributing to land build-up [66]. When analyzed on a yearly basis, the area is marked by a high average wind speed (4.5 m/s) and a very small number of days with calm weather (4.9 days) [67]. Days extremely windy, including those with storms, occur mainly in the autumn and winter. During an average storm, waves of 4–5 m high and 60–80 m long develop. The direct reason for the development of storm surges along the Polish coast is low-pressure systems with accompanying thermal fronts, moving to the east or south-east, from the Norwegian Sea towards Scandinavia [68]. Significant causative factors for changes occurring within the coastal zone system are played by systems of coastal currents, which take part in, among other things, the transport of deposits [65].

2.2. Data

In this study, airborne photos taken in 1951, 1964, 1975, 1984, 1995, 2004, 2010, and 2016 were used. The detailed characteristics of the images and their sources are shown in Table 1. The images were taken at various scales. The images from 1951–1984 were taken in diapositives and scanned into digital format with a precision scanner (PHOTOSCAN TD Intergraph-Zeiss, resolution 14 μm). Digital images for 1995, 2004 and an orthophotomap for 2010 were obtained from the Polish Centre of Surveying and Cartographic Documentation. Data from 1995 were recorded under the EU program PHARE. Moreover, an orthophotomap of 2016, published as WMS (Web Map Service) by the Polish Main Office of Surveying and Cartography, was used.

Table 1. The characteristics of source aerial images (1951–2004) and source orthophotomaps (2010, 2016).

Year	1951	1964	1975	1984	1995	2004	2010	2016
Camera type	RC-5	RC-8	RC-5/8	RC-10	RC-20	RC-20	-	-
Format (cm)	18 × 18	23 × 23	23 × 23	23 × 23	23 × 23	23 × 23	-	-
Scale	1:25,000	1:23,000	1:29,500	1:25,000	1:26,000	1:26,000	-	-
Pixel size (μm)	14	14	14	14	21	14	-	-
Source	diapositive	diapositive	diapositive	diapositive	digital	digital	digital	WMS

2.3. Processing of Aerial Data

The data from 1951–2004 were subjected to geometric rectification. The aerial images were blocks of four strips (years: 1951, 1964, 1975, 1984) and eight strips (years: 1995, 2004). The photographs from 1951 and 1964 were characterized by slightly lower quality and a poorer photo interpretation potential due to scratches and damages of the emulsion on the diapositives. It was necessary to improve their quality by using, among others, smoothing and noise removal filters. In the aerotriangulation process, the number of used Ground Control Points (GCPs) depended on the photointerpretation potential of the images in the given year, related to land cover changes. The GCPs were obtained from GPS measurements and based on topographic maps. Totally 45–69 GCP points, depending on the image, were used. The number of GCPs and the results of the aerotriangulation are presented in the table (Table 2). The RMS error for individual coordinates did not exceed the value of 0.5 m.

Table 2. Results of the aerotriangulation process for images for subsequent years.

Year	1951	1964	1975	1984	1995	2004
Number of photos in block	52	36	24	39	46	36
Number of used GCPs	45	59	61	69	61	59
RMS X (m)	0.36	0.45	0.36	0.41	0.11	0.37
RMS Y (m)	0.46	0.44	0.31	0.42	0.10	0.37
RMS Z (m)	0.46	0.45	0.43	0.36	0.22	0.28

The photogrammetric materials in the form of aligned blocks of aerial photographs with a stereoscopic coverage from six years: 1951, 1964, 1975, 1984, 1995, 2004 were used to create different Digital Elevation Model (DEM) for each period. For this purpose, measurement of points and vectors representing the terrain surface was performed using a Dephos digital photogrammetric station. For each stereo model, a grid of points with dimensions of 15 m × 15 m (points with a set of coordinates—x, y, z) was measured. Data were supplemented with characteristic details of the terrain (hills, slopes, watercourses, water bodies, depressions, areas excluded from development, and others).

As a result of the orthorectification process, orthophotomaps with a 0.5 m ground pixel resolution in the EPSG 2173 coordinate system were obtained. The orthophotomaps fulfilled the accuracy standards for cartographic materials.

2.4. Analysis of Changes in the Baltic Coastal Zone

Analyses of changes in the coastal zone of the Baltic Sea were carried out on the 188–218 km section of the coast related to seven time intervals (Table 2) covering the years from 1951–2016 (Figure 2).

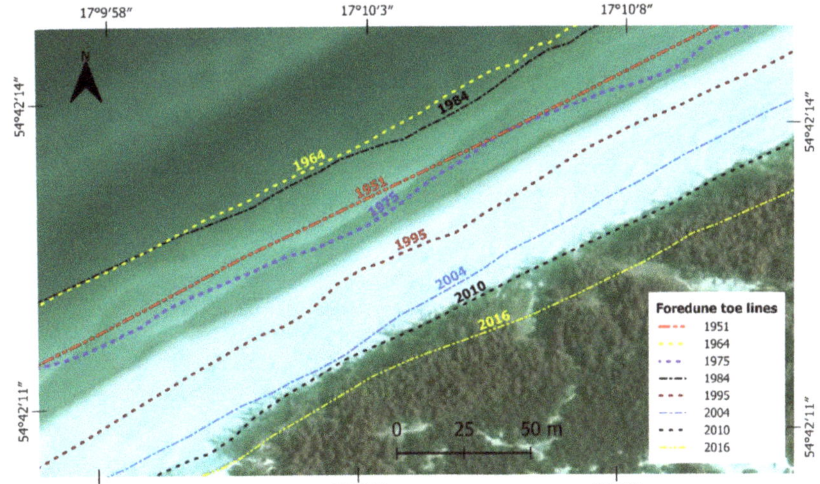

Figure 2. Foredune toe lines in following years on the orthophotomap from 2010. The position of the foredune toe line reflects the course of the shoreline in the given years.

The foredune toe line (FTL) overlaps with the natural line formed by vegetation on the beach is an important indicator of shoreline movement [46–49]. For each data set, vegetation lines were identified and mapped based on the visual interpretation of the orthophotomaps (Figures 2–4). That allowed the course of the line along the base of the front dunes to be vectorized. Additionally, for the 1951–2004 years, stereoscopic observations of aerial photographs were used to enhance the accuracy of foredune toe line vectorization. This allowed for avoiding errors in cases where the vegetation line was difficult to interpret. In sections of coast lacking a foredune, the lines of the base of the main dune were vectorized.

A reference line (RL) was determined along the coast according to the Polish Maritime Office's Kilometer line (MOK) (Figure 3). According to the methodology adopted in studies of shoreline changes in the Baltic Sea [43], the analysis of the magnitude of coastline changes was carried out based on transects set at short intervals—every 10 m. The transects were perpendicular to the reference line (Figure 4). It made it possible to obtain more precise measurement results (quasi-continuous) compared with measurements utilizing 100–500 m profiles as applied in the state monitoring of the Baltic shore [69]. The distance from the reference line to the coastline was calculated for each year. As examples, 1975 and 2010 are presented in Figure 4. The obtained RL-FTL distances made it possible to determine differences in the shoreline course in the following years and the extent of shifts/changes of the coastline in each period.

The vectors of intersecting lines create polygons, i.e., closed areas on which accumulation and erosion occur. Based on the geometry (coordinates) of the obtained polygons (Figure 5), their surface areas were calculated. The transformation of linear objects into 2D and calculation of the size of accumulated areas and eroded areas for respective time intervals were carried out with GIS software (GeoMedia, QGIS)).

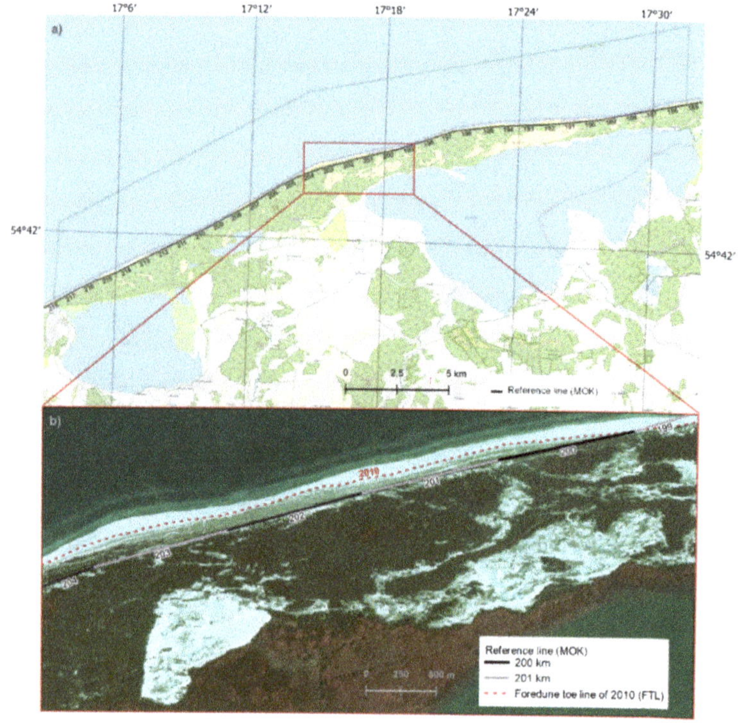

Figure 3. The route of the reference line (RL)—Polish Maritime Office's Kilometer line (MOK), (**a**) along the examined coast section of the Baltic Sea, the base map—OpenStreetMap, (**b**) The routes of the selected part of RL and FTL line in year 2010 (red color)—the base map—orthophotomap dated 2010.

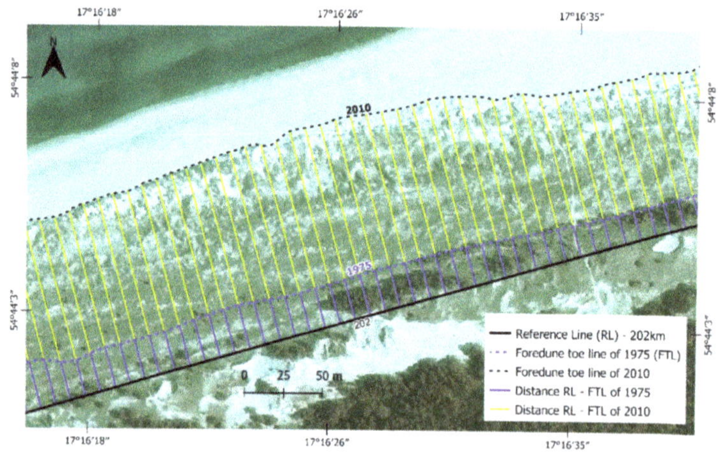

Figure 4. Transects perpendicular to the reference line (MOK) determined at intervals of 10 m. The distance was defined as the reference line—foredune toe lines in a given year: RL-FTL 2010; RL-FTL 1975. The base map—orthophotomap dated 2010.

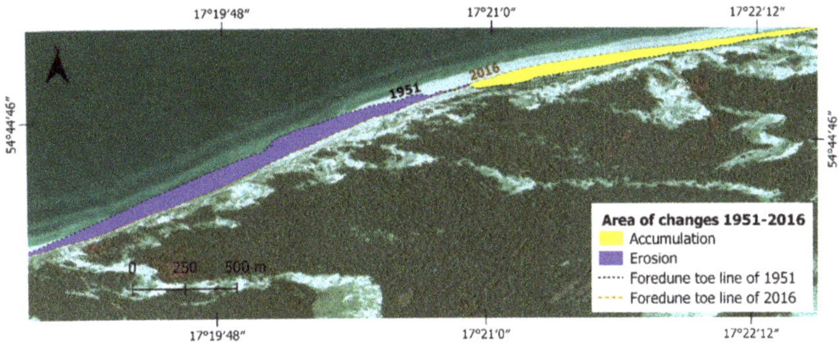

Figure 5. The vectors of intersecting foredune toe lines from 1951 and 2016 create polygons (closed areas), which were used to calculate the size of the accumulated and eroded areas. The base map/orthophotomap dated 2010.

It should be noted that the last two periods were shorter (6 years) than the previous ones (from 9–13 years). To carry out analysis, they should be combined into a 12 year period, i.e., 2004–2016. For each time interval, the coast area at which the morphodynamic processes took place was calculated. Then, we calculated the total area of changes in individual periods (1) and the average value of the change in area in a given period (2).

$$A_T = \Sigma a_{i,} \quad (1)$$

$$A_{TY} = A_T/n, \quad (2)$$

where:

- A_{TY}—average area for a period of time
- A_T—total area in the time period
- a—partial area in the time period
- T—period of time
- n—number of years

3. Results

Based on the obtained results, an analysis and determination of changes in the coastal zone of the southern Baltic over 65 years (1951–2016) were carried out. Graphs and tables (Figures 6–10, Table 3) provide an overview of the surface area of erosion/accumulation changes over the decades and their spatial-temporal distribution of those changes.

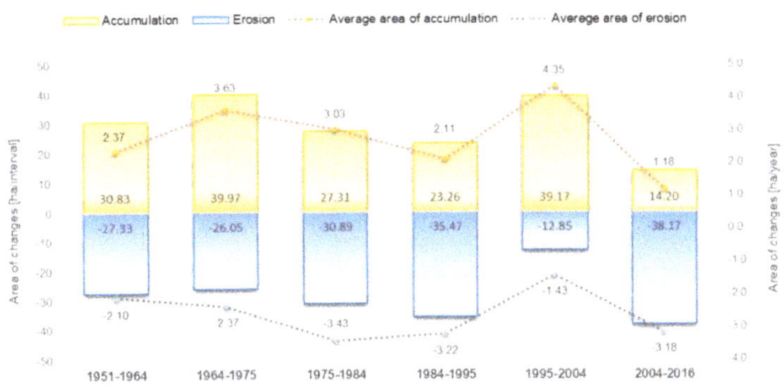

Figure 6. Graph of the changes in accumulated and eroded areas [ha] from 1951–2016 and the average area [ha] per year (188–218 km).

Graphs of the total area of change in subsequent intervals and the average value for each year were prepared for all periods (Figure 6). The results are also presented as a graph and map of the total change that took place over 65 years (Figure 10).

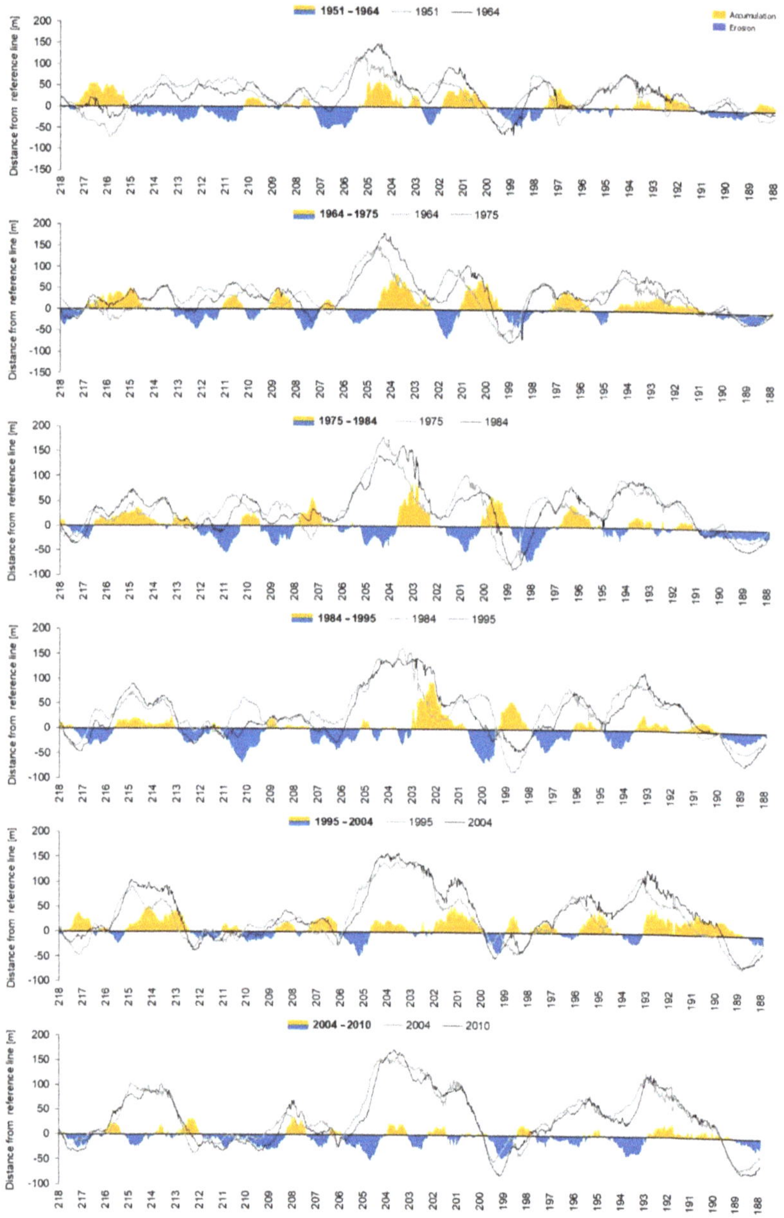

Figure 7. Graphs of the changes in the accumulated and eroded areas in subsequent time intervals (periods close to a decade) on the studied part of the coast (188–218 km according to MOK).

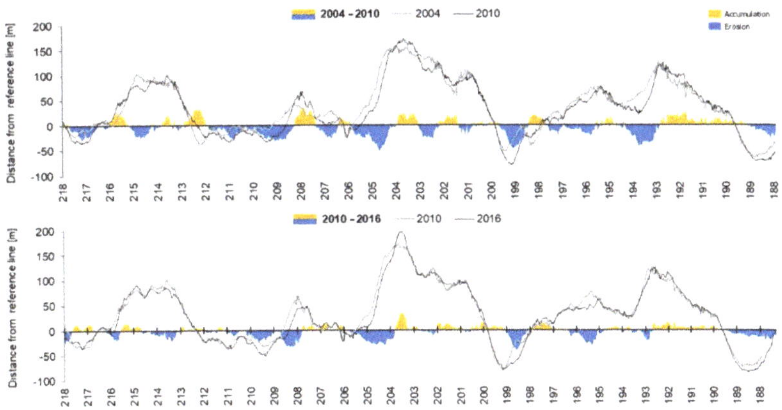

Figure 8. Graphs of the changes in the accumulated and eroded areas in two 6 year time intervals (2004–2010 and 2010–2016) on the studied part of the coast (188–218 km according to MOK).

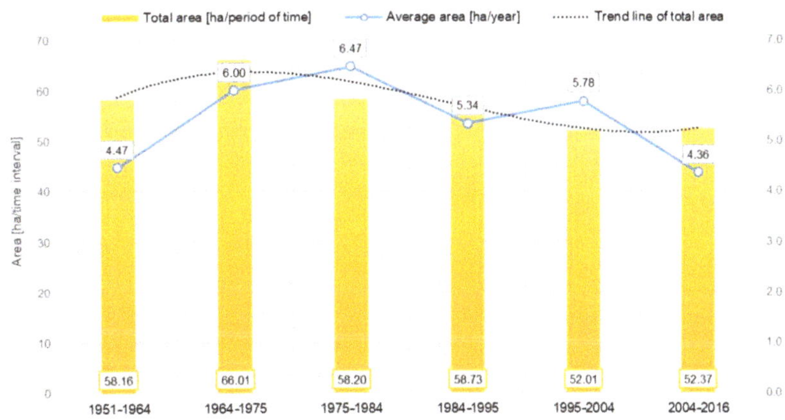

Figure 9. Total area of coastal change in subsequent time intervals.

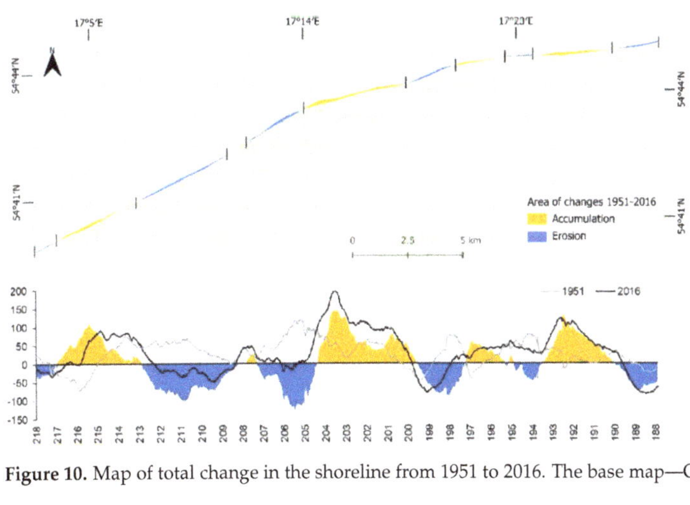

Figure 10. Map of total change in the shoreline from 1951 to 2016. The base map—OpenStreetMap.

Table 3. Area of coastal changes in subsequent time series: value of accumulated and eroded area, total area of changes, and the balance of the impact of morphodynamic processes on the studied part of the coast.

Nature of Changes		Area of Coastal Changes in Subsequent Time Series							
		1951–1964	1964–1975	1975–1984	1984–1995	1995–2004	2004–2016	2004–2010	2010–2016
Accumulation	(ha)	30.83	39.97	27.31	23.26	39.17	14.20	11.96	7.66
	(%)	53	61	47	40	75	27	31	31
Erosion	(ha)	−27.33	−26.05	−30.89	−35.47	−12.85	−38.17	−26.94	−17.41
	(%)	47	39	53	60	25	73	69	69
Total area	(ha)	58.16	66.01	58.20	58.73	52.01	52.37	38.90	25.07
Budget	(m)	3.5	13.92	−3.59	−12.21	26.32	−23.97	−14.98	−9.75

3.1. Analysis of the Areas Eroded and Accumulated in the Examined Period

The analysis of the changes in accumulated and eroded areas showed that in the examined section of the coast in 1951–1964, the accumulation process slightly dominated, covering 30.83 ha, which represented a change amounting to 53% of the total area. The time interval 1964–1975 was characterized by clear intensification of the accumulation processes (Figures 6 and 7, Table 3). The accumulated area increased by more than 9 ha. That means 61% of the total surface changed, representing accretion, which prevailed during this interval. At the same time, there was a slight decrease in eroded areas from 27 to 26 ha (Figure 6 and Table 3), which, in the studied period, represented only 39% of the total area. A comparison with the previous period led to an observation that along the western part of the studied coast (i.e., 214.3 to 205.5 km, according to MOK), eroded sections were shortened, the trend of erosive fragments was reversed for accumulative areas (i.e., 215.2–214.3 km and 211.3–210.3 km MOK), and vice versa (Figure 7). Along the entire examined coast, an extension of the shoreline sections with accumulation characteristics was observed.

In the subsequent two intervals, 1975–1984 and 1984–1995, the accumulation area decreased accordingly from 40 to 27.3 ha, and down to 23.3 ha. The eroded area increased from 26 to 30.89 ha in the 1975–1984 interval, while during the next, the growth was up to 35.47 ha. It was the biggest area subjected to shore erosion during the studied period of 65 years. A visible elongation of eroded sections was also noticed (Figure 7).

The interval of 1995–2004 was characterized by the most extended accretionary processes, which determined the reduction of the coastlines/areas affected by erosion to only 25% of the total (Figure 6 and Table 3). The elongation of accumulation sections was observed. On the coast sections where shore reconstruction processes took place, significant shifts in the shoreline were observed in the range of 20–40 m, sometimes exceeding 50 m (Figure 7). The significant predominance of long sections of accumulation and shorter fragments of shore erosion indicate an essential intensification of accumulation trends and a decrease in erosion trends.

The analysis for 2004–2016 showed a clear increasing trend of coastal destruction processes. Compared to the previous period, the eroded area increased three times from 12.85 ha to 38.17 ha (Figures 7 and 8, Table 3). The 2004–2016 period was the most erosive interval. These results align with those obtained by analyzing data from the shorter 6 year periods. It can be seen that the time interval of 2004–2010 was characterized by a significant lowering of the intensity of the accumulation processes from an area of 39.17 ha to 11.96 ha (24.97 ha). At the same time, there was an intensification of shore erosion processes. The eroded area increased by more than double; from 12.85 to 26.94 ha. In the most recent period (2010–2016), there was a permanent trend for the area of coastal fragments to decrease where beach build-up was taking place (from 11.96 to 7.66 ha). At the same time, there was a significant decrease in the size of the eroded area (from 26.94 to 17.41 ha).

Balance of the Impact of Morphodynamic Processes on the Studied Part of the Coast

The balance between the sums of accumulated and eroded areas in subsequent time intervals was analyzed (Figures 6–8 and Table 3). Comparable amounts of accumulated and eroded areas were observed in the first time interval, and their participation was 53% and 47%, respectively (Table 3). In subsequent years (1964–1975), clear intensification of accumulation changes was observed, which manifested itself in a positive balance (+13.92 ha) of the area of accumulation; the proportion of the eroded area stayed on a similar level, i.e., 26.05 ha, as compared to the previous period. A negative balance occurred in two subsequent decades, equalling 3.59 ha in 1975–84 and 12.21 ha in the following decade (1984–1995). This demonstrates an intensification of erosion and dominance of the shore destruction processes. The following time interval (1995–2004) was characterized by a definite weakening of erosion compared to previous decades. The eroded area decreased by almost three times (35.47 to 12.85 ha). The retreat of the shore occurred on almost one-quarter of the length of the examined coast. The positive balance of +26.32 ha of accumulation area provides evidence of the reversal of the erosion trend and the decisive predominance of the shore accumulation processes over the interval.

In the period from 2004–2016, a negative balance of the area of change was observed (−23.97 ha) along with a significant dominance of erosion processes (Figures 6–8 and Table 3). It should be noted that compared to the previous period, the area on which coastal destruction processes occurred increased by 48%. In contrast, the size of the area of the accumulative nature decreased three times. The detailed analysis of the shorter six-year intervals indicated a negative balance in both periods, i.e., 14.98 and 9.75 ha. The coast erosion occurred in 69% of the examined area.

Based on the analysis of changes in coastal areas over the whole examined period (1951–2016), a slight predominance of the accumulated area over the eroded area, +78.19 (51%) and 74.22 (49%), respectively, was identified (Figure 6, Table 3). Moreover, in a comparison of the areas subjected to changes, in subsequent periods, the area of shore accumulation was +174.74 ha (51%), while the area of shore subjected to erosion was 170.76 ha (49%). The obtained results of the study showed an equilibrium between the processes of accumulation and erosion, proving that the development of the studied coastline has been balanced. There were only changes in the trend from erosive to accumulative processes and vice versa.

The total area of shore change in subsequent years ranged from 52.01 to 66,01 ha (Figure 9 and Table 3). A detailed analysis of the total areas of accumulation and erosion in subsequent time intervals indicated the intensification of coastal processes in 1964–1975, and the increase in the area of changes in the coastal area by 7.85 ha was observed, as compared to the previous interval (Table 3). The years 1975–1984 were marked by a decrease in the area of change down to 58.2 ha, which corresponds to the change value in the 1951–1964 period. In the following time interval (1984–1995), the change area reached a comparable value, i.e., 58.73 ha. In 1995–2004 and 2004–2016, there was a noticeable decrease in the area subjected to changes, about 52 ha, thus indicating a weakening of dynamic coastal processes within the examined part of the coastline.

4. Discussion

Sea level fluctuations exert a great impact on the functioning of the coastal zone system and the course of processes occurring therein. Changes in water level occurring in the coastal zone may have the nature of periodic oscillations (tides) or non-periodic changes, such as storm surges [65]. Sandy beaches act naturally as shoreline buffers, absorbing the energy of waves and adjusting dynamically to the seasonal or long-term wave climate. During extreme wave phenomena, major changes in the coast morphology may occur, which may significantly affect the coast stability [70]. The frequency of storm surges on the Polish coast has intensified over the last fifty years. Also, the maximum levels and duration of surges have increased. The increase in the number of low-pressure fronts moving above the Baltic area resulted in the growth of the number of storm days [71]. The clear increasing tendency results from an increased number of extreme weather phenomena due to climate changes. That process also impacts distinct changes in sea levels over the last hundred years [72].

Following the increase in sea levels and frequency of high storm surges, the Polish coast is threatened with erosion [73–75]. Visible on the larger part of the dune coast is the shore withdrawal inland [73,76–78]. The rate of the Polish coast retraction in the 20th century has been estimated at 0.8–1.0 m per year [76,79]. However, after each storm surge, the dune toe or its top (ridge) is recorded to have moved as much as by 3–4 m per year [80,81]. This is confirmed by the analysis of eroded/accumulated areas for the examined Baltic coast section as presented herein (Figures 6–8 and, Table 3) and previous research [82].

Analyses of the Baltic coast changes based on 1951–2016 aerial photos take into account the accumulative effect of the impact of processes occurring in the coastal zone, including, among other things, changes in sea levels, coastal jet systems, damages resulting from storm surges, etc., [52]. Due to the strict protection of the examined part of the coast, located within the National Park limits, one can exclude negative effects of anthropogenic activity, and the same about the insignificant impact of the Lupawa River estuary. Research [83] has analyzed the variability of wave motion parameters for 1950–2010. There was no significant diversity in wave parameters, which could justify the variability of the coastal zone.

Based on the analysis of accumulated and eroded areas, the fragment of the examined Baltic coast in 1951–1964 is marked by a balanced inventory of coastal areas, on which processes of shore building up/damaging take place (Table 3). The analysis of the frequency of storm surge occurrences [84] and research results [75,85] denote a moderate influence of extreme weather phenomena in the period from 1947–1970 on changes within the coastal zone. In the subsequent years of 1964–1975, intensification of coast building processes (aggregation of the coast) and an increase in surfaces of accumulated areas were noted (Figures 6 and 7, Table 3). This is reflected in the results of the analysis of storm surge occurrence frequency at the southern Baltic Sea [75,84,85]. According to the analysis performed by [68], the period of 1960–1969 was marked by the smallest number of storms on the Polish part of the Baltic Sea coast. This is confirmed by studies undertaken by [86] and conducted in 1963–1964, which proved that 60% of the Lebsko Spit area was subject to the process of accumulation.

The prevalence of coast destruction activity, and larger eroded areas, was recorded in multiannual periods, during which severe and frequent storm surges occurred. The negative balance of the size of accumulated and of eroded areas was recorded in 1975–1984 and again in 1984–1995. Those periods overlap with dates of increased storm activity on the southern Baltic coast [84,87]. Research [88,89] conducted on other sections of the Polish coast confirms that. The strongest effects of the activity of the coast destruction processes were noticed in 1984–1995 (Figures 6 and 7, Table 3). The increased surface area of erosion zones in the period under discussion may result from an increased storm activity on the southern Baltic (207 storms recorded) [71,84]. The length of the coast sections under destruction was extended (Figure 7) compared to the previous periods [82,88].

After 1995, the number of stormy days sharply decreased [71,84], reflected in results obtained for the multi-annual period of 1995–2004. The noticeable prevalence of areas that are accumulative in nature (75% of the surface area of the examined coast section) denotes a decrease in the impact of destructive factors) (Figures 6 and 7, Table 3).

The years 2004–2016 were dominated by destructive shore processes regarding over 70% of the examined coast surface area (Figures 7 and 8, Table 3). This confirms the occurrence of intensive morphodynamic processes resulting from heavy storm surges in 2004, 2009, and 2012 [21,77,80,90]. During the 6-year periods of 2004–2010 and 2010–2016, a demonstrable retraction of land towards the south was recorded. An intensification and domination of shore destruction processes was observed (Figures 7 and 8, Table 3). Considerable land losses along the whole coast were recorded after heavy storms in 2004, 2006, and 2009 [77,80,90]. Following the heavy storm in November 2006, the dunes retracted 2 to 5 m [73]. The highest number of storm surges over all analyzed mareographic stations was recorded in the last period of 2010–2016 [68]. Two storm surges following each other in January 2012, with high seawater levels, caused massive losses along the entire Polish coast, even on the hitherto accumulative part of the coast. Regarding the examined section (the Lebsko Spit), the dune toe retracted up to 3 m [21].

5. Conclusions

The use of historic/current aerial images and orthophotomaps for the analysis of changes in the coastline of the southern coast of the Baltic Sea made it possible to determine the size and intensity of morphodynamic changes (erosion and accumulation) in the time interval of 1951–2016, i.e., over 65 years. The total area subjected to coastal morphodynamic processes in subsequent time intervals revealed an intensification of coastal processes in the 1960s and an increase in the area subjected to changes compared to the first studied period (1951–1964). In the time interval 1975–1984 there was a decrease in the area subjected to changes, slightly below the level of change observed in the first period. In the following decade, the area subjected to transformations corresponded with the size of the area of change from the 1950s. In 1995–2004, accumulation processes dominated. At the same time, there was a further decrease in the area subjected to deformation, which indicates a slight weakening of dynamic coastal processes in the studied area of the coast as studied by the paper authors. Compared to all previously studied periods, the last interval (2004–2016) was the most erosive. A threefold increase in the eroded surface area was due to the growth of the shore destruction processes in this period, mainly due to strong storms in 2004, 2006, 2009, and 2012. The results obtained for the 65-year period indicate a balance between the processes of accumulation and erosion, proving the development of the examined coastline was balanced.

In summary, results obtained from the analysis of orthophotos from 1951–2016 indicate that severe storms have a significant impact on the magnitude of changes (accumulation/erosion) and the formation of the studied coastline. Due to the strict protection of the examined part of the coast (it is located within the borders of the National Park), negative anthropogenic impact can be excluded.

Author Contributions: Conceptualization, K.M.; methodology, K.M. and E.G.; software, K.M. and E.G.; validation, K.M. and E.G.; formal analysis, K.M. and E.G.; investigation, K.M. and E.G.; resources, K.M. and E.G.; writing—original draft preparation, K.M. and E.G.; writing—review and editing, K.M. and E.G.; visualization, K.M. and E.G.; supervision, K.M.; funding acquisition, K.M. All authors have read and agreed to the published version of the manuscript.

Funding: This research was funded by the Gdansk University of Technology, Faculty of Civil and Environmental Engineering.

Institutional Review Board Statement: Not applicable.

Informed Consent Statement: Not applicable.

Data Availability Statement: Not applicable.

Conflicts of Interest: The authors declare no conflict of interest.

References

1. Brown, A.G.; Tooth, S.; Bullard, J.E.; Thomas, D.S.G.; Chiverrell, R.C.; Plater, A.J.; Murton, J.; Thorndycraft, V.R.; Tarolli, P.; Rose, J.; et al. The geomorphology of the Anthropocene: Emergence, status and implications. *Earth Surf. Process. Landf.* **2017**, *42*, 71–90. [CrossRef]
2. Stive, M.J.; Aarninkhof, S.G.; Hamm, L.; Hanson, H.; Larson, M.; Wijnberg, K.M.; Nicholls, R.J.; Capobianco, M. Variability of shore and shoreline evolution. *Coast. Eng.* **2002**, *47*, 211–235. [CrossRef]
3. Suursaar, Ü.; Kullas, T.; Otsmann, M.; Kõuts, T. Extreme sea level events in the coastal waters of western Estonia. *J. Sea Res.* **2003**, *49*, 295–303. [CrossRef]
4. Manca, E.; Pascucci, V.; Deluca, M.; Cossu, A.; Andreucci, S. Shoreline evolution related to coastal development of a managed beach in Alghero, Sardinia, Italy. *Ocean Coast. Manag.* **2013**, *85*, 65–76. [CrossRef]
5. Pradhan, B.; Rizeei, H.M.; Abdulle, A. Quantitative assessment for detection and monitoring of coastline dynamics with temporal RADARSAT images. *Remote Sens.* **2018**, *10*, 1705. [CrossRef]

6. Chaaban, F.; Darwishe, H.; Louche, B.; Battiau-Queney, Y.; Masson, E.; Khattabi, J.E.; Carlier, E. Geographical information system approach for environmental management in coastal area (Hardelot-Plage, France). *Environ. Earth Sci.* **2011**, *65*, 183–193. [CrossRef]
7. Natesan, U.; Parthasarathy, A.; Vishnunath, R.; Kumar, G.E.J.; Ferrer, V.A. Monitoring Longterm Shoreline Changes along Tamil Nadu, India Using Geospatial Techniques. *Aquat. Procedia* **2015**, *4*, 325–332. [CrossRef]
8. Mukhopadhyay, A.; Mukherjee, S.; Mukherjee, S.; Ghosh, S.; Hazra, S.; Mitra, D. Automatic shoreline detection and future prediction: A case study on Puri coast, Bay of Bengal, India. *Eur. J. Remote Sens.* **2012**, *45*, 201–213. [CrossRef]
9. Al Fugura, A.; Billa, L.; Pradhan, B. Semi-automated procedures for shoreline extraction using single RADARSAT-1 SAR image. *Estuar. Coast. Shelf Sci.* **2011**, *95*, 395–400. [CrossRef]
10. Liu, H.; Jezek, K.C. Automated extraction of coastline from satellite imagery by integrating Canny edge detection and locally adaptive thresholding methods. *Int. J. Remote Sens.* **2004**, *25*, 937–958. [CrossRef]
11. Nassar, K.; Mahmod, W.E.; Fath, H.; Masria, A.; Nadaoka, K.; Negm, A. Shoreline change detection using DSAS technique: Case of North Sinai coast, Egypt. *Mar. Georesour. Geotechnol.* **2018**, *37*, 81–95. [CrossRef]
12. Oyedotun, T.D.T.; Ruiz-Luna, A.; Navarro-Hernández, A.G. Contemporary shoreline changes and consequences at a tropical coastal domain. *Geol. Ecol. Landsc.* **2018**, *2*, 104–114. [CrossRef]
13. Dunn, R.A.; Genz, A.S.; Fletcher, C.H.; Frazer, L.N.; Rooney, J.J. The Predictive Accuracy of Shoreline Change Rate Methods and Alongshore Beach Variation on Maui, Hawaii. *J. Coast. Res.* **2007**, *231*, 87–105. [CrossRef]
14. Angnuureng, D.B.; Almar, R.; Senechal, N.; Castelle, B.; Addo, K.A.; Marieu, V.; Ranasinghe, R. Shoreline resilience to individual storms and storm clusters on a meso-macrotidal barred beach. *Geomorphology* **2017**, *290*, 265–276. [CrossRef]
15. Jiménez, J.A.; Sancho-García, A.; Bosom, E.; Valdemoro, H.I.; Guillén, J. Storm-induced damages along the Catalan coast (NW Mediterranean) during the period 1958–2008. *Geomorphology* **2012**, *143*, 24–33. [CrossRef]
16. Sallenger, A.H.; Stockdon, H.F.; Fauver, L.; Hansen, M.; Thompson, D.; Wright, C.W.; Lillycrop, J. Hurricanes 2004: An overview of their characteristics and coastal change. *Estuaries Coasts* **2006**, *29*, 880–888. [CrossRef]
17. Sherman, D.J.; Hales, B.U.; Potts, M.K.; Ellis, J.T.; Liu, H.; Houser, C. Impacts of Hurricane Ike on the beaches of the Bolivar Peninsula, TX, USA. *Geomorphology* **2013**, *199*, 62–81. [CrossRef]
18. Zhang, K.; Douglas, B.; Leatherman, S. Do Storms Cause Long-Term Beach Erosion along the U.S. East Barrier Coast? *J. Geol.* **2002**, *110*, 493–502. [CrossRef]
19. Reeve, D.E.; Fleming, C.A. A statistical-dynamical method for predicting long term coastal evolution. *Coast. Eng.* **1997**, *30*, 259–280. [CrossRef]
20. Langley, S.K.; Alexander, C.R.; Bush, D.M.; Jackson, C.W. Modernizing Shoreline Change Analysis in Georgia Using Topographic Survey Sheets in a GIS Environment. *J. Coast. Res.* **2003**, *38*, 168–177.
21. Łabuz, T.A. Erosion and its rate on an accumulative Polish dune coast: The effects of the January 2012 storm surge. *Oceanologia* **2014**, *56*, 307–326. [CrossRef]
22. Harley, M.D.; Turner, I.L.; Short, A.D.; Ranasinghe, R. Assessment and integration of conventional, RTK-GPS and image-derived beach survey methods for daily to decadal coastal monitoring. *Coast. Eng.* **2011**, *58*, 194–205. [CrossRef]
23. Gonçalves, J.A.; Henriques, R. UAV photogrammetry for topographic monitoring of coastal areas. *ISPRS J. Photogramm. Remote Sens.* **2015**, *104*, 101–111. [CrossRef]
24. Casella, E.; Rovere, A.; Pedroncini, A.; Stark, C.P.; Casella, M.; Ferrari, M.; Firpo, M. Drones as tools for monitoring beach topography changes in the Ligurian Sea (NW Mediterranean). *Geo-Mar. Lett.* **2016**, *36*, 151–163. [CrossRef]
25. Barlow, J.; Gilham, J.; Ibarra Cofrã, I. Kinematic analysis of sea cliff stability using UAV photogrammetry. *Int. J. Remote Sens.* **2017**, *38*, 2464–2479. [CrossRef]
26. Talavera, L.; Del Río, L.; Benavente, J.; Barbero, L.; López-Ramírez, J.A. UAS as tools for rapid detection of storm-induced morphodynamic changes at Camposoto beach, SW Spain. *Int. J. Remote Sens.* **2018**, *39*, 5550–5567. [CrossRef]
27. Fabbri, S.; Grottoli, E.; Armaroli, C.; Ciavola, P. Using high-spatial resolution UAV-derived data to evaluate vegetation and geomorphological changes on a dune field involved in a restoration endeavour. *Remote Sens.* **2021**, *13*, 1987. [CrossRef]
28. Priest, G.R. Coastal Shoreline Change Study Northern and Central Lincoln County, Oregon. *J. Coast. Res.* **1999**, *28*, 140–157.
29. Duo, E.; Fabbri, S.; Grottoli, E.; Ciavola, P. Uncertainty of Drone-Derived DEMs and Significance of Detected Morphodynamics in Artificially Scraped Dunes. *Remote Sens.* **2021**, *13*, 1823. [CrossRef]
30. Fisher, J.S.; Overton, M.F. Interpretation of Shoreline Position from Aerial Photographs. In Proceedings of the 24th International Conference on Coastal Engineering, Kobe, Japan, 23–28 October 1994.
31. Fletcher, C.; Richmond, B.; Rooney, J.; Barbee, M.; Lim, S.C. Mapping Shoreline Change Using Digital Orthophotogrammetry on Maui, Hawaii. *J. Coast. Res.* **2003**, *38*, 106–124.
32. Ford, M. Shoreline changes interpreted from multi-temporal aerial photographs and high resolution satellite images: Wotje Atoll, Marshall Islands. *Remote Sens. Environ.* **2013**, *135*, 130–140. [CrossRef]
33. Novikova, A.; Belova, N.; Baranskaya, A.; Aleksyutina, D.; Maslakov, A.; Zelenin, E.; Shabanova, N.; Ogorodov, S. Dynamics of permafrost coasts of Baydaratskaya Bay (Kara Sea) based on multi-temporal remote sensing data. *Remote Sens.* **2018**, *10*, 1481. [CrossRef]

34. Gairin, E.; Collin, A.; James, D.; Maueau, T.; Roncin, Y.; Lefort, L.; Dolique, F.; Jeanson, M.; Lecchini, D. Spatiotemporal Trends of Bora Bora's Shoreline Classification and Movement Using High-Resolution Imagery from 1955 to 2019. *Remote Sens.* **2021**, *13*, 4692. [CrossRef]
35. Pardo-Pascual, J.E.; Sánchez-García, E.; Almonacid-Caballer, J.; Palomar-Vázquez, J.M.; Priego de los Santos, E.; Fernández-Sarría, A.; Balaguer-Beser, Á. Assessing the accuracy of automatically extracted shorelines on microtidal beaches from landsat 7, landsat 8 and sentinel-2 imagery. *Remote Sens.* **2018**, *10*, 326. [CrossRef]
36. Maglione, P.; Parente, C.; Vallario, A. Coastline extraction using high resolution WorldView-2 satellite imagery. *Eur. J. Remote Sens.* **2014**, *47*, 685–699. [CrossRef]
37. Sekovski, I.; Stecchi, F.; Mancini, F.; Del Rio, L. Image classification methods applied to shoreline extraction on very high-resolution multispectral imagery. *Int. J. Remote Sens.* **2014**, *35*, 3556–3578. [CrossRef]
38. García-Rubio, G.; Huntley, D.; Russell, P. Evaluating shoreline identification using optical satellite images. *Mar. Geol.* **2015**, *359*, 96–105. [CrossRef]
39. Bini, M.; Casarosa, N.; Luppichini, M. Exploring the relationship between river discharge and coastal erosion: An integrated approach applied to the Pisa coastal plain (Italy). *Remote Sens.* **2021**, *13*, 226. [CrossRef]
40. Turner, I.L.; Harley, M.D.; Almar, R.; Bergsma, E.W.J. Satellite optical imagery in Coastal Engineering. *Coast. Eng.* **2021**, *167*, 103919. [CrossRef]
41. Bishop-Taylor, R.; Nanson, R.; Sagar, S.; Lymburner, L. Mapping Australia's dynamic coastline at mean sea level using three decades of Landsat imagery. *Remote Sens. Environ.* **2021**, *267*, 112734. [CrossRef]
42. Castelle, B.; Masselink, G.; Scott, T.; Stokes, C.; Konstantinou, A.; Marieu, V.; Bujan, S. Satellite-derived shoreline detection at a high-energy meso-macrotidal beach. *Geomorphology* **2021**, *383*, 107707. [CrossRef]
43. White, S.A.; Wang, Y. Utilizing DEMs derived from LIDAR data to analyze morphologic change in the North Carolina coastline. *Remote Sens. Environ.* **2003**, *85*, 39–47. [CrossRef]
44. Shrestha, R.L.; Carter, W.E.; Sartori, M.; Luzum, B.J.; Slatton, K.C. Airborne Laser Swath Mapping: Quantifying changes in sandy beaches over time scales of weeks to years. *ISPRS J. Photogramm. Remote Sens.* **2005**, *59*, 222–232. [CrossRef]
45. Hobbs, P.; Gibson, A.; Jones, L.; Poulton, C.; Jenkins, G.; Pearson, S.; Freeborough, K. Monitoring coastal change using terrestrial LiDAR. *Geol. Soc. Lond. Spec. Publ.* **2010**, *345*, 117–127. [CrossRef]
46. Kuhn, D.; Prüfer, S. Coastal cliff monitoring and analysis of mass wasting processes with the application of terrestrial laser scanning: A case study of Rügen, Germany. *Geomorphology* **2014**, *213*, 153–165. [CrossRef]
47. Pye, K.; Blott, S.J. Assessment of beach and dune erosion and accretion using LiDAR: Impact of the stormy 2013-14 winter and longer term trends on the Sefton Coast, UK. *Geomorphology* **2016**, *266*, 146–167. [CrossRef]
48. Fuse, T.; Ohkura, T. Development of shoreline extraction method based on spatial pattern analysis of satellite SAR images. *Remote Sens.* **2018**, *10*, 1361. [CrossRef]
49. Loiotine, L.; Andriani, G.F.; Jaboyedoff, M.; Parise, M.; Derron, M.H. Comparison of remote sensing techniques for geostructural analysis and cliff monitoring in coastal areas of high tourist attraction: The case study of polignano a mare (southern Italy). *Remote Sens.* **2021**, *13*, 5045. [CrossRef]
50. Toure, S.; Diop, O.; Kpalma, K.; Maiga, A. Shoreline Detection using Optical Remote Sensing: A Review. *ISPRS Int. J. Geo-Inf.* **2019**, *8*, 75. [CrossRef]
51. De Sanjosé Blasco, J.J.; Gómez-Lende, M.; Sánchez-Fernández, M.; Serrano-Cañadas, E. Monitoring retreat of coastal sandy systems using geomatics techniques: Somo Beach (Cantabrian Coast, Spain, 1875–2017). *Remote Sens.* **2018**, *10*, 1500. [CrossRef]
52. Furmańczyk, K.; Dudzinska-Nowak, J. Przewidywanie wielkości erozji brzegu w rejonie miejscowości Trzęsacz [The prediction of shoreline erosion in the area of Trzęsacz]. In *Zintegrowane Zarzadzanie Obszarami Przybrzeznymi w Polsce—Stan Obecny i Perspektywy. Morze—Lad Wzajemne Relacje*; Uniwersytet Szczeciński: Szczecin, Poland, 2006.
53. Furmanczyk, K.; Musielak, S.; Prajs, J. Remote sensing characteristics of dynamics of Hel's Penisula fragment of shore line. In Proceedings of the 11th EARSel Symposium, Graz, Austria, 1–5 July 1991; pp. 208–215.
54. Leatherman, S.P. Shoreline Change Mapping and Management Along the U.S. East Coast. *J. Coast. Res.* **2003**, *38*, 5–13.
55. Coyne, A.M.A.; Fletcher, C.H.; Richmond, B.M. Mapping Coastal Erosion Hazard Areas in Hawaii: Observations and Errors. *J. Coast. Res.* **1999**, *28*, 171–184. [CrossRef]
56. Orviku, K.; Jaagus, J.; Kont, A.; Ratas, U.; Rivis, R. Increasing activity of coastal processes associated with climate change in Estonia. *J. Coast. Res.* **2003**, *19*, 364–375.
57. Dailidienė, I.; Davulienė, L.; Tilickis, B.; Stankevičius, A.; Myrberg, K. Sea level variability at the Lithuanian coast of the Baltic Sea. *Boreal Environ. Res.* **2006**, *11*, 109–121.
58. Reis, H.Ç.; Bozkurt, S.; Bayram, B.; Oğurlu, M.; Şeker, D.Z.; Janpaule, I. Shoreline Extraction and Change Detection using 1:5000 Scale Orthophoto Maps: A Case Study of Latvia-Riga. *Int. J. Environ. Geoinformatics* **2018**, *2*, 1–6. [CrossRef]
59. Samuelsson, M.; Stigebrandt, A. Main characteristics of the long-term sea level variability in the Baltic sea. *Tellus Ser. A Dyn. Meteorol. Oceanogr.* **1996**, *48*, 672–683. [CrossRef]
60. Ryabchuk, D.; Kolesov, A.; Chubarenko, B.; Spiridonov, M.; Kurennoy, D.; Soomere, T. Coastal erosion processes in the eastern Gulf of Finland and their links with geological and hydrometeorological factors. *Boreal Environ. Res.* **2011**, *16*, 117–137.
61. Ato Armah, F. GIS-based Assessment of Short Term Shoreline Changes in the Coastal Erosion-Sensitive Zone of Accra, Ghana. *Res. J. Environ. Sci.* **2011**, *5*, 643–654. [CrossRef]

62. Nicholls, R.J.; Townend, I.H.; Bradbury, A.P.; Ramsbottom, D.; Day, S.A. Planning for long-term coastal change: Experiences from England and Wales. *Ocean Eng.* **2013**, *71*, 3–16. [CrossRef]
63. Miszalski, J. *Współczesne Procesy Eoliczne na Pobrzeżu Słowińskim. Studium Fotointerpretacyjne [Present-Day Aeolian Processes on the Slovonian Coastline. A Photointerpretation Study]*; IGiPZ PAN: Warszwa, Poland, 1973.
64. Borówka, R.K. Present Day Dune Processes and Dune Morphology on the Łeba Barrier, Polish Coast of the Baltic. *Geogr. Ann. Ser. A Phys. Geogr.* **1980**, *62*, 75–82. [CrossRef]
65. Musielak, S. Geneza i funkcjonowanie systemu przyrodniczego morskiej strefy brzegowej [Genesis and functioning of the natural system of the marine coastal zone]. In *ZZOP w Polsce—Stan Obecny i Perspektywy, Brzeg Morski Zrównoważony, cz. 2*; Uniwersytet Szczeciński: Szczecin, Poland, 2006.
66. Borowka, M. Przebieg procesow deflacji I akumulacji na poweierzchni nadbrzeznych walow wydmowych [Deflation and accumulation processes on the surface of coastal dune ridges]. *Badania Fizjogr. Pol. Zachodnia, Ser. A Geogr. Fiz.* **1979**, *32*, 31–48.
67. Rabski, K. Mezoklimatyczne tło obszaru Słowińskiego Parku Narodowego [The mesoclimatic background of the area of the Slovinski National Park]. *Park. Nar. Rezerw. Przyr.* **1992**, *11*, 37–54.
68. Przygrodzki, P.; Letkiewicz, B. Charakterystyka wezbrań sztormowych wzdłuż polskiego wybrzeża Morza Bałtyckiego [Characteristics of storm surges along the Polish coast of the Baltic Sea]. *Inżynieria Morska Geotech.* **2015**, *3*, 158–165.
69. Boniecka, H.; Gajda, A.; Gawlik, W.; Marcinkowsk, T.; Olszewski, T.; Szmytkiewicz, M.; Skaja, M.; Szmytkiewicz, P.; Chrząstkowska, N.; Piotrowska, D. *Monitoring i Badania Dotyczące Aktualnego Stanu Brzegu Morskiego—Ocena Skuteczności Systemów Ochrony Brzegu Morskiego Zrealizowanych w Okresie Obowiązywania Wieloletniego, Programu Ochrony Brzegów MORSKICH" [Monitoring and research Concerning the Current State of the Sea Shore—Assessment of the Effectiveness of the Sea Shore Protection Systems Relized in the Period of Validity of the Long-Term "Programme of the Protection of Sea Shores"]*; IBW PAN: Gdansk, Poland, 2013.
70. Scott, T.; Masselink, G.; O'Hare, T.; Saulter, A.; Poate, T.; Russell, P.; Davidson, M.; Conley, D. The extreme 2013/2014 winter storms: Beach recovery along the southwest coast of England. *Mar. Geol.* **2016**, *382*, 224–241. [CrossRef]
71. Formela, K.; Marsz, A.A. Zmienność liczby dni ze sztormem nad Bałtykiem (1971–2009) [Changeability in the number of days with gale over the Baltic Sea (1971–2009)]. *Pr. Stud. Geogr.* **2011**, *47*, 189–196.
72. Miętus, M. Zagrożenia Wybrzeży i Pobrzeży Południowego Bałtyku WYNIKAJCE ze współczesnej Zmiany Klimatu [Threats to the coasts and shores of the southern Baltic Sea from contemporary climate change]—Uniwersytet Gdański. In *Krajoznawstwo Pomorskie Wobec Wyzwań Współczesności*; Wrzosek, R., Ed.; PTTK: Gdansk, Poland, 2020; ISBN 978-83-957233-0-8.
73. Łabuz, T.A. Polish coastal dunes-affecting factors and morphology. *Landf. Anal.* **2013**, *22*, 33–59. [CrossRef]
74. Uścinowicz, G.; Kramarska, R.; Kaulbarsz, D.; Jurys, L.; Frydel, J.; Przezdziecki, P.; Jegliński, W. Baltic Sea coastal erosion; a case study from the Jastrzębia Góra region. *Geologos* **2014**, *20*, 259–268. [CrossRef]
75. Wolski, T.; Wiśniewski, B. Changes of maximum sea levels at selected gauge stations on the Polish and Swedish baltic coast. *Stud. Pr. Wydz. Nauk Ekon. Zarządzania* **2012**, *29*, 209–227.
76. Zawadzka-Kahlau, E. *Tendencje Rozwojowe Polskich Brzegów Bałtyku Południowego [Development Tendencies of the Polish Coasts of the Southern Baltic Sea]*; Gdanskie Towarzystwo Naukowe: Gdansk, Poland, 1999.
77. Zawadzka-Kahlau, E. *Morphodynamics of the Southern Baltic Dune Coasts*; Wydawnictwo Uniwersytetu Gdanskiego: Gdansk, Poland, 2012.
78. Labuz, T.A. Brzegi wydmowe polskiego wybrzeza Baltyku. *Czas. Geogr.* **2005**, *76*, 19–47.
79. Dubrawski, R.; Zawadzka-Kahlau, E. *Przyszłość Ochrony Polskich Brzegów Morskich [The Future of Sea Shore Protection]*; Zakład Wydawnictw Naukowych Instytutu Morskiego: Gdansk, Poland, 2006.
80. Łabuz, T.A.; Kowalewska-Kalkowska, H. Coastal erosion caused by the heavy storm surge of November 2004 in the southern Baltic sea. *Clim. Res.* **2011**, *48*, 93–101. [CrossRef]
81. Łabuz, T.A. Coastal Response to Climatic Changes: Discussion with Emphasis on Southern Baltic Sea. *Landf. Anal.* **2012**, *21*, 43–55.
82. Michalowska, K.; Glowienka, E.; Pekala, A. Spatial-Temporal Detection of Changes on the Southern Coast of the Baltic Sea Based on Multitemporal Aerial Photographs. *Int. Arch. Photogramm. Remote Sens. Spat. Inf. Sci.* **2016**, *XLI-B2*, 49–53. [CrossRef]
83. Zhang, W.; Schneider, R.; Kolb, J.; Teichmann, T.; Dudzinska-Nowak, J.; Harff, J.; Hanebuth, T.J.J. Land-sea interaction and morphogenesis of coastal foredunes—A modeling case study from the southern Baltic Sea coast. *Coast. Eng.* **2015**, *99*, 148–166. [CrossRef]
84. Wiśniewski, B.; Wolski, T. *Katalogi Wezbrań i Obniżeń Sztormowych Poziomów Morza Oraz Ekstremalne Poziomy Wód na Polskim Wybrzeżu [Catalogues of SEA level Storm Surges and Falls and Extreme Water Levels on the Polish Coast]*; Wyd. Nauk. Akad. Mors.; Maritime University of Szczecin: Szczecin, Poland, 2009.
85. Rotnicki, K. Problem holoceńskiej transgresji Bałtyku Południowego na wybrzeżu środkowym Polski w świetle nowych danych z obszaru Niziny Gardzieńsko-Łebskiej [Problem of the Holocene transgressions along the Polish Baltic middle coast in the light of new data from Gardno–Łeba]. In *Ewolucja Geosystemów Nadmorskich Południowego Bałtyku*; Bogucki Wydawnictwo Naukowe: Poznan, Poland, 1999; pp. 121–140.
86. Marsz, A. Charakterystyka geomorfologiczna Mierzei Lebskiej i Niziny Gardensko-Lebskiej [Geomorphological characteristics of the Łeba Barrier and the Gardens-Lebsk Lowland]. In *Materiały Konferencji Terenowej: Gleby Słowińskiego Parku Narodowego*; PTGleb.: Poznan, Poland, 1975; pp. 7–25.

87. Sztobryn, M.; Stigge, H.J.; Wielbińska, D.; Weidig, B.; Stanisławczyk, I.; Kańska, A.; Krzysztofik, K.; Kowalska, B.; Letkiewicz, B.; Mykita, M. Storm surges in the southern Baltic (western and central parts). *Bund. Seesch. Hydrogr. Rep.* **2004**, *39*, 74.
88. Furmanczyk, K.; Dudzinska-Nowak, J. Effects of extreme storms on coastline changes: A southern Baltic example. *J. Coast. Res.* **2009**, *56*, 1637–1640.
89. Harff, J.; Deng, J.; Dudzińska-Nowak, J.; Fröhle, P.; Groh, A.; Hünicke, B.; Soomere, T.; Zhang, W. What determines the change of coastlines in the baltic sea? In *Coastal Research Library*; Springer: Cham, Switzerland, 2017; Volume 19, pp. 15–35.
90. Łabuz, T.A.; Kowalewska-Kalkowska, H. Coastal abrasion of the Świna Gate Sandbar (Pomeranian Bay coast) caused by the heavy storm surge on 15 October 2009. In Proceedings of the Storm Surges Congress 2010, Hamburg, Germany, 13–17 September 2010.

remote sensing

MDPI

Article

Using Augmented and Virtual Reality (AR/VR) to Support Safe Navigation on Inland and Coastal Water Zones

Tomasz Templin [1], Dariusz Popielarczyk [1,*] and Marcin Gryszko [2]

1 Department of Geodesy, Institute of Geodesy and Civil Engineering, Faculty of Geoengineering, University of Warmia and Mazury in Olsztyn, Oczapowskiego 2, 10-719 Olsztyn, Poland; tomasz.templin@uwm.edu.pl
2 INFEO Marcin Marek Gryszko, Królewny Śnieżki 18, 10-696 Olsztyn, Poland; marcin.gryszko@infeo.pl
* Correspondence: dariusz.popielarczyk@uwm.edu.pl

Abstract: The aim of this research is to propose a new solution to assist sailors in safe navigation on inland shallow waters by using Augmented and Virtual Reality. Despite continuous progress in the methodology of displaying bathymetric data and 3D models of the bottoms, there is still a lack of solutions promoting these data and their widespread use. Most existing products present navigation content on 2D/3D maps onscreen. Augmented Reality (AR) technology revolutionises the way digital content is displayed. This paper presents the solution for the use of AR on inland and coastal waterways to increase the safety of sailing and other activities on the water (diving, fishing, etc.). The real-time capability of AR in the proposed mobile application also allows other users to be observed on the water in limited visibility and even at night. The architecture and the prototype Mobile Augmented Reality (MAR) applications are presented. The required AR, including the preparation methodology supported by the Virtual Reality Geographic Information System (VRGIS), is also shown. The prototype's performance has been validated in water navigation, specifically for exemplary lakes of Warmia and Mazury in Poland. The performed tests showed the great usefulness of AR in the field of content presentation during the navigation process.

Keywords: Digital Elevation Model (DEM); geo-information services; Mobile Augmented Reality (MAR); inland and coastal water zones; navigation; spatial databases and GIS; Virtual Reality Geographic Information System (VRGIS)

Citation: Templin, T.; Popielarczyk, D.; Gryszko, M. Using Augmented and Virtual Reality (AR/VR) to Support Safe Navigation on Inland and Coastal Water Zones. *Remote Sens.* **2022**, *14*, 1520. https://doi.org/10.3390/rs14061520

Academic Editors: Marta Wlodarczyk-Sielicka, Katarzyna Bradtke, Paweł Terefenko and Jacek Lubczonek

Received: 20 February 2022
Accepted: 16 March 2022
Published: 21 March 2022

1. Introduction

The water environment is an unexplored world, which usually remains hidden. It forms part of the natural environment widely used for recreation, leisure, and tourism. Each year, countless people actively spend time on inland and coastal waterways using sailboats, motorboats, houseboats, or other transportation facilities.

Nautical tourism combines sailing and boating with vacation and holiday activities. It merges a variety of activities such as: travelling from port to port on a cruise ship, participating in sailing events such as regattas, chartering a boat and spending time on the water using the facilities (ports, marinas, restaurants, or entertainment venues). Inland water tourism supervisors do not require a license to operate a sailing yacht with a hull length of up to 7.5 m or a motor yacht with an engine power of up to 10kW for water tourism purposes.

People enjoying recreational water bodies support themselves with traditional paper maps or Personal Navigation Devices (PNDs) to navigate the water. Electronic devices with digital maps or dedicated nautical chart plotters are a barrier that makes navigating on shallow water reservoirs difficult for many users. Similar to most conventional navigation systems, they present navigation information in an abstract form, with arrows indicating the intended direction or as a "bird's eye view" map with the intended path [1]. However, navigation on the water is not as easy as car navigation and requires a good understanding of the terrain. If this understanding is inadequate, the safety of sailors/motorboats is at risk and the vessel can be damaged, which may also have an adverse effect on the environment.

Augmented Reality (AR) technology has revolutionised the way digital content is displayed. By superimposing digital information directly on real objects, environments or maps, AR allows people to process digital and physical information simultaneously, improving their ability to absorb information, make decisions, and execute tasks quickly.

In recent years, AR technology has been popularised in various applications [2]. In geography, these include geovisualisation [3], spatial cognition [4], civil engineering [5], surveying and mapping engineering [6], environmental simulation [7], or education [8,9]. In tourism, AR is used to visualise places and activities before and during travel. It allows travellers to visit places of interest in a more meaningful way through digital overlays that contain interactive information about a place's culture or history [10–12]. In the world of cultural heritage, applications include virtual reconstructions of sites with the relocation of environment and objects [13,14].

Implementing AR for navigation is one of the most popular augmented reality applications in everyday life [15]. Using a mobile device camera combined with a Global Navigation Satellite System (GNSS), users can see their route against a real-world view. They can safely move from point A to point B based on GNSS position (autonomous or further improved by object recognition methods or Artificial Intelligence (AI) algorithms). In this process, the methodology of displaying navigation content to the user is becoming an important aspect. The use of AR has now become a subject of intensive research. In addition, this technology has had commercial implementations in the automotive sector. There is still work to be done to achieve an optimal interface and methodology that does not distract the user during navigation. As some authors have indicated [16], navigation in real environments with digital maps and augmented reality in mobile devices has caused some problems and is still challenging.

Safe navigation on the water requires access to up-to-date, reliable information on the shape of the bottom of the water body, location of dangerous places (shallows, stones, overhead power lines), navigable routes and information on the current situation on the water. They are obtained from various data sources and are usually stored in a digital database [17]. Geographic Information System (GIS) software is used to update and add new contents to the database, allowing for the analysis of the bottom shape, considering the current water level, identifying safe navigable routes, and marking dangerous places. Although the results are three-dimensional and digital, they are still visualised using traditional maps or bathymetric charts or displayed on two-dimensional (2D) screens. Although the 2D digital map can be very effective for navigational purposes, bathymetric presentations based on AR technology can be useful additionally for diving or fishing purpose.

Another element affecting the navigation process on shallow reservoirs is water traffic and the relationship between boaters. The real-time capability of AR can be exploited to present information about moving objects. In contrast to maritime applications, information identifying moving objects, the Automatic Identification System (AIS), is missing on inland reservoirs. Inshore, some systems monitor the situation on the water (for example—https://zegluj.mobi/mapa.html/, accessed on 10 March 2022) and keep their users' real-time position, which enables to use them for traffic management and socialising among the sailing community.

There are many applications available in the market for maritime and inland navigation. However, there is a lack of AR-based mobile applications dedicated to the general public. AR can add the advantage of true mobility and location awareness. It allows creating only those virtual objects necessary to supplement what a camera perceives (reality) [16]. This type of application is defined as Mobile Augmented Reality (MAR). It is an augmented reality-based technical solution dedicated to mobile devices (smartphone/tablet), Head-Up Displays (HUD), or optical Head-Mounted Display (HMD) [18]. It extends and enhances the experience of the mobile device user.

Only a small part of AR research focuses on water application. Bartolini et al. [19] describe the architecture of a novel infrastructure for coastal management with an innovative visualisation tool based on Augmented Virtuality. Mirauda et al. [20] presented a prototype Augmented Reality application specifically designed for monitoring water resources. The main goals were to provide an option to visualise 3D objects and real-time textual environmental data with an AR interface directly in the field.

Other authors concentrated on preparing the methodology of integrating and preparing products for VR and AR purposes using optical and acoustic data for underwater sites. An extensive review of the sensors and the methodologies used in archaeological underwater 3D recording and mapping is presented by Menna et al. [21].

Bruno et al. [22] present the methodology for preparing 3D semi-immersive or fully immersive underwater diving simulations. They use numerical products to prepare augmented diving services supporting divers in the navigation process inside an archaeological site. The tablet position and orientation are displayed like a 3D map representing the environment around the diver, information about underwater artefacts and structures.

There is a lack of research results in the literature on the feasibility of using AR to support users in inland waters. Most existing studies concentrate on coastal areas requiring advanced equipment and dedicated software.

Some research groups have started to employ augmented reality in water navigation on the sea. Many focus on implementing AR as innovative IT solutions to improve shipping safety, avoiding marine accidents caused by human error [23]. The primary goal of this research was to add new virtual contents to increase the number of observed vessels and add objects that are invisible but can affect the safety of navigation. For example, Bandara et al. [24] considered the feasibility of using AR to enhance visualisation in maritime operations to avoid collision in different environmental conditions. It showed that AR navigation lights, geo-locked to real physical hazards, can present navigational information in compromised visibility.

Hugues et al. [25] showed one of the first integrated solutions: a vision system and a thermal camera with augmentation. They conclude that the functionalities provided by AR must differ depending on people and weather conditions and require up-to-date contextual information. Oh et al. [23] proposed a navigation aid system based on AR technology that displays various overlaid navigation information on images from cameras to support swift and accurate decision making by officers. They verified their interface in the field and emphasised that further research on the AR interface to include a method for the efficient display and handling of information is needed.

The theoretical aspect of implementing AR for navigation was presented by Grabowski [26]. He presented several important research questions to introduce Wearable Immersive Augmented Reality (WIAR) systems in ship navigation. Procee et al. [27] focused on the user support aspect by conducting cognitive work analysis to derive a scientific base for a functional interface that best supports navigators in their work. To do this, they used Head Mounted Display (HMD).

Some authors analysed practical techniques used to visualise AR content in outdoor environments. For example, Hertel et al. [28] investigate the perceived egocentric distance of virtual objects in an open outdoor environment and how the visual attributes of coloration, shape, and relation to the floor influence the perceived depth.

The most compressive review of using AR for maritime navigation data visualisation was conducted by Laera et al. in two publications [29,30]. The first presents research and methodology for developing AR that assists sailing purposes. Their research concluded that current solutions for visualising sailing data have serious limitations and are technically outdated. In the second study, the authors investigate augmented reality technology in the field of maritime navigation. They assume that AR can be an evolutionary step to improve safety and reduce stress on board. However, the application of AR in the nautical field, specifically the maritime one, has not been sufficiently investigated. There is still much to be done regarding interface proposals.

Despite the rapid development of AR technology, the number of commercial implementations remains relatively small. Many Research and Development (R&D) projects show that due to the specificity of created products (handling only a limited number of tasks) and their various requirements (e.g., lighting conditions, techniques for superimposing objects), it is difficult to transfer the applied solutions to other applications directly. This is especially true for implementing AR technology outdoors in demanding, powerful, and hostile natural environment conditions.

The use of AR on water causes many technical difficulties. When navigating on water reservoirs, we usually encounter:

- Harsh environment in which the mobile application has to consider the water as an environment; weather conditions including extreme temperatures, intense sunlight, precipitation, and visibility obstructions; and strong movements due to waves (onboard ship) and in the case of motorboats' high speeds and accelerations.
- Obstructed conditions associated with the use of a positioning system. In many cases (especially near steep, wooded shore, narrow channels), some obstacles make it difficult to determine the GNSS position.
- The use of wireless networks for data transmission is often problematic due to existing bandwidth limitations or being completely blocked by a lack of signal coverage. It causes problems in real-time database access and affects the architecture of the mobile application.
- A ship on water is not static but moves, and its movements are heave, roll, and pitch (vessel motions). Therefore, navigation on water requires locating the mobile device, determining the ship's movement relative to the ground, and locating/moving the person (or device) to the ship.
- As in marine navigation, coastal navigation often lacks easy-to-interpret navigation marks or other visual reference points (problems for less experienced operators).
- A problem with handling electronic navigation devices is that they often have a complicated Graphical User Interface (GUI) that provides too much information. In addition, this information is not well organised, making users experience it as complicated rather than helpful.
- Optical viewpoint problem. For small water boats, the visibility of the water surface is relatively small because the optical viewpoint in the water body field is 1-2 m above the water level.

Uwe von Lukas et al. [31] additionally highlight issues arising from the use of AR in a maritime environment. They point out several factors contributing to the relatively weak development of technology in this area, highlighting relatively low R&D intensity, many small and medium-sized companies, and a conservative attitude of users.

However, the rapid advancement of technology results in today's mobile devices boasting fast, powerful Central Processing Units (CPUs) supported by Graphical Processing Units (GPUs). They have high-resolution screens, megapixel cameras, and various sensors such as powerful wireless data transmission methods, Inertial Measurement Units (IMUs), compass, and GNSS [32]. It enables them to provide a universal platform to prepare outdoor MAR apps dedicated to a wide range of users.

2. Materials and Methods

The survey of users' opinions (Section 2.1) shows that boaters and other water sports practitioners on inland and coastal waterways expect a simple, easy-to-use mobile app to improve navigation safety. It should offer a user-friendly interface that supports them without requiring knowledge of regulations, technology, and specialised equipment.

This requirement is satisfied by AR solutions. However, currently there is no representative standard of navigation information elements related to AR technology. This issue is the subject of many studies, and some research has investigated and tested end-users to define representative guidelines. The presented research is consistent with the above trend. It involved designing and testing an initial version of the visual AR interface for the prototype MAR application. The following subsections demonstrate the results of a survey on user needs in this area, the concept proposal for an interface using AR on water, and its verification as the AR interface for a prototype MAR application.

2.1. The AR Interface User Needs Analysis

The authors aimed to determine the potential of AR technology for navigation purposes on the water by developing a navigational MAR application for inland and coastal water bodies and determine the usefulness of traditional navigation data sources such as bathymetric datasets, remote-sensing products, and 3D models as AR content. We also propose and implement a universal methodology for preparing AR content using its own VR solution.

The following assumptions were made as the starting point:

- The augmented contents will describe the part of the water reservoir where the user is positioned, help him reach the destination safely and more efficiently by augmenting the real world with navigation information, indicating the route, enabling efficient movement, and avoiding dangerous places.
- The app will offer technical opportunities to improve travel safety and help avoid dangerous navigation obstacles by alerting when approaching a dangerous area.
- GNSS will be the source of locations (track the user's location, geolocate other users or objects in the AR (e.g., obstacles a ship may encounter)).
- The application will present a navigation chart and user location/speed/direction, danger areas, other users, and visual appearance of 3D bottom models.

The MAR application should be widely available for mainstream phones and tablets. The MAR market has developed significantly in recent times. However, simple MAR apps available in application stores are still not widely adopted by consumers. Users consider these apps to be fun but not usable, and they quickly stop using them. That causes a need to prepare an easy-to-use, intuitive GUI based on UX design principles.

The authors conducted a survey addressing people active on the water to determine user awareness of AR in mobile navigational applications in 2021. The survey was implemented using an electoral form on 120 users of the Zegluj app. Answers are presented in Figures 1 and 2.

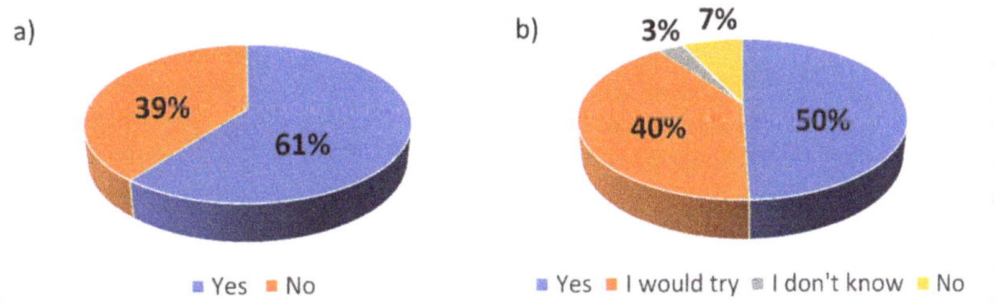

Figure 1. The evaluation of the need to implement AR functionality in a mobile navigation application: (**a**) Do you use an application supporting spending time by the water? (**b**) Would you use AR function in navigation application on the water?

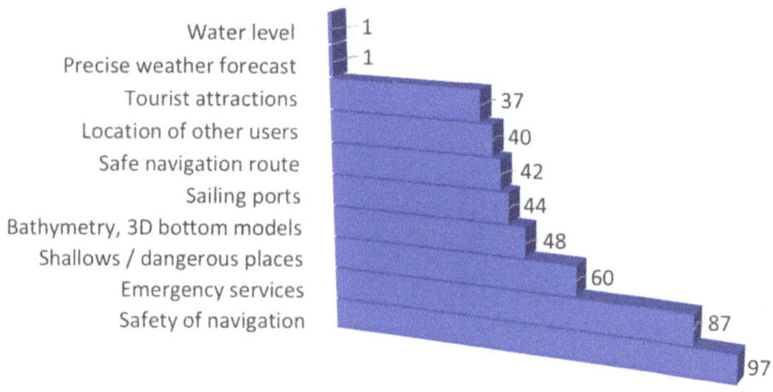

Figure 2. Identification of prioritised functionalities from the perspective of the mobile application user.

The analysis showed that people are aware of modern technology and use it to support their water activities (over 60%). Almost half of the respondents declared they will use the app with AR functionality, 40.2% would try it, and 3.1% had no opinion. Only 7.2% declared no need to use AR.

Safety of navigation is the most wanted feature where almost 97% of respondents consider information about shallow water and dangerous areas as important and 64% of them as very important. In addition, 87% recognise reporting a threat to emergency services as a needed function. Nevertheless, additional information such as bathymetry (48%), sailing ports (44%), other users (40%), and tourist attractions (37%) are also important.

In the following step, a list of potential elements to be displayed was analysed to determine the AR interface requirements of an on-water navigation application. Variables affecting the boat guidance process were defined. Of the information considered for display, the following were highlighted:

- Compass;
- Obstacles;
- Course;
- Boat route and speed;
- Position (longitude and latitude);
- Wind direction;
- Depth;
- Waypoints;
- Distance to waypoint;
- Estimated time of arrival (eta);
- Traffic information;
- Non-navigation zone;
- Course over ground.

This is the essential information for route calculation, monitoring, and primary checks of the boat status and navigation area. Due to the limitations of mobile device screens, the minimum set of parameters presented on the screen was selected from the complete list.

The preliminary selection of parameters intended to be displayed in the interface was made based on the authors' experience and a survey conducted among expert navigators and sailors. Three thematic categories relevant to amateur water sports practitioners were considered. This selection assigned an importance rate for every parameter displayed on the screen. The results are shown in Figure 3.

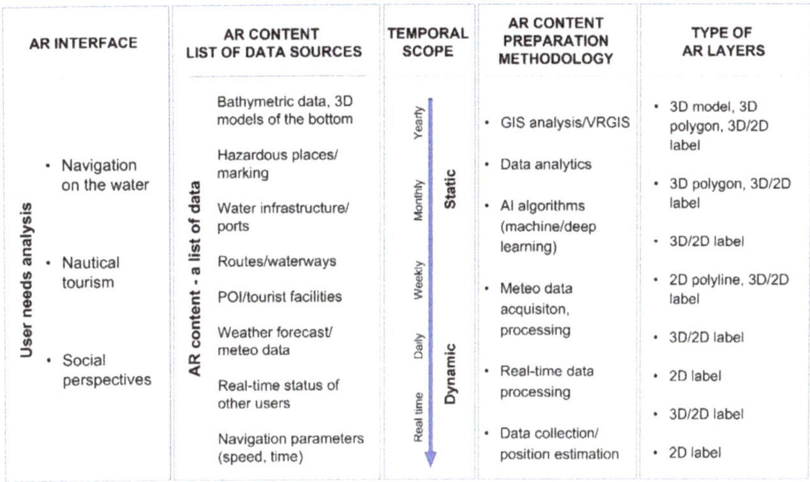

Figure 3. User needs analysis for visual AR interface.

Three application categories for AR interface were considered for ranking the required data: navigation on the water, nautical tourism, and social perspective. Potential data sources, temporal scope (needed frequency of update), geometric type of data, and data processing methodology for each category were also defined. The category of AR information is closely related to the type of activity on the water. In fact, these parameter sets may vary in importance depending on the specific task of the application.

The vast majority of the presented data relates to the navigation task on the water. It requires access to bathymetric data, dangerous places, markings of paths, and user travel routes. Nautical tourism involves accessing context-related information concerning the current position, including water infrastructure, tourist facilities, points of interest and weather data. The last category includes up-to-date information about boats to establish contacts, organise regattas, or use crowdsourcing data (e.g., monitor free moorings).

Analysing the scope of current research related to the use of AR on the water, it can be pointed out that the vast majority of it focuses on an interface designed for Head Mounted Devices (HMDs) used by professionals (trained individuals with experience in ship/boat driving). The authors' research concentrated on interfaces dedicated to individual consumers. They have little experience but are authorised to navigate through bodies of water on various vessels. The focus should be on identifying those elements that are important to them. The four types of AR functionality proposed to be implemented in the application are shown in Figure 4.

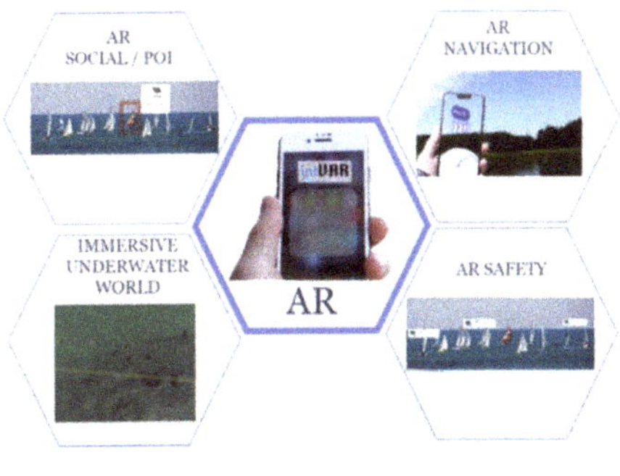

Figure 4. Types of AR functionality supported by the proposed architecture.

2.2. Architecture of the System

Traditionally, mobile geo-application refers to popular applications and frontiers, including Location-Based Services (LBS), social media, and Augmented Reality (Fu 2015). LBS refers to information services that integrate the location of mobile devices to provide added value to the user. The social media features offer solutions that can help users communicate easily. AR combines data from different sources with data from the human senses.

LBS is an up-and-coming technology to deliver valuable services. It is the foundation for developing outdoor MAR applications. An LBS requires five essential components: the service provider's software application, a mobile network to transmit data and requests for service, a content provider to supply the end-user with geo-specific information, a positioning component, and the end-user's mobile device. The most crucial element is the positioning module that provides information about the current location. Nowadays, GPS, Galileo, GLONASS, BeiDou satellite positioning systems, and EGNOS are used. They provide a submeter accuracy of position determination [33,34].

The integration of GNSS and AR technologies offers new technical opportunities to improve travel safety and to help avoid dangerous navigation obstacles. In order to display AR layers, an appropriate system architecture is needed to deliver the content to MAR app users.

The proposed architecture is based on several interconnected building blocks. Each of them is responsible for different tasks. The typical architecture of a location-based mobile augmented reality system is used to acquire data about the observed world, acquire AR content, and overlay it on the viewed scene. The principle of operation and overall architecture of the proposed AR inland and coastal water application highlighting the major modules and their relationships are shown in Figure 5. The diagram consists of the following modules:

- Tracking module—module responsible for obtaining geographical position and orientation of the mobile device based on the sensors (GNSS receiver, accelerometer, gyroscope).
- Interaction module—the subsystem that allows users to interact with electronic devices (AR interface).
- Presentation module—the subsystem responsible for displaying content using AR layers (3D objects, polylines, 2D/3D labels, pictures).
- A world model—AR content, a database containing information describing the real world, including data about presented objects and methods.
- Communication module—communication channel between a client and server, the component responsible for sending/receiving data into a database.
- Filter module (data filtering)—a module responsible for selecting AR content based on preferences set by the user.
- Search engine—the module that selects content based on a query generated by the client.

In the proposed solution, the architecture is based on a 2-tier client–server architecture. Client–server architecture is a computing model in which the server hosts, delivers and manages most of the resources and services to be consumed by the client. This type of architecture has one or more client computers connected to a server over a network or internet connection.

In this model, tiers realised the following functions:

1. Client-side—responsible for gathering information from the user, sending the user information to the business/data services for processing, receiving the server processing results, and presenting those results to the user.
2. Server-side—responsible for receiving input from the client. It interacts with the data server to perform operations that the application was designed to automate (AR

content, weather forecast, DEM's of the bottom). It sends the processed results to the client side.

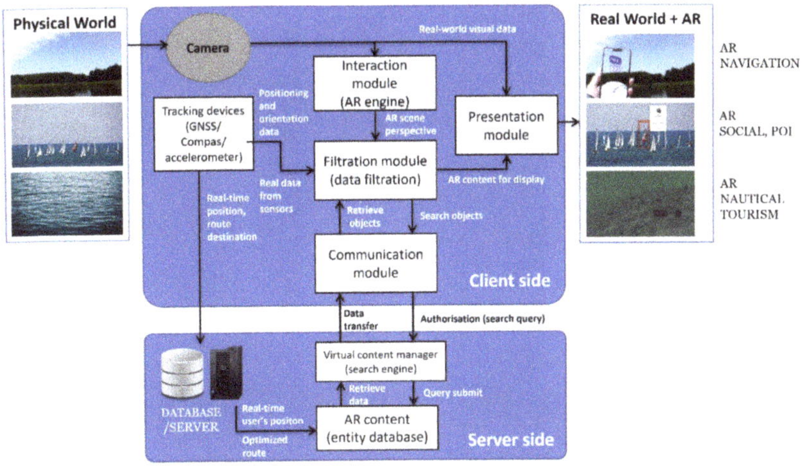

Figure 5. The overall architecture of the proposed AR shallow water application highlighting the major modules and their relationships.

In a typical situation, the communication is initiated by the client. The client obtains the position and orientation of a mobile device based on the tracking device module. Based on the parameters set by the user interface (interaction module) and calculated relative pose, the database query is generated, and objects captured by the mobile device are identified. After recognising objects, the subsystem returns object IDs, camera position, and orientation for each object. The client downloads this data into the presentation module and displays it.

2.3. *AR Content*

Augmented reality content is computer-generated input used to enhance parts of a user's physical world via smartphones, tablets, or smart glasses. Users can generate this content directly or use previously prepared materials. AR content is sometimes delivered as 3D models or visual, video, or audio content.

Section 2.1 identifies the sets that constitute the visual elements of the AR interface (AR layers). A potential data source and a methodology for processing it to deliver the required AR content has been identified for each of them. A detailed list of AR layers is provided in Figure 6.

A critical source of data for users on the water is bathymetry. In reality, the helmsman cannot see the shape of the bottom. He can only see the water's surface. Bathymetry information is not just a curiosity, but a critical element affecting the safety of navigation. Based on the shape and morphometry of the bottom, it is possible to determine areas that are dangerous to navigate (shallows), rocky reefs, single boulders that can damage yachts, motorboats, and houseboats, as well as sudden dips and shallows, which are dangerous for swimming, especially for children and teenagers, but attractive for anglers and divers.

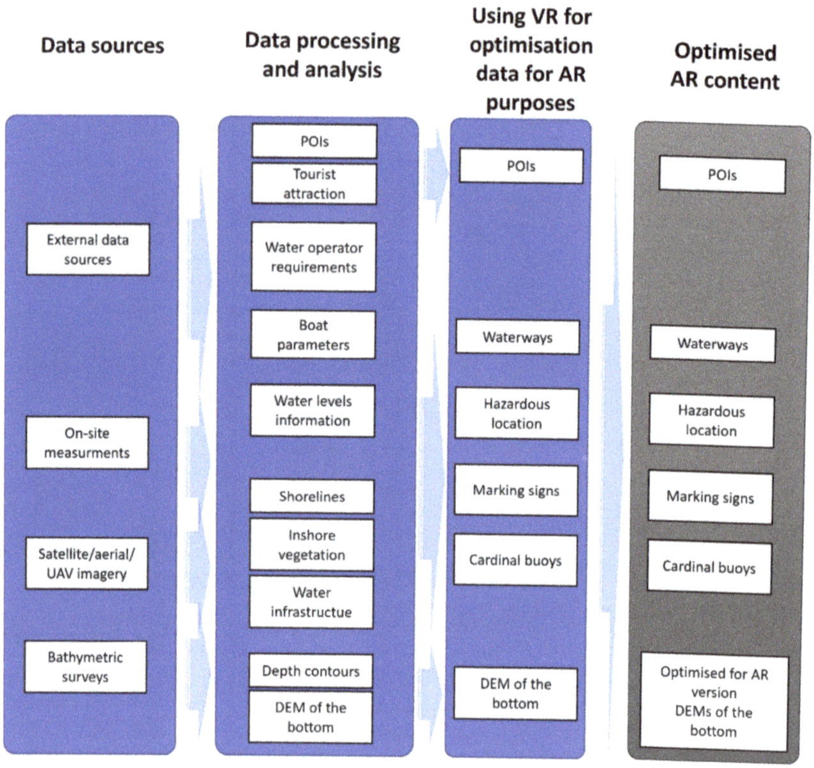

Figure 6. Data sources and processing methodology for preparing AR-optimised layers.

3. Measurements and Validation

3.1. Study Area

Our chosen test area, the Great Masurian Lakes, is a beautiful region often called the Land of Thousand Lakes. The selected area is located in the northeastern region of Poland, known as the Masurian Lake District (Masurian Lakeland). It is the most popular region among sailors. This fashionable region is visited by about 1 million Poles and 200 thousand foreigners every year. In the summer, only in Masuria, about 50 thousand sailors and motorboaters (about 10 thousand yachts) spend their free time actively every day.

Generally, water covers more than 70 percent of the earth, but more than 80 percent of its surface remains unexplored. Unfortunately, many reservoirs in Poland have shallows with rocks, reefs, and fallen trees, dangerous for sailors. Dangerous places make marine navigation very difficult. Therefore, all shallow water areas should have up-to-date bathymetric charts to ensure the safety of shipping lanes [17,35–37]. Nowadays, high-resolution multibeam echosounders are used more commonly. They allow not only the obtaining of a very accurate bathymetry of the shape of the tank bottom but also of its characteristics and structure. Multibeam systems also identify underwater objects and obstacles (rocks, wrecks, sunken trees).

For the feasibility tests of the prototype version of the MAR app on water reservoirs, two lakes have been chosen: the biggest lake in Poland, Śniardwy, and one of the longest, Bełdany. The first step to building the database for the MAR app was to perform bathymetric measurements.

3.2. Bathymetric Surveys/Data Analysis

Experiments were realised on two sites (Figure 7) with different morphology: the largest Polish lake, Śniardwy (July 2019), and post-glacial lake, Bełdany, one of the most popular among sailors (November/December 2019). These lakes were measured using various hydroacoustic technologies. Lake Śniardwy was measured with a single-beam echosounder (SBES) based on profiles spaced 10–50 m apart. Lake Bełdany was measured with a high-resolution multibeam echosounder (MBES), giving several points per square meter.

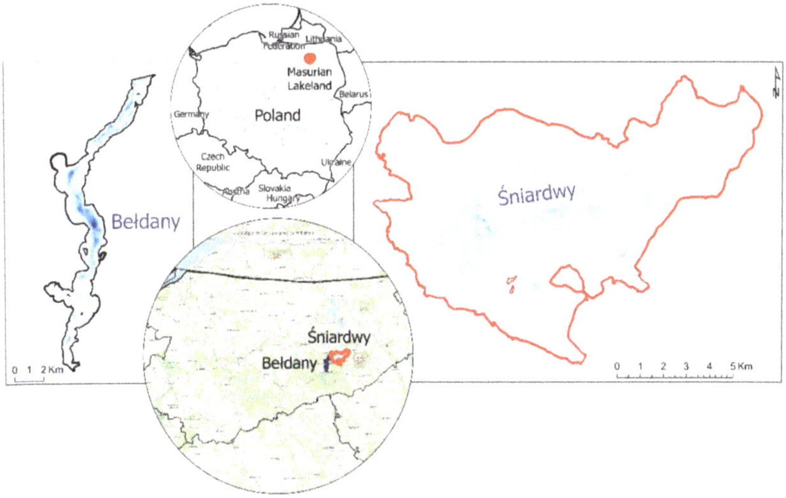

Figure 7. Study area.

AR content for the MAR app was prepared by carrying out the following task. First, an on-site survey of hazardous areas was made for this area, their horizontal location was identified, and bathymetry was analysed. As a part of the survey work, GNSS and hydroacoustic techniques were used to verify the shape of the lake bottom and identify shallow, hazardous areas. The work was then continued using GIS software (ESRI ArcGIS Pro) and self-made VRGIS software for bathymetric data analysis.

Professional hydrographic surveys of selected parts of both lakes were carried out to identify dangerous places (shallows with rocks). Lake Śniardwy was measured with an SBES Simrad EA501p single-beam echo sounder, while Lake Bełdany was measured with a Reson T50P multibeam system. In the next step, based on a numerical terrain model of the bottom, shallow places dangerous for navigation were determined. Computer simulation of water level lowering was used, taking into account bathymetry and long-term average water levels in the lake (Figure 8).

In the next step, we elaborated a design to place cardinal buoys (IALA system: International Association of Marine Aids to Navigation and Lighthouse Authorities) to assist sailors in avoiding obstacles. The marking was implemented after direct verification of correctness and water rescue services (Figure 9). Finally, a navigation lanes layer was developed using the developed bottom models and identified hazardous areas (safe and reliable). These routes indicate safe waterways using side marker buoys.

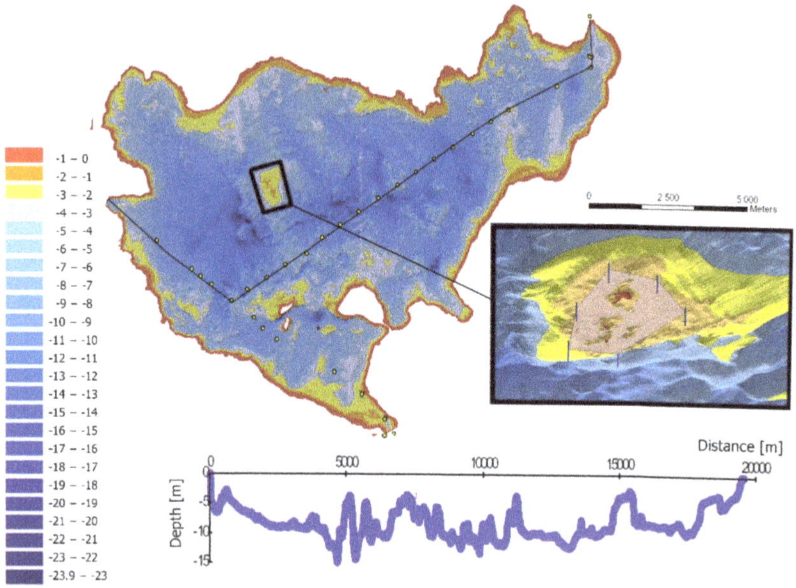

Figure 8. Marking the navigable route on the largest lake in Poland, Śniardwy Lake, with the depth profile along the route. Inset map presents a sample of cardinal buoy location (IALA system) used to mark hazardous locations obtained from GIS analyses.

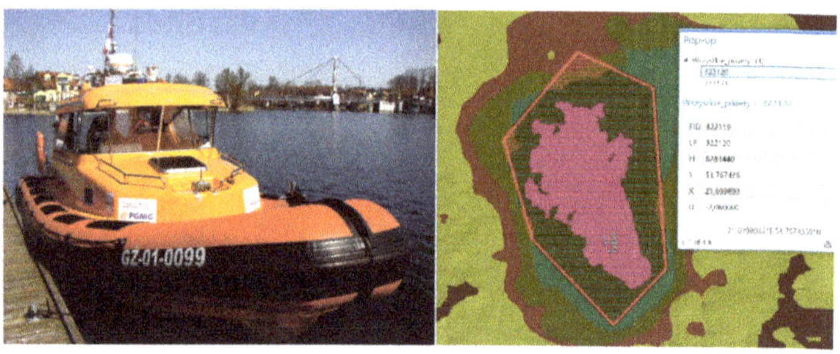

Figure 9. On-site verification of cardinal buoys (measuring platform—Water Rescue Service boat (**left**), the model with depth readings (**right**)).

GIS analyses applied in the data processing allowed identifying dangerous places, cardinal signs, and buoys for their marking. GIS software allows for 3D visualisation of the obtained results on the monitor screen. This visualisation method offers the lowest immersion level, resulting in limitations to the visual assessment of the validity of the obtained findings. It also does not control the optimisation procedure for further use of prepared content in the MAR application.

In order to optimise the procedure of data adaptation to the needs of the prepared mobile application, additional software was used to increase the level of immersion through the use of VR. The dedicated, self-made VRGIS software was used. It is based on the Unity game engine (https://unity.com/, accessed on 10 March 2022) and allows importing GIS data, visualising 3D models (TIN, GRID, textures), performing analysis, and adding and verifying hazardous areas. In order to verify the already-prepared data, it was further analysed using VR Oculus Rift glasses with motion controllers and gesture manoeuvres. The methodology of verifying hazardous area signs based on DEM of the bottom is presented in Figure 10.

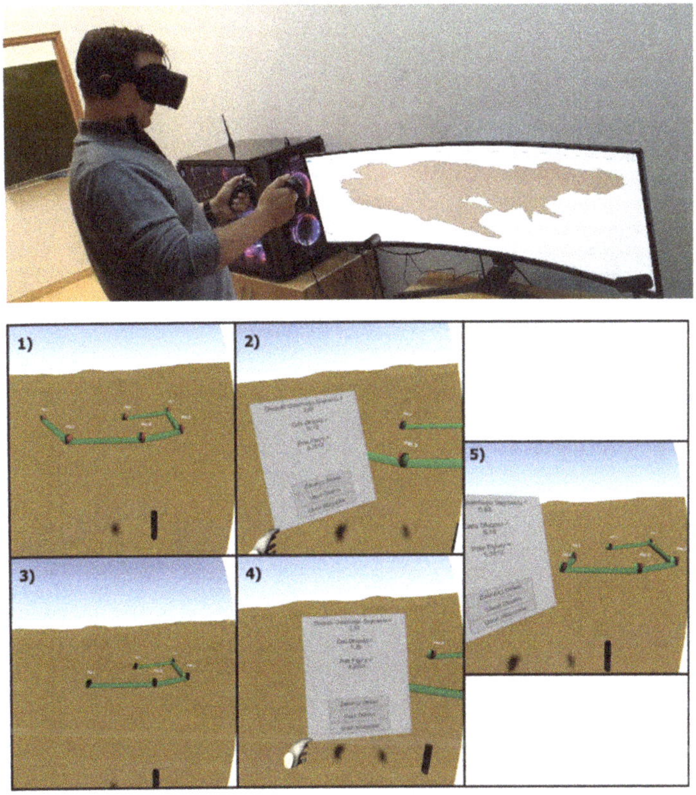

Figure 10. Optimisation of marking location and verification of hazardous area signs based on DEM of the bottom using self-developed VRGIS app. Screens from Oculus Rift with motion controllers: (**1**) The new point creation. (**2**) Settings new attributes for created point. (**3**) Removing existing points. (**4**) Display all values after deleting one point. (**5**) Display all values after adding one point and refresh the display.

The proposed methodology optimised AR and prepared the Śniardwy and Bełdany Lake DEM bottom model by adjusting the number of vertices, spatial distribution, and texture overlay verification. In addition, the marking of waterways and dangerous areas layer was also loaded as a text file and then it verified the spatial distribution of the markings and the correctness of the mapped waterways.

3.3. *Prototype of MAR App*

3.3.1. MAR App Functionality

The main element of the MAR mobile application is a unique functionality using AR technology. The key functionality of MAR is a visualisation of a route on the water, the identification of obstacles which are difficult to perceive, or the display of information on the water reservoir directly where the phenomenon is taking place.

The application will add computer-generated content to the observed (through a camera built into the device) real world. The augmented contents will describe the part of the water reservoir showing the route, enabling efficient navigation, and avoiding dangerous areas. Augmented reality in water navigation will support safe travel on the water and guide the user to the specified destination. The integration of GNSS and AR technologies offers new technical opportunities to improve travel safety and to help avoid dangerous navigation obstacles.

In order to guide the skipper along a specific path, it was decided to use an interface similar to AR car navigation. The research tested different abstract shapes such as lines and planes and different 3D objects such as arrows and 3D object models.

3.3.2. MAR App Implementation

To realise the assumptions made and test the effectiveness of displaying objects on water, a prototype MAR app for iOS was implemented and performed all the necessary tests. Finally, for the app dedicated to navigating in a water environment, we decided to use:

- 2D labels (descriptions);
- 3D labels;
- Abstract shapes (boats moving on water);
- Watermarking symbols;
- Realistic 3D objects (such as virtual buoys rendered on the water surface);
- DEMs of the bottom.

Examples of implementation of the algorithm for superimposing selected objects (2D labels, 3D objects) based on location and data defining the orientation of the smartphone are shown in Figure 11.

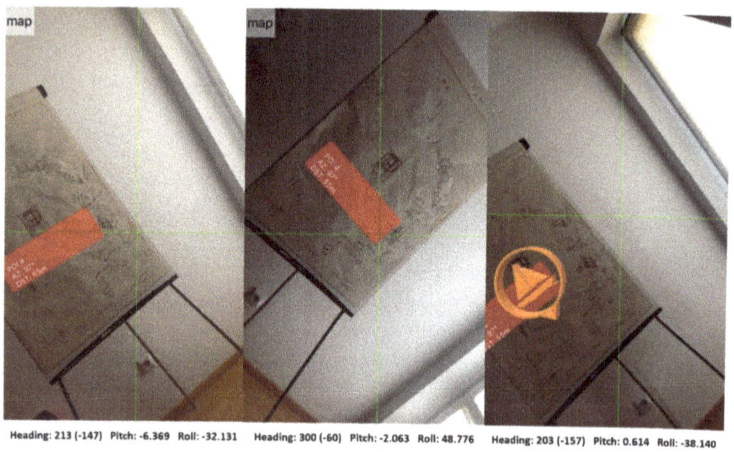

Figure 11. Test application presenting labels and 3D models of spatial objects based on sensor readings.

The proposed solution is based on a location-based approach to AR. In this solution, it is essential to define rules for displaying objects' visibility priorities in the recorded real-world view. Using latitude, longitude, and altitude data, the app computes the relative location of target Points of Interest (POIs) and displays them correctly on screen. After determining the location of the object, graphical symbols with interactive annotations are overlaid on top of the captured image.

Due to the potentially increasing number of objects to display, a unique mechanism is introduced to consider the spatial relationships. In the prototype of the MAR app, an algorithm was proposed to optimise the displayed content. It utilises distance threshold and direction from the camera sensor to the objects to determine how many will be overlaid on top of the captured image. It allows the filtering of objects and visualising of only part of them, not too far from the device and inside the active field of view.

The implementation requires combining mobile device position and orientation (determined with GNSS receiver, IMU, and magnetometer) with geographical objects using spatial relationships. The approach considers three aspects of the problem: distance, angle, and the assigned priority determined by these parameters (order) (Figure 12).

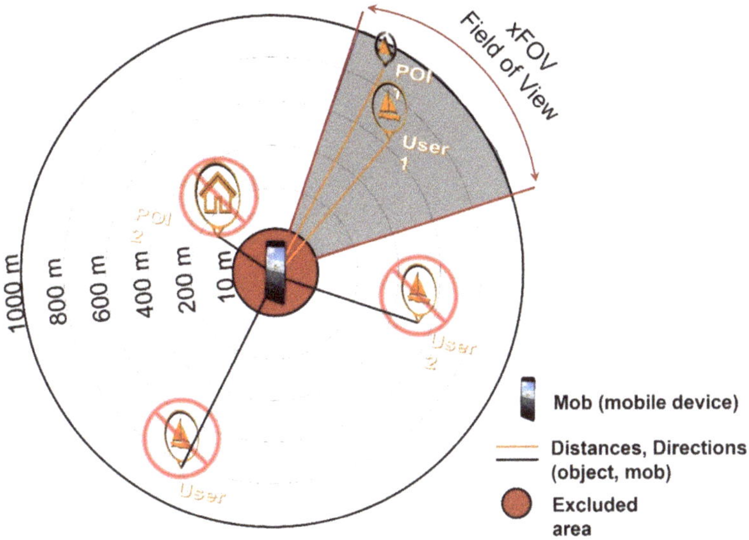

Figure 12. Concept of AR content matching in dynamic water conditions. Geographic objects are defined by distance criterion (5 distance thresholds), horizontal directions, and actual horizontal field of view.

When the mobile device sensors determine the position and orientation, a database query is generated to check the objects of interest and retrieve the GNSS coordinates of features meeting the distance criterion (up to 1000 m). In the next step, the list is further filtered based on the distance assumptions calculated between the visual sensor and the geographical objects and the orientation of the mobile device. The methodology is shown in Figure 13.

The acceptable distance criterion represents the value of the distance between the visual sensor and the geographical object (minimum and maximum). The spherical distance is calculated using the haversine formula Equation (1).

$$distance_{mob\_user1} = 2Rarcsin\sqrt{\begin{matrix}\left(sin\frac{Lat_{mob}-Lat_{user1}}{2}\right)^2 + cos(Lat_{mob})\times \\ cos(Lat_{user1})\times\left(sin\frac{Lon_{mob}-Lon_{user1}}{2}\right)^2\end{matrix}} \quad (1)$$

where:

- R is the radius of the Earth (km);
- Lat_{mob}, Lat_{user1} are the latitude of mobile device and latitude of point1 (user position);
- Lon_{mob}, Lon_{user1} are the longitude of mobile device and latitude of point1 (user position).

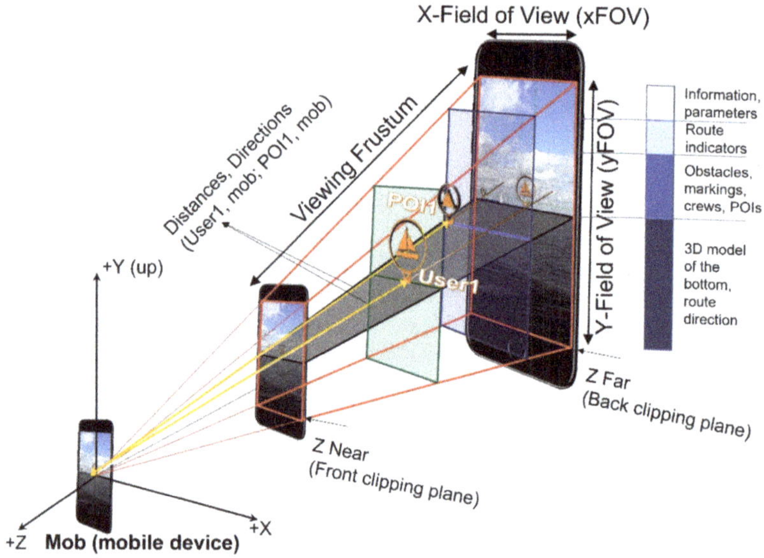

Figure 13. A visualisation of a real-world 3D coordinate system, determined by the viewing frustum created by the visual sensor, with the origin at the centre of the visual sensor. The X and Y axes are parallel to the screen. The Z-axis, which corresponds to the negative orientation direction of the visual sensor, is perpendicular to the screen. The AR layer layout concept for the MAR app prototype is on the right.

In the next step, the aspect of orientation is analysed. A direction is calculated to determine whether the object is within the actual horizontal field of view. It determines whether the object is inside the field of view and should be displayed as AR content. The direction value is calculated according to Equation (2).

$$direction_{mob\_user1} = arctan\left(\frac{\begin{array}{c} cos(Lat_{mob}) * sin(Lat_{user1}) - sin(Lat_{mob}) \\ *cos(Lat_{user1}) * cos(Lon_{user1} - Lon_{mob}) \end{array}}{sin(Lon_{user1} - Lon_{mob}) * cos(Lat_{user1})}\right) \times \frac{180}{\pi} \quad (2)$$

where:

- Lat_{mob}, Lat_{user1} are the latitude of mobile device and latitude of point1 (user position);
- Lon_{mob}, Lon_{user1} are the longitude of mobile device and latitude of point1 (user position).

The calculated distances and directions are compared with the corresponding thresholds and any geographic object that does not meet the requirements is not displayed. Additionally, each object has been assigned a priority (the category of the object). They define the label/3D model's size and colour of the object depending on its type. In the prototype version of the application, five categories based on distance (200 m, 400 m, 600 m, 800 m, 1000 m) have been proposed.

4. Tests and Results

An AR inland and coastal water navigation system prototype was developed based on the architecture described in Section 2.2. Our solution is intended for the entire community of shallow-water navigators, mainly for the inexperienced people whose ability to note relevant information on the water called 'seaman's eye' is insufficient [38]. A simple smartphone app could simply serve in providing added certainty for novice sailors.

Using the prototype MAR app, simulated and real application experiments were conducted. In the experiment, each AR functionality of the prototype was tested (on the water), the applicability in varying locations on the boat was evaluated, and the suitability as a navigation device was assessed. The ability to display selected AR layers and their usefulness in navigation was tested.

There are several categories of augmented reality technology, each with different objectives and applicational use cases. For the prototype of the MAR app, the most popular types: marker-based AR, markerless AR (sometimes called location-based or sensor-based AR), superimposition-based AR, and projection-based AR were explored.

Marker-based augmented reality (also called Image Recognition) uses a camera and a visual marker, such as a QR/2D code, to produce a result only when a reader senses the marker. The markerless AR uses a GNSS, digital compass, gyroscope, or accelerometer embedded in the device to provide data based on position.

Projection-based augmented reality works by projecting artificial light onto real-world surfaces. Superimposition-based augmented reality either partially or fully replaces the original view of an object with a newly augmented view of that same object.

The first tests were performed to explore the suitability of different augmented reality approaches in the water environment. The recent research shows significant problems with using native AR Software Development Kits (SDKs) for water surface detecting. For example, Nowacki et al. [39] tested the potential of the ARCore and ARKit platforms on different surfaces and under different conditions. They concluded that the water mapping sheet was disappointing in both cases (ARCore, ARKit), but ARCore detected many more characteristic points.

The first test investigated the feasibility of using water surface recognition with the popular ARKit and ARCore solutions. The tests were conducted during experiments on two Great Masurian Lakes—Śniardwy and Bełdany. Unfortunately, the mobile device could not calculate the world origin (centre and orientation of the reference system where other objects are located) or find any tracking surface. Another test on Lake Bełdany under different weather conditions was conducted. The lake was calm, and the water surface was flat that day. The device calculates the world origin and tracks the surface only when the shore is partially in the camera's field of view. When the device's camera moved so that only water was in the field of view, the origin and all 3D objects placed in the scene began to float due to the lack of tracking points.

The results showed that a different approach should be taken, omitting the tracking of the world and focusing only on the device's sensors. With more accurate sensors integrated into new smartphones, this approach (location-based) has become increasingly important and gained popularity, especially in the application area of personal navigation; storytelling; Architecture, Engineering and Construction (AEC); cultural heritage; and tourism.

Further work assumed that AR objects were at the same height as the device on the water. This assumption eliminated the problem of low accuracy in determining altitude using GNSS methods. A test of application that presented labels and 3D objects in the device view based on a horizontal position, compass, and accelerometer was developed.

Due to technical capabilities, it was developed on the iOS platform using the native ARKit platform. The AR content is superimposed as a 3D object, 2D/3D labels, polylines, or polygons on the real world obtained from the smartphone camera. All AR objects are displayed on a smartphone screen. There are three information elements represented simultaneously in this solution: DEM of the bottom, markings, and traffic information.

The solution presents three groups of AR information elements. The route is indicated by a line between buoys/cardinal marks to provide the user with the necessary information about navigable routes and dangerous places. The DEM of the bottom shows depth and shallowing information as a 3D model. 3D objects show other users' locations with information about them (safety and social aspect) and selected POIs.

The information elements change size and colour depending on the distance from the boat to improve distance perception. Some of the described AR objects are shown in Figure 14.

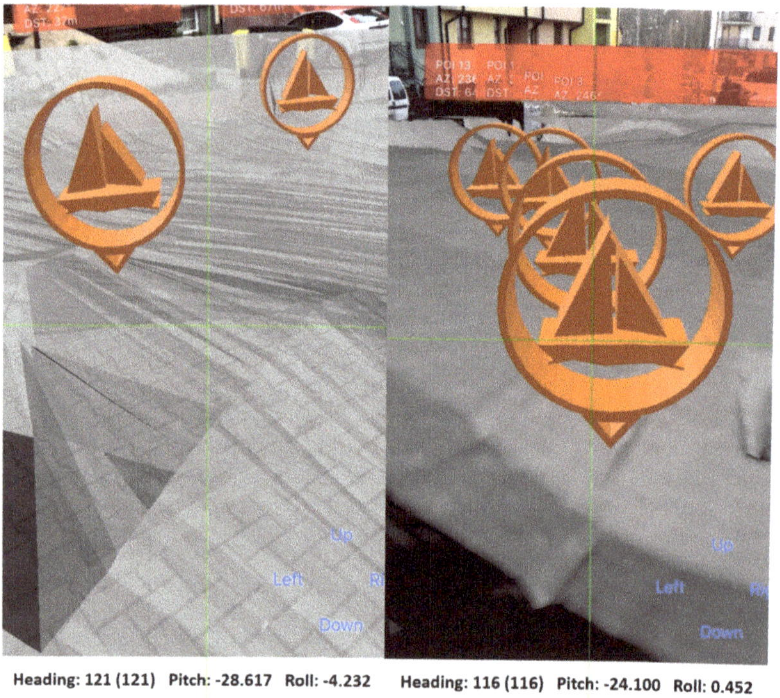

Figure 14. Examples of Augmented Reality applied to the navigation process—data fusion of the virtual location of sailors and Digital Elevation Model of the bottom of the water reservoirs.

In the following step, tests were conducted using the prototype MAR application on real data. The results are shown in Figure 15, which shows other users' locations whose position comes from the Zegluj navigation system (http://zegluj.mobi, accessed on 10 March 2022) and selected POIs. The tests showed that location accuracy did not matter when presenting distant objects. Some problems occurred with objects located at distances lower than 200 m. AR 3D objects "shifted" into the inaccurate position.

Further tests were conducted to reduce the impact of the inaccurate position on AR display. Changes were made to the code to optimise the frequency of refreshing the position depending on the travelled distance. In addition, changes have been made to improve the accuracy of the user device's location, and new filters have been introduced to optimise the determined position.

Tests were conducted on an iPhone 8. Approximately 50 points with 3D icons and the DTM model of Lake Śniardwy were rendered. The results were auspicious, and there were no rendering problems. The application worked very smoothly.

The movement through the digital terrain model while driving was also tested. When the device was in motion, location inaccuracies did not affect the presentation or immersion in using the application. The impact of inaccuracies in location determination was noticeable only if the device did not change its location at all. However, the frequency of position and orientation refreshing should depend on the distance, eliminating this problem.

Figure 15. The screenshot of the prototype of the MAR application (**right**). The AR visual interface presents real-time AR content about other boats and POIs (**left**).

5. Discussion

One of the main goals of this work was to verify the usefulness of AR technology for navigation in inland and coastal waters. In this research, the authors explored the feasibility of mobile devices as a hardware platform for building an AR-based mobile application. A visual AR interface was proposed for a general user consistent with the concept of nautical tourism, with thematic categories defining the range of content presented on the application screen. Due to the issue's complexity, it is rather a voice in the discussion, showing a methodology in preparing data for an AR application, a potential system architecture supporting a mobile AR application, and the proposition of an AR interface dedicated to this issue.

Existing solutions focus on implementing AR within professional, costly, and dedicated marine vessel solutions. At the same time there is a large group of sailors with less experience navigating on challenging sea and shallow areas. There are also professionals who sail recreationally in shallow waters. Despite much research, there is a lack of unification within a universal approach. However, the authors agree that the critical element is to work on an optimal AR interface and to produce AR context-related data dedicated for onscreen display. This paper examines these issues from the perspective of the user, who expects to be supported in the process of safe navigation on the water as well as contents that facilitates activities on the water. The importance of the presented results using AR for inland and shallow zones navigation thus lies both in their generality and relative ease of application to new areas.

The presented prototype AR application is the first approach that will be further developed. The work identified several issues and difficulties in implementing AR on mobile devices. However, the obtained results correspond with those by authors of other studies, who indicate that AR can be a key technology supporting water users. The experiments show that this technology has the potential to show a model of a body of water, to help efficient navigation on the water by augmenting the real world with navigational information, indicating the route, and avoiding dangerous places. Additionally, it can provide social information about other users, places worth visiting, and tourist infrastructure. Therefore, it can significantly impact the development of nautical tourism, becoming a source of new innovative tourism services.

One of the main limitations of the proposed solution is the hardware itself. For a mobile device to work efficiently, it must be placed appropriately on the boat, providing a camera view in front of the subject and a clear view of the screen by the captain. To simplify the research, the authors did not address the topic related to the safety of using this solution, the boat user's control, and the simultaneous observation of the screen of the mobile device. These topics are the subject of many studies. Several authors point to the need for HMDs in this area. It seems that a big opportunity is the introduction of HUD technology, which is especially evident in the automotive industry. There, the display of AR content on the windshield (HUD) is becoming more common in navigation systems in cars to minimise visual interference for the driver while providing relevant information while driving. Work is also currently underway to apply similar solutions to water navigation. They aim to display information for the skipper, steering the motorboat efficiently.

It is challenging to implement augmented reality on water. For most algorithms, it is difficult to achieve both good accuracy and high efficiency. Sensor-based methods can achieve reasonable efficiency, but their performance is often limited by the low precision of sensors used in mobile devices. Vision methods typically require significant computational and memory space to process an image consisting of many pixels. They require matching the image to a large database and estimating the geometric transformation between the captured image and the recognised object from the database.

A hybrid approach that integrates vision-based and sensor-based methods can potentially combine their complementary advantages, but its implementation is still a complex task and requires great knowledge. The authors have already conducted several experiments in optimising position and orientation estimation, position filtering, and optimising its display.

The performance of mobile devices is one of the most critical factors in implementing AR technology. Other important factors are low power consumption and memory management. Although the performance of mobile Central Processing Units (CPUs) has increased significantly and Graphics Processing Units (GPUs) can speed up calculations, the memory of today's smartphones is still not enough for performing some advanced operations (such as object recognising). Implementing the client–server model is a possible solution. It allows realising some operation on the server-side or cloud via the Internet. However, the network connection is not accessible anywhere, especially in rural areas with low population density. Due to these reasons, it is necessary to consider the functional requirement for the system.

A system architecture is a conceptual model that defines the structural behaviour of a system. It could be described as a set of significant decisions about the organisation and components of a system solution.

The proposed system architecture was defined based on the recent methodologies. Nevertheless, the complexity of the proposed architecture prevents the proposed solution from fully meeting all of the requirements.

The most popular mobile systems are installed on various hardware platforms provided by different vendors. The MAR app should work on devices with different screen sizes and resolutions, which brings on technical risks for the app. It requires determining different testing strategies and analysing the performance of individual modules.

6. Conclusions

This paper presents and describes the architecture of a novel application for AR navigation in inland and coastal waters. While the proposed architecture partially relies on existing, well-known technological solutions, the overall concept constitutes remarkable advancements with respect to existing systems. This research aimed to show the current state of the art of existing AR solutions for nautical applications and propose a methodology to introduce AR for mainstream mobile devices for shallow water navigation and nautical tourism development. The proposed solution is the first step to test AR implementation's feasibility and propose future AR evolution in this field.

The presented solution also proposes a participative approach where sailors can become part of the data acquisition process. They are a valuable resource providing information on hazard locations, traffic information, current weather conditions and much more. This allows increasing up-to-date AR contents dynamically. Due to validation issues with data provided using crowdsourcing, it needs to be verified using strict validation procedures before displaying.

The concept of the proposed AR solution has already been set up as a working prototype of the AR inland water app of the Masurian Lakeland region in Poland with several operative modules. The deployment of the whole vision will require long-term work due to the need to integrate a wide range of different technical solutions. The authors will continue this work to support safety in the navigation process on inland and coastal water with AR technology.

Currently, there is no representative standard of navigation information elements related to AR technology. This issue has yet to be investigated and tested on end-users to define representative guidelines. As research and development progresses, various navigation aid systems based on different displays will be further developed and utilised to increase the immersion level of the provided AR content. It is expected that this application of augmented technology will prevent future accidents on the water and provide new content.

In conclusion, the presentation of Augmented Reality in an aquatic environment is a complicated and challenging task. Compared to other industry sectors, this AR application is still underdeveloped. This shows that additional research is needed to leverage this technology's potential fully. Additionally, tools and processes need to be optimised for efficiency.

The study showed that further research is needed on the technical side of the platform. There are options to extend the platform with external sensors, cameras, and miniaturised scanners or radars to provide real-time data that can be analysed using artificial intelligence, object detection, collision avoidance, and even autonomous movement.

Author Contributions: Conceptualization, T.T., D.P., and M.G.; methodology, T.T., D.P., and M.G.; validation, T.T. and M.G.; formal analysis, T.T. and M.G.; investigation, D.P.; resources, T.T. and D.P.; writing—original draft preparation, T.T.; writing—review and editing, T.T. and D.P.; visualisation, T.T. and M.G.; funding acquisition, D.P. All authors have read and agreed to the published version of the manuscript.

Funding: This research was founded by: University of Warmia and Mazury, Olsztyn, Poland, and ESA Contract No. 4000127827/19/NL/MM/ff.

Institutional Review Board Statement: Not applicable.

Informed Consent Statement: Not applicable.

Data Availability Statement: Not applicable.

Acknowledgments: The authors wish to thank INFEO company for support and giving us access to the system data.

Conflicts of Interest: The authors declare no conflict of interest.

References

1. Narzt, W.; Pomberger, G.; Ferscha, A.; Kolb, D.; Müller, R.; Wieghardt, J.; Hörtner, H.; Lindinger, C. Augmented reality navigation systems. *Univ. Access Inf. Soc.* **2006**, *4*, 177–187. [CrossRef]
2. Kamel Boulos, M.N.; Lu, Z.; Guerrero, P.; Jennett, C.; Steed, A. From urban planning and emergency training to Pokémon Go: Applications of Virtual Reality GIS (VRGIS) and Augmented Reality GIS (ARGIS) in personal, public and environmental health. *Int. J. Health Geogr.* **2017**, *16*, 7. [CrossRef] [PubMed]
3. Hruby, F.; Ressl, R.; Valle, G.D.L.B.D. Geovisualization with immersive virtual environments in theory and practice. *Int. J. Digit. Earth* **2019**, *12*, 123–136. [CrossRef]
4. Schrom-Feiertag, H.; Settgast, V.; Seer, S. Evaluation of indoor guidance systems using eye tracking in an immersive virtual environment. *Spat. Cogn. Comput.* **2017**, *17*, 163–183. [CrossRef]
5. Behzadan, A.H.; Dong, S.; Kamat, V. Augmented reality visualization: A review of civil infrastructure system applications. *Adv. Eng. Inform.* **2015**, *29*, 252–267. [CrossRef]
6. Lu, W.; Zhao, L.; Xu, R. Remote sensing image processing technology based on mobile augmented reality technology in surveying and mapping engineering. *Soft Comput.* **2021**. [CrossRef]

7. Lütjens, M.; Kersten, T.P.; Dorschel, B.; Tschirschwitz, F. Virtual reality in cartography: Immersive 3D visualization of the Arctic Clyde Inlet (Canada) using digital elevation models and bathymetric data. *Multimodal Technol. Interact.* **2019**, *3*, 9. [CrossRef]
8. Saidin, N.F.; Halim, N.D.A.; Yahaya, N. A review of research on augmented reality in education: Advantages and applications. *Int. Educ. Stud.* **2015**, *8*, 1–8. [CrossRef]
9. Pereira, G.H.D.A.; Stock, K.; Delazari, L.; Centeno, J.A.S. Augmented reality and maps: New possibilities for engaging with geographic data. *Cartogr. J.* **2018**, *54*, 313–321. [CrossRef]
10. Kaji, S.; Kolivand, H.; Madani, R.; Salehinia, M.; Shafaie, M. Augmented Reality in Smart Cities: A Multimedia Approach. Available online: https://researchonline.ljmu.ac.uk/id/eprint/8035/1/Augmented%20Reality%20in%20Smart%20Cities%20A%20Multimedia%20Approach.pdf (accessed on 13 March 2020).
11. Serravalle, F.; Ferraris, A.; Vrontis, D.; Thrassou, A.; Christofi, M. Augmented reality in the tourism industry: A multi-stakeholder analysis of museums. *Tour. Manag. Perspect.* **2019**, *32*, 100549. [CrossRef]
12. Loureiro, S.M.C.; Guerreiro, J.; Ali, F. 20 years of research on virtual reality and augmented reality in tourism context: A text-mining approach. *Tour. Manag.* **2020**, *77*, 104028. [CrossRef]
13. Barrile, V.; Fotia, A.; Bilotta, G.; De Carlo, D. Integration of geomatics methodologies and creation of a cultural heritage app using augmented reality. *Virtual Archaeol. Rev.* **2019**, *10*, 40–51. [CrossRef]
14. Templin, T.; Popielarczyk, D. The use of low-cost unmanned aerial vehicles in the process of building models for cultural tourism, 3D web and augmented/mixed reality applications. *Sensors* **2020**, *20*, 5457. [CrossRef] [PubMed]
15. Dünser, A.; Billinghurst, M.; Wen, J.; Lehtinen, V.; Nurminen, A. Exploring the use of handheld AR for outdoor navigation. *Comput. Graph.* **2012**, *36*, 1084–1095. [CrossRef]
16. Schmidt, M.A.R.; de Mendonca, A.L.A. Challenges in Personal Mobile Navigation. 2018. Available online: https://use.icaci.org/wp-content/uploads/2018/11/SchmidtMendonca.pdf (accessed on 3 March 2021).
17. Popielarczyk, D.; Templin, T. Application of integrated GNSS/hydroacoustic measurements and GIS geodatabase models for bottom analysis of Lake Hancza: The deepest inland reservoir in Poland. *Pure Appl. Geophys.* **2014**, *171*, 997–1011. [CrossRef]
18. Yu, J.; Fang, L.; Lu, C. Key technology and application research on mobile augmented reality. In Proceedings of the 2016 7th IEEE International Conference on Software Engineering and Service Sciences (ICSESS), Beijing, China, 26–28 August 2016; pp. 547–550.
19. Bartolini, S.; Mecocci, A.; Pozzebon, A.; Zoppetti, C.; Bertoni, D.; Sarti, G.; Caiti, A.; Costanzi, R.; Catani, F.; Ciampalini, A.; et al. Augmented virtuality for coastal management: A holistic use of in situ and remote sensing for large scale definition of coastal dynamics. *ISPRS Int. J. Geo-Inf.* **2018**, *7*, 92. [CrossRef]
20. Mirauda, D.; Erra, U.; Agatiello, R.; Cerverizzo, M. Applications of mobile augmented reality to water resources management. *Water* **2017**, *9*, 699. [CrossRef]
21. Menna, F.; Agrafiotis, P.; Georgopoulos, A. State of the art and applications in archaeological underwater 3D re-cording and mapping. *J. Cult. Herit.* **2018**, *33*, 231–248. [CrossRef]
22. Bruno, F.; Barbieri, L.; Mangeruga, M.; Cozza, M.; Lagudi, A.; Čejka, J.; Liarokapis, F.; Skarlatos, D. Underwater augmented reality for improving the diving experience in submerged archaeological sites. *Ocean Eng.* **2019**, *190*, 106487. [CrossRef]
23. Oh, J.; Park, S.; Kwon, O.-S. Advanced navigation aids system based on augmented reality. *Int. J. e-Navig. Marit. Econ.* **2016**, *5*, 21–31. [CrossRef]
24. Bandara, D.; Woodward, M.; Chin, C.; Jiang, D. Augmented reality lights for compromised visibility navigation. *J. Mar. Sci. Eng.* **2020**, *8*, 1014. [CrossRef]
25. Hugues, O.; Cieutat, J.-M.; Guitton, P. An experimental augmented reality platform for assisted maritime navigation. In Proceedings of the 1st Augmented Human International Conference, Megève, France, 2–3 April 2010. [CrossRef]
26. Grabowski, M. Research on wearable, immersive augmented reality (WIAR) adoption in maritime navigation. *J. Navig.* **2015**, *68*, 453–464. [CrossRef]
27. Procee, S.; Borst, C.; van Paassen, R.; Mulder, M. Toward functional augmented reality in marine navigation: A cognitive work analysis. In *Proceedings of COMPIT*; Technical University Hamburg: Hamburg, Germany, 2017.
28. Hertel, J.; Steinicke, F. Augmented reality for maritime navigation assistance—Egocentric depth perception in large distance outdoor environments. In Proceedings of the 2021 IEEE Virtual Reality and 3D User Interfaces Conference, Lisbon, Portugal, 27 March–1 April 2021. [CrossRef]
29. Laera, F.; Foglia, M.M.; Evangelista, A.; Boccaccio, A.; Gattullo, M.; Manghisi, V.M.; Gabbard, J.L.; Uva, A.E.; Fiorentino, M. Towards sailing supported by augmented reality: Motivation, methodology and perspectives. In Proceedings of the 2020 IEEE International Symposium on Mixed and Augmented Reality Adjunct (ISMAR-Adjunct), Recife, Brazil, 9–13 November 2020; pp. 269–274. [CrossRef]
30. Laera, F.; Fiorentino, M.; Evangelista, A.; Boccaccio, A.; Manghisi, V.M.; Gabbard, J.; Gattullo, M.; Uva, A.E.; Foglia, M.M. Augmented reality for maritime navigation data visualisation: A systematic review, issues and perspectives. *J. Navig.* **2021**, *74*, 1073–1090. [CrossRef]
31. Von Lukas, U.; Vahl, M.; Mesing, B. Maritime applications of augmented reality—Experiences and challenges. In *Virtual, Augmented and Mixed Reality. Applications of Virtual and Augmented Reality*; Lecture Notes in Computer Science; Springer: Cham, Switzerland, 2014; Volume 8526.
32. Daponte, P.; De Vito, L.; Picariello, F.; Riccio, M. State of the art and future developments of the Augmented Reality for measurement applications. *Measurement* **2014**, *57*, 53–70. [CrossRef]

33. Paziewski, J.; Kealy, A.; Gikas, V.; Geng, J. Recent advances in ubiquitous positioning systems for mobility applications. *Meas. Sci. Technol.* **2021**, *32*, 090201. [CrossRef]
34. Ciećko, A.; Bakuła, M.; Grunwald, G.; Ćwiklak, J. Examination of multi-receiver GPS/EGNOS positioning with Kalman filtering and validation based on CORS stations. *Sensors* **2020**, *20*, 2732. [CrossRef]
35. Specht, M.; Stateczny, A.; Specht, C.; Widźgowski, S.; Lewicka, O.; Wiśniewska, M. Concept of an innovative autonomous unmanned system for bathymetric monitoring of shallow waterbodies (INNOBAT system). *Energies* **2021**, *14*, 5370. [CrossRef]
36. Makar, A.; Specht, C.; Specht, M.; Dąbrowski, P.; Szafran, M. Integrated geodetic and hydrographic measurements of the yacht port for nautical charts and dynamic spatial presentation. *Geosciences* **2020**, *10*, 203. [CrossRef]
37. Specht, M.; Specht, C.; Szafran, M.; Makar, A.; Dąbrowski, P.; Lasota, H.; Cywiński, P. The use of USV to develop navigational and bathymetric charts of yacht ports on the example of national sailing centre in Gdańsk. *Remote Sens.* **2020**, *12*, 2585. [CrossRef]
38. Crenshaw, R.S. *Naval Shiphandling*; US Naval Institute: Annapolis, MD, USA, 1965.
39. Nowacki, P.; Woda, M. Capabilities of ARCore and ARKit platforms for AR/VR applications. *Adv. Intell. Syst. Comput.* **2019**, *987*, 358–370. [CrossRef]

remote sensing

MDPI

Article

Assessment of Adjacency Correction over Inland Waters Using Sentinel-2 MSI Images

Rejane S. Paulino [1,2,*], Vitor S. Martins [3], Evlyn M. L. M. Novo [1,2], Claudio C. F. Barbosa [1,2], Lino A. S. de Carvalho [4] and Felipe N. Begliomini [1,2]

1 Earth Observation and Geoinformatics Division, National Institute for Space Research, São José dos Campos 12227-010, SP, Brazil; evlyn.novo@inpe.br (E.M.L.M.N.); claudio.barbosa@inpe.br (C.C.F.B.); fnincao@hotmail.com (F.N.B.)
2 Instrumentation Laboratory for Aquatic Systems (LabISA), Earth Observation Coordination of National Institute for Space Research (INPE), São José dos Campos 12227-010, SP, Brazil
3 Department of Agricultural & Biological Engineering, Mississippi State University, Starkville, MS 39762, USA; vmartins@abe.msstate.edu
4 Departamento de Meteorologia, Universidade do Rio de Janeiro (UFRJ), Rio de Janeiro 21941-916, RJ, Brazil; lino.sander@igeo.ufrj.br
* Correspondence: rejane.paulino@inpe.br or rejane.dspaulino@gmail.com

Abstract: Satellite remote sensing data have been used for water quality mapping, but accurate water reflectance retrieval is dependent on multiple procedures, such as atmospheric and adjacency corrections. For the latter, physical-based methods are used to minimize the adjacency effects caused by neighboring land targets close to water pixels, and implementation requires atmospheric and environmental parameters, such as aerosol optical depth and horizontal range (i.e., distance in meters) of the adjacency effect (H_{Adj}). Generally, the H_{Adj} is empirically defined by users and can lead to substantial errors in water reflectance when incorrectly used. In this research, a physical-based approach with three empirical methods to determine the H_{Adj} (fixed, SIMilarity Environment Correction—SIMEC, and Adaptative Window by Proportion—AWP-Inland Water) were used to correct and characterize the adjacency effects in Sentinel-2 images over Brazilian inland waters. An interactive inversion method of the deep blue waveband estimated the aerosol loading for the atmospheric correction procedure. The results of atmospheric and adjacency corrections were validated against in-situ reflectance data. The inverted aerosol loading achieved a good agreement with in-situ measurements, especially at visible wavelengths (Mean Absolute Percentage Error—MAPE for eutrophic (~56%), bright (~80%), and dark (~288%) waters). The adjacency correction performance was near similar between the SIMEC and AWP-Inland Water methods in eutrophic and bright waters (MAPE difference < 3%). However, only the AWP Inland Water method provided a smaller error (MAPE ~53%) for dark waters compared to the fixed (~108%) and SIMEC (~289%) methods, which shows how critical H_{Adj} parametrization is for low water reflectance values. Simulations of different atmospheric and adjacency effects were performed, and they highlighted the importance of adjacency correction under aerosol loading higher 0.1, which is a typical aerosol loading in a dry climate season, and over extremely dark, low-reflectance waters. This paper contributes to further understanding adjacency effects in medium spatial resolution imagery of inland waters using a physical-based approach including the uncertainties in H_{Adj} determination.

Keywords: adjacency effects; surface reflectance; atmospheric correction; aerosol; 6SV; radiative transfer

Citation: Paulino, R.S.; Martins, V.S.; Novo, E.M.L.M.; Barbosa, C.C.F.; de Carvalho, L.A.S.; Begliomini, F.N. Assessment of Adjacency Correction over Inland Waters Using Sentinel-2 MSI Images. *Remote Sens.* **2022**, *14*, 1829. https://doi.org/10.3390/rs14081829

Academic Editors: Marta Wlodarczyk-Sielicka, Katarzyna Bradtke, Paweł Terefenko and Jacek Lubczonek

Received: 24 February 2022
Accepted: 4 April 2022
Published: 11 April 2022

1. Introduction

Inland waters are essential ecosystems for life on Earth [1]. They are sources of drinking water, socio-economic services (e.g., power generation, fisheries, recreation, and irrigation), and ecosystem services (e.g., biodiversity maintenance) [2]. By 2050, scientific projections indicate that most of the world's population (*52%) will be living in water-stressed areas [3], predicting that water demand will surpass its availability in the following decades. Thus, monitoring these environments is crucial to ensure water availability for its multiple uses. Biophysical monitoring of inland waters at large space–time scales is only feasible with satellite remote sensing data sources. The advance of new satellite sensors with better radiometric quality and higher spectral, spatial, and temporal resolutions, such as Landsat-8 OLI and Sentinel-2 MSI [4–7], has sparked the water remote sensing community's interest in using these systems for the mapping of inland waters' Optically Active Components—OACs [8–11]. However, atmospheric and adjacency effect corrections [12] are still a challenge in achieving accurate water reflectance estimates for high- and medium-resolution satellite imagery.

The adjacency effect is caused by reflected photons from surrounding land targets that are scattered by atmosphere components into the sensor's field of view [13–15]. This complex phenomenon reduces the contrast between the high-reflectivity (land) and the low-reflectivity (water) surfaces. As a result, scattered photons from the land targets close to the water bodies can distort water spectral reflectance, mainly affecting small water bodies [12]. The magnitude of these effects depends on several factors, such as atmosphere composition (e.g., aerosol particle properties), land-cover type, viewing and illumination geometry, sensor characteristics (e.g., spatial resolution), and the shape and size of water bodies [16–18]. When neglected, the adjacency effect can introduce significant errors in retrieving the water-leaving signal [19], limiting the use of satellite imagery for estimating OACs.

Adjacency correction in satellite imagery uses physical approaches that allow the application of radiative transfer theory and Atmospheric Point Spread Function—APSF to quantify and remove the adjacency effects [15,20,21]. In this context, APSF describes the contribution of reflected photons from the neighboring targets into the interest target [22,23]. Tanré et al. [15] and Vermote et al. [21] have suggested a robust formulation for APSF recovery based on Monte Carlo simulations, where the APSF results from the contribution of molecular (or Rayleigh) and aerosol scattering. The inherent challenge in applying this method is estimating the range of the adjacency effect (H_{Adj}), which refers to the maximum extent of the adjacency effects around the interest target, because it depends on often unknown factors [24]. Alternatively, the H_{Adj} has been chosen arbitrarily [25,26] or defined iteratively by comparing the corrected image reflectance with the spectral responses of targets [18,27,28]. An improper choice of H_{Adj} can lead to over- or under-correction of water reflectance due to the mischaracterization of environmental conditions around the water target.

Few studies have assessed the accuracy of the correction of adjacency effects on water reflectance [12,29,30]. The methodology complexity and limited open-source software for adjacency effect correction can partially explain this lacune. In this way, the assessment of water reflectance products has been generally performed in terms of (i) intercomparison between different atmospheric correction methods assuming a uniform surface (i.e., without adjacent correction) [31,32], (ii) validation of adjacency effects for large lakes (surface area greater than 370 km^2) or coastal waters [29,33], or (iii) theoretical water reflectance simulations for a variety of atmospheric conditions and coastal water types [16,17,34]. Over inland waters, adjacency effects are under-discussed, and a recent study showed the importance of correcting adjacency problems when estimating OACs in these environments [19]. For this reason, a physical-based method to fix adjacency effects with a practical empirical assumption for H_{Adj} estimation remains necessary research for inland waters.

This study explores the adjacency effects over Brazilian inland waters in medium spatial resolution Sentinel-2 (A and B) MSI images. The paper's focus is to assess the satellite imagery reflectance uncertainties caused by adjacency effects using a physical-based approach [15,21] on small lakes surrounded by dense forest cover and on a large water urban reservoir considering variable aerosol loadings. Our assessment of satellite-derived water reflectance contributes to understanding (i) the effect of optical water types on adjacency correction, (ii) the performance of a physical method in complex environments (e.g., atmospheric scattering conditions and shape and size of water bodies), (iii) the proper definition of the size of the H_{Adj}, (iv) the impact of external factors on modeling the adjacency effect for inland waters, and (v) the conditions under which this effect is negligible. Here, we analyze two different methods to retrieve the H_{Adj}: (i) considering the size of the H_{Adj} constant along the water body (hereafter called fixed), and (ii) varying the H_{Adj} pixel-by-pixel (hereafter called adaptive). The adaptive methods tested were SIMilarity Environment Correction—SIMEC [33] and Adaptative Window by Proportion applied to Inland Water—AWP-Inland Water. AWP-Inland Water is a proto-algorithm developed in this study based on the proportion of non-water targets within the window. The atmospheric and adjacency correction validation used field data acquired in a time interval close to the satellite overpasses. Moreover, this study further investigates the magnitude of the adjacency effect using the Second Simulation of a Satellite Signal in the Solar Spectrum (6SV) to simulate its impact under different water optical types, aerosol properties (e.g., aerosol loading and model), and land-cover types.

2. Materials and Methods

2.1. Study Area

Five small lakes (Mamirauá—MAM, Pirarara—PIR, Concordia—CON, Branco—BRA, and Mutum—MUT) and one large reservoir (Billings—BIL) were selected for the analysis (Figure 1). These water bodies represent different sceneries regarding land cover, shape and size, and optically active water components in Brazil's north and southeast regions. Billings reservoir (127 km^2) is one of the largest water systems in the metropolitan region of São Paulo State, serving multiple purposes including public water supply, energy generation, fisheries, and recreation [35,36]. Frequent algae blooms and potentially toxic cyanobacteria dominate Billings' optical properties and impair its water quality [35,37–39]. The reservoir is divided into eight narrow arms that contribute to a broader and elongated central body [36]. In general, the arms' average width is approximately 500 m, but sometimes it is smaller than 100 m. These narrow arms and the proximity to urban targets make the Billings reservoir a suitable study site to investigate the adjacency effect.

The other selected water bodies encompass Amazon floodplain lakes with surface areas smaller than 3 km^2. Two of them, Mamirauá and Pirarara, are located inside the Mamirauá Sustainable Development Reserve—MSDR close to the confluence of the Solimões and Japurá rivers. The remaining three lakes, Concordia, Branco, and Mutum, belong to the Juruá River floodplain. These lakes are located in well-preserved areas under low human influence and are surrounded by a dense flooded forest cover. The water flow exchange with the fluvial systems [40] and the land coverage around the lakes [41] strongly influence the bio-optical compositions of these ecosystems. For example, Pirarara lake has brighter waters because it is connected to the Japurá River and receives a high inflow from its temporarily sediment-laden waters. Four of these lakes have dark waters with reflectance values smaller than 4%. In these cases, they are perennial lakes and suffer influence from the surrounding forest, which during the rising and flooding season, washes into the lakes large amounts of dissolved organic matter accumulated year-round [41,42].

The water bodies chosen in this study are potential environments for bio-optical modeling applied to the water quality monitoring of urban reservoirs and sustainable management of small lakes using remote sensing data [10,37,43]. For that reason, the investigation of factors that influence the accurate recovery of OACs, including the adjacency effect, becomes essential in these regions.

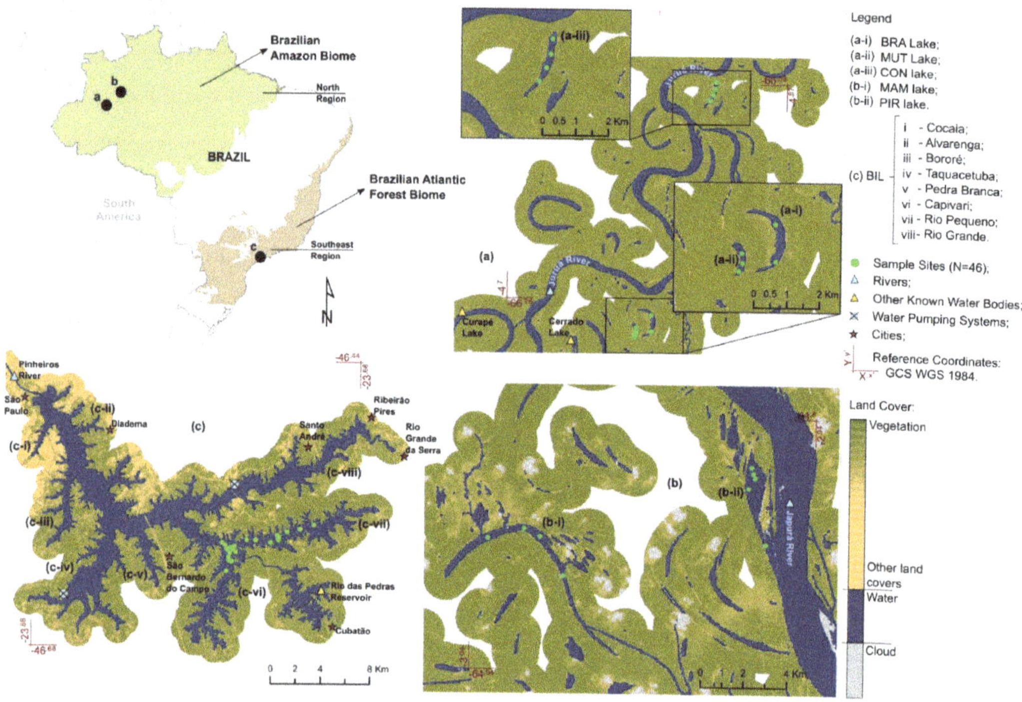

Figure 1. Overview of the study area: (**a**,**b**) Amazon floodplain lakes and (**c**) urban reservoir. The sampling points are illustrated by green dots. The green–pale orange gradient represents the land cover (e.g., vegetation and other land covers), and the blue and gray colors refer to water and clouds, respectively.

2.2. Dataset

2.2.1. MSI/Sentinel-2 Data

Multi-Spectral Instrument—MSI sensors onboard Sentinel-2 (A and B) satellites were used to assess adjacency effect correction in this study. The MSI Earth observation data are acquired with fine spatial resolution (10, 20, and 60 m depending on the band) and radiometric resolution (12-bit) in 13 spectral bands localized in the visible near-infrared and short-wave-infrared regions. The Sentinel-2 mission consists of two satellites that carry identical sensors that allow a revisit time of five days at the equator [44]. Although this instrument did not aim at remote sensing aquatic environments, it has accurately performed inland water applications [8,10,45]. Additionally, it increases the possibility for monitoring small water bodies (area smaller than 0.002 km^2; see [46]). In addition to water quality monitoring applications, the MSI configuration also helps to investigate the adjacency effect, since: (i) on surfaces with high-contrast among land covers (e.g., water and earth interface), the contamination of the spectral information may be more disrupting in images of medium- and high-spatial resolution; (ii) there is a trend of the adjacency effect to be greater in small and narrow water bodies; and (iii) the increase in time-frequency of image acquisition enables a more significant number of field samplings concurrent to satellite overpass, increasing the number of samples for calibration and validation of the satellite sensor's measurements.

Images were acquired over the water body regions on 27 August 2017 (MAM and PIR), 20 August 2019 (CON, BRA, and MUT), and 9 August 2021 (BIL). The scenes were downloaded from the Copernicus Open Access Hub website (https://scihub.copernicus.

eu/) as L1C products that were corrected for radiometric and geometric distortions with pixel values referring to Top of Atmosphere—TOA reflectance [47]. The selected images had less than 10% cloud cover and cloud-free conditions over all the field sampling sites (Section 2.2.2). After that, the atmosphere and adjacency effect corrections were applied to the TOA reflectance data. Details are discussed in Sections 2.4. and 2.5.

2.2.2. Field Data

This study used in-situ radiometric data collected to validate MSI surface reflectance imagery after atmospheric and adjacency corrections. Field campaigns in the water bodies occurred during 25–28 August 2017 (MAM and PIR), 19–20 August 2019 (CON, BRA, and MUT), and 8–9 August 2021 (BIL). The time difference between the in-situ reflectance measurements and satellite images can reduce the data correlation due to water composition variability [48,49]. Therefore, we used a time window of ±48 h for the match-up analysis. Over inland waters, commonly size time windows of 2 h—7 days are applied to validate the atmospheric and adjacency corrections [12,29,31,50]. In addition, we also highlight that the water bodies selected did not present fast changes during the field campaigns because both systems, the reservoir and small lakes, were under conditions that minimized abrupt variations in the optical components of the water in a short period. For instance, Billings reservoir has a hydraulic residence time (392 days) and a maximum depth (18 m) [51] that both decrease the water column turbulence and bottom sediment resuspension. Likewise, the small Amazon floodplain lakes, although their water level and circulation are influenced by the flood pulse, these processes occur gradually throughout the hydrological year [40,52,53] with a minimal influence in short time windows during the rising season in meandering rivers, such as Juruá [54].

A total of 46 in-situ samples from BIL (N = 28) and other lakes (N = 18) were collected. A suitable distance from the water body edge was adopted for all sample sites to avoid or reduce the effect of pixel mixture at the land–water interface, bottom reflectance, and adjacency contamination from the surrounding targets (e.g., forest). The sampling sites were placed on the lakes' central regions, and at BIL, they were split into two categories: sites placed at distances smaller than 100 m (N = 14) and larger than 100 m (N = 14). Due to BIL's large area, collecting in-situ samples close to the reservoir's edge was fundamental to validating the adjacency effect because this problem is much larger near land targets around water bodies [17]. In general, the depth of these sites (>3 m) in comparison with the in-situ Secchi depth (<1 m) measured indicates that shallow water effects did not influence the data collected near the water body' boundary; that is, this study assumes that these regions are optically deep. We had difficulties sampling throughout BIL due to its extension and the need for feasible illumination conditions during the measurements. Thus, the samples were collected mainly along the Rio Pequeno arm and at the beginning of the Capivari arm (see Figure 1).

At each sample site, the remote sensing reflectance (R_{rs}) was estimated using the radiometric quantities obtained by three intercalibrated spectroradiometers (TriOS-RAMSES). All sensors operated simultaneously and measured the total water-leaving radiance ($L_t(\lambda, \theta_v, \phi_v)$), the downwelling sky radiance ($L_{sky}(\lambda, \theta'_v, \phi'_v)$), and the total irradiance incident on the water surface ($E_s(\lambda)$) within 350–950 nm wavelengths (at ~3.3 nm increments). The measurements were performed between 10:00 a.m. and 1:00 p.m., and the sensors were positioned at selected angles to minimize sun glint effects following the recommendations of [55]: with the Sun as a reference, the zenith (θ_v) and azimuth (ϕ_v) angles of the L_t term were $(45^\circ, 90^\circ - 135^\circ)$ and $(\theta'_v, \phi'_v) = (\theta_v + 90^\circ, \phi_v = \phi'_v)$ for the L_{sky} term. With these data, each radiometric record was resampled at 1 nm, and the spectral R_{rs} was calculated utilizing the following equation:

$$R_{rs}(\lambda) = \frac{L_t(\lambda, \theta_v, \phi_v) - Rho_{sky}(\theta'_v, \phi'_v, \theta_0, W)\, L_{sky}(\lambda, \theta'_v, \phi'_v)}{E_s(\lambda)} \left(sr^{-1}\right) \quad (1)$$

where Rho_{sky} is a coefficient that corrects the skylight reflection effects and depends on the wind speed (W), view geometry (θ'_v, ϕ'_v), and Sun zenith angle (θ_0); it can be obtained in [56]. Afterward, the water reflectance (ρ_w) was obtained by multiplying R_{rs} by the value of π. The ρ_w was used in the simulation of the Sentinel-2 MSI bands (visible to near-infrared) using their Spectral Response Function—SRF to generate multispectral data comparable to the corrected image reflectance values in this study:

$$\rho^*_w(B_i) = \frac{\int_a^b \rho_w(\lambda) \times SRF(\lambda)\, d\lambda}{\int_a^b SRF(\lambda)} \text{ (Unitless)} \quad (2)$$

where ρ^*_w is the MSI reflectance simulated from in-situ data, B_i is the MSI spectral band, [a, b] is the range of the spectral band, and λ is the wavelength.

2.3. The Selection of Water Types

Water composition affects the shape and magnitude of the spectra (Figure 2). In general, the signals of eutrophic and bright waters are up to ~2 times greater than that of dark waters at visible wavelengths (~400–700 nm), and this difference decreases along the spectrum. When observed by the orbital sensors, water bodies with different optical types present distinct behaviors regarding atmospheric and adjacency noise [5,17,42]. Thus, the grouping of water bodies according to their spectral features helps to understand the adjacency effect on inland waters. In this way, we classified the water bodies into artificial eutrophic urban reservoirs (BIL) and natural lakes surrounded by forest, classifying them afterward into bright- (PIR) and dark- (MAM, CON, BRA, and MUT) water lakes, following the criteria defined in [12]. They are named eutrophic, bright, and dark waters throughout this study. It is essential to highlight that the water bodies inserted in these clusters were under different conditions (e.g., atmospheric scattering, land cover, illumination geometry, etc.) that affect the adjacency effect magnitude. These different contexts limited a direct comparison among the water types. Thus, for a proper analysis, the three groups were observed individually.

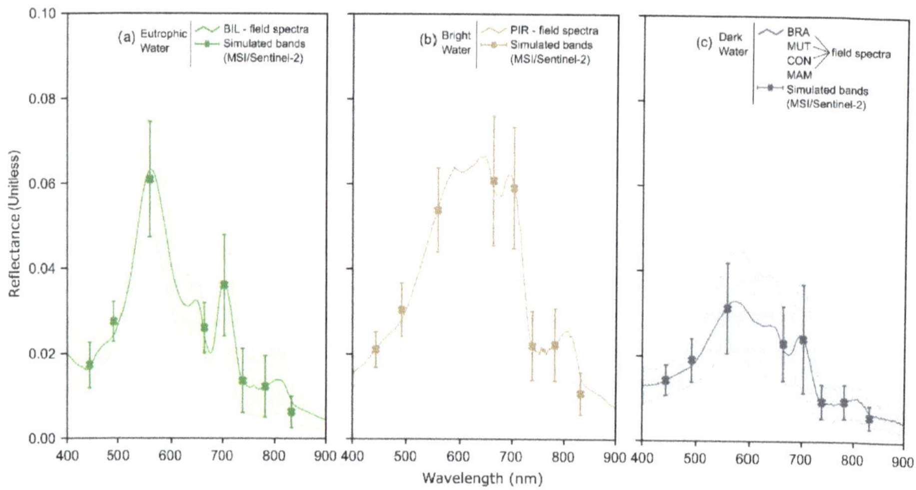

Figure 2. Spectral values of water reflectance measured in-situ for selected water bodies. The water bodies were grouped into three types: (**a**) eutrophic, (**b**) bright, and (**c**) dark waters. Solid lines and shaded areas indicate the average and standard deviation of water reflectance, respectively. The markers and error bars indicate the reflectance values simulated for the MSI sensor bands.

2.4. Atmospheric Correction Procedure

Atmospheric Correction—AC is essential for using remote sensing images to monitor aquatic environments [12,31,50]. In the water, the goal of AC is to remove from TOA reflectance the atmospheric scattering and absorption effects caused by the aerosol and molecules and the gases, respectively. A 6SV model was applied for the AC of the MSI images. It is a radiative transfer code that simulates the TOA reflectance through atmosphere conditions and has been widely adopted in water applications [8,23,25,57]. According to the 6SV model under the assumption of surfaces with Lambertian characteristics and adjacency effect, the target (water) reflectance at the sensor level ($\hat{\rho}_w$) is derived as follows:

$$\hat{\rho}_w(B_i, i_0, j_0) = B - C \cdot \rho_{env}(B_i, i_0, j_0) \tag{3}$$

where ρ_{env} is the average reflectance of the environment (this factor is described in Section 2.5), B_i is the MSI spectral band, and (i_0, j_0) are the Cartesian coordinates of the pixel. Factors B and C include the atmospheric content, and they are obtained by simplifying the radiative transfer equation:

$$A = \left(\frac{\rho_{TOA}}{T_{gOG} \cdot T_{gO_3}} - \rho^{atm} \right) \cdot \frac{1}{T^{\downarrow}_{(\mu_s)} \cdot T_{gH_2O}} \tag{4}$$

$$B = \frac{A}{t^{\uparrow}_{dir}(\mu_v)} \tag{5}$$

$$C = \frac{\left(t^{\uparrow}_{dif}(\mu_v) + A \cdot S_{atm}\right)}{t^{\uparrow}_{dir}(\mu_v)} \tag{6}$$

where ρ_{TOA} is the reflectance at the TOA, ρ^{atm} is the atmosphere intrinsic reflectance, $T^{\downarrow}_{(\mu_s)}$ is the total atmosphere transmission (downward), and $t^{\uparrow}_{dir}(\mu_v)$ and $t^{\uparrow}_{dif}(\mu_v)$ are the transmissions direct and diffuse of the atmosphere (upward), respectively. μ_s and μ_v are the geometric conditions (cosine of the zenith angle) of illumination and viewing, respectively; T_g is the gaseous transmission by water vapor (T_{gH_2O}), ozone (T_{gO_3}), and other gases (T_{gOG}), which include dioxide (CO_2) and monoxide (CO) of carbon, oxygen (O_2), nitrogen dioxide (NO_2), and methane (CH_4); and S_{atm} designates the atmosphere spherical albedo. The initial approximation to retrieve $\hat{\rho}_w$ was performed using the assumption of uniform surfaces ($\hat{\rho}^{**}_w$), i.e., considering surfaces without the adjacency effect $\left(\rho_{env}(B_i, i_0, j_0) - \hat{\rho}_w(B_i, i_0, j_0)\right)$. The ($\hat{\rho}^{**}_w$) can be written as:

$$\hat{\rho}^{**}_w(B_i, i_0, j_0) = \frac{B}{1 + C} \tag{7}$$

The 6SV model requires predefined knowledge of atmospheric conditions for the regions where the water bodies are localized. The necessary inputs include the geometry of illumination and viewing, the amount of water vapor and ozone in a vertical path through the atmosphere, the water body altitude, and the aerosol characteristics, which comprise the aerosol model and Aerosol Optical Depth at 550 nm—AOD_{550} [21,23]. These parameters were acquired using the average value recovered within the defined buffer around BIL (10 km) and the other water bodies (5 km), except for the geometric conditions obtained from the information contained in the MSI image metadata. Total columnar ozone and water vapor were extracted from the Level-3 MODIS daily global product MOD08 (MOD08_D3) and MODIS atmospheric products derived from the Multi-Angle Implementation of an Atmospheric Correction—MAIAC algorithm (MCD19A2 Collection 6), respectively. The amounts of ozone and water vapor were calculated as the average value within a time window of ±24 h of the MSI sensor overpass to reduce uncertainties related to the MODIS sensor. The altitude was obtained using an SRTM (30 m) digital elevation model for each water body.

Among all the information needed to run the 6SV model, the aerosol contribution is the most challenging, since its properties (e.g., extinction and scattering coefficients, asymmetry factor, and phase function) are largely unknown. In this study, a continental aerosol model was used, and the AOD_{550} parameter was estimated through an interactive inversion method with a radiative transfer equation using the field data as a reference (Appendix A). We evaluated the AC performance for the water bodies using the AOD_{550} obtained from both the inversion method and the satellite aerosol product (MCD19A2 Collection 6). The AOD at 550 nm obtained with the MODIS sensor followed the same criteria defined for the other parameters used to model the atmospheric quantities. A summary of the input data used in each water body for AC is shown in Appendix B (Table A1).

2.5. Adjacency Effect Correction Procedure

The adjacency effect is a critical factor for inland waters and an essential step in processing remote sensing images applied in mapping OACs. This paper analyzed the adjacency effect in the selected water bodies according to the physical method suggested in [15,21]. The method is based on the APSF, which describes the dispersal of photons as they leave the surfaces and are propagated through the atmospheric layer. In this method, the adjacency contribution to the target pixel is calculated as the weighted average of the reflectance values of its surrounding pixels. The weight factor, referred to as APSF or environment function, denotes the probability that the atmosphere components scatter the reflected photons from the pixels within an array towards the sensor. In principle, the weight value decreases with increasing distance from the target pixel. The adjacency contribution is defined as:

$$\rho_{env}(B_i, i_0, j_0) = \left(\sum_{i=1}^{N}\sum_{j=1}^{N} F\left(B_i, \bar{r}(i,j)\right)\right)^{-1} \cdot \sum_{i=1}^{N}\sum_{j=1}^{N} \hat{\rho}_w^{**}(B_i, i, j) \cdot F\left(B_i, \bar{r}(i,j)\right) \quad (8)$$

where:

$$F\left(B_i, \bar{r}\right) = \frac{\uparrow t_{dif}^{R} \cdot F_R(\bar{r}) + \uparrow t_{dif}^{A} \cdot F_A(\bar{r})}{\uparrow t_{dif}^{R+A}} \quad (9)$$

where $\bar{r}(i,j)$ is the position of an array of pixels centered on the target pixel (i_0, j_0), $(i_0, j_0) \cdot \hat{\rho}_w^{**}$ is the surface reflectance calculated considering a uniform surface (Equation (8)), B_i is the MSI spectral band, N is the number of pixels within of an array (we considered a contribution window of m × m pixels as the range of the adjacency effect), F is the weight or APSF, and $F_R(\bar{r})$ and $F_A(\bar{r})$ refer to the environment function for molecular (or Rayleigh) and aerosol scattering, respectively. $\uparrow t_{dif}^{R}$ and $\uparrow t_{dif}^{A}$ are the diffuse transmittances (upward) for molecular and aerosol scattering, respectively, and $\uparrow t_{dif}^{R+A}$ is the total diffuse transmittance from target to sensor $\left(\uparrow t_{dif}^{R+A} = \uparrow t_{dif}^{R} + \uparrow t_{dif}^{A}\right)$. F is obtained through the atmosphere scattering characteristics (content of molecules and aerosol particles) and can be expressed as a sum of the contribution of these spreads weighted by their respective transmittances (Equation (9)) [15,21]. To recover the atmospheric scattering information, we applied the functions $F_R(\bar{r})$ and $F_A(\bar{r})$ specified in [23]. These functions were calculated assuming an average atmosphere condition using a continental aerosol model [21,23]. The generic expression of these functions is given by:

$$F_x(\bar{r}) = a \cdot e^{-\alpha \cdot \bar{r}} + b \cdot e^{-\beta \cdot \bar{r}} \quad (10)$$

For $F_R(\bar{r})$, the coefficients correspond to a = 0.930, α = 0.08, b = 0.070, and β = 1.10. For $F_A(\bar{r})$, they are a = 0.448, α = 0.270, b = 0.552, and β = 2.83. $\bar{r}$ denotes the distance from the surrounding pixel to the target pixel (in km).

An essential question when applying this method is how to define the size of the H_{Adj}, which indicates the maximum distance that a target pixel can be affected by its surrounding pixels. Ideally, the window size of the adjacency effect is mainly determined by the aerosol vertical distribution, aerosol optical depth, satellite spatial resolution, geometry of observation, and type of surrounding target [14,24,58]. However, information about the actual aerosol vertical distribution is often unknown, and as the extent of influence of the surrounding pixels depends on many factors, it is not easy to calculate the size of the H_{Adj} exactly. Alternative methods include empirical approaches to determine the H_{Adj} using fixed and adaptative windows [18,25,28]. Three methods to estimate the size of the H_{Adj} for the selected water bodies were analyzed: (i) fixed window. It was defined using atmospheric correction and validated with the data collected in-situ. For each water body, we applied different fixed window sizes ranging from 100 × 100 m to 1500 × 1500 m and observed which presented the better AC result, assuming a no-uniform surface (Equation (3)). (ii) Adaptative window using SIMEC proposed to multispectral sensors by [33]. And (iii) the AWP-Inland Water adaptative window method. Methods (ii) and (iii) are described in the next section. All these approaches were used to retrieve the component ρ_{env} to solve for the adjacency effect in Equation (3).

2.5.1. SIMEC

SIMEC describes the recovery of the H_{Adj} from the NIR similarity spectrum. The method was developed for hyperspectral airborne data, being applied in aquatic environments only to multispectral sensors, such as the MSI data [18,29,33]. Its central assumption is that the water spectrum shape in the near-infrared region is known and invariant. The SIMEC method suggests a simple ratio in two near-infrared MSI bands with central wavelengths at 705 nm and 783 nm [29]. For each water pixel, the result is then iteratively compared to the water spectrum invariant shape range at 780 nm, as defined in [59]. The window size is determined when the ratio value satisfies the water invariant spectrum condition. This method has some restrictions on the near-infrared region's water signal that limit its extensive use. Highly turbid waters, macrophyte growth, intense algae blooms, or optically shallow waters have a near-infrared signal which differs from the water invariant spectrum. Under these conditions, the SIMEC method cannot be applied [18,33]. This study assumed that the water body regions used in the validation were free from these effects.

2.5.2. AWP-Inland Water

The Adaptative Window by Proportion applied to Inland Water, or AWP-Inland Water, is an empirical algorithm based on the occurrence of non-water targets within the window. It is expected that adjacency effect magnitude increases with increasing non-water targets around the target pixel [17,24] and that the window size needs to adapt to local conditions across the water body. This occurs because higher APSF weights are associated with non-water targets. For example, if larger window sizes are used for water pixels close to land, unrealistic adjacency effect magnitude may occur due to overestimating the adjacency contribution. On the other hand, smaller window sizes attributed to water pixels located far from the land can underestimate the adjacency effect. AWP-Inland Water minimizes these uncertainties by controlling the relationship between the distance of the water pixel from the land and the weight distribution of the APSF through the proportion of the targets within the H_{Adj}.

The AWP-Inland Water algorithm has three stages: (i) definition and calculation of the proportion of non-water targets within the window, (ii) building of the $W_{(0,1)}$ factor, and (iii) calculation of the adjacency effect contribution. Targets' proportions within the window were chosen empirically by defining different proportion ranges, which were then applied to estimate the adjacency effect in the water bodies (0–10%, 10–20%, 20–30%, 30–40%, and

40–50%). The best range was selected for each water body by comparing the adjacency corrected water reflectance with the in-situ reflectance data. These proportion values were calculated using the Modified Normalized Difference Water Index—MNDWI [60], employing a simple threshold less than or equal to 0.20 to mask the non-water targets. In the method, different window sizes were attributed for each pixel inside the water body. The spectral index was calculated to indicate the window size value (m × m pixels) referring to the desired proportion of non-water targets. From this value, the $W_{(0,1)}$ factor was generated. The binary factor $W_{(0,1)}$ is a matrix with elements 0 and 1, having the same size of the APSF weight matrix. Both the size of the $W_{(0,1)}$ factor matrix and the size of the weight matrix were fixed at 5 km, i.e., the maximum H_{Adj} was defined by default at this value. For the $W_{(0,1)}$ factor, the element equal to 1 occupied the matrix center up to the window size value referent to the proportion of non-water targets (m − 1 pixels) defined in stage (i). The goal was to control the APSF weight matrix growth through element-wise multiplication. The AWP-Inland Water method computed the contribution of the adjacency effect as follows:

$$\rho_{env}\left(B_i, i_0, j_0\right) = \left(\sum_{i=1}^{N}\sum_{j=1}^{N} U\left(B_i, \bar{r}(i,j), t\right)\right)^{-1} \cdot \sum_{i=1}^{N}\sum_{j=1}^{N} \hat{\rho}_w^{**}\left(B_i, i, j\right) \cdot U\left(B_i, \bar{r}(i,j), t\right) \quad (11)$$

where:

$$U\left(B_i, \bar{r}(i,j), t\right) = F\left(B_i, \bar{r}(i,j)\right) \cdot W_{(0,1)}(t) \quad (12)$$

Note that we added another weight U to the equation. It results from the iteration between the APSF weight array (F) and the $W_{(0,1)}$ factor array, and it depends on the proportion of non-water targets (t).

2.6. Statistical Analysis

The performance of the atmospheric correction and the adjacency effect correction in the remote sensing images was assessed using coefficient of determination—R^2, Mean Absolute Percentage Error—MAPE, Root Mean Square Error—RMSE, and average ratio analyses. The computed data were compared with ground measurements.

3. Results

3.1. Inversion Model (AOD_{550}^{**}) versus MODIS Aerosol in the Atmospheric Correction

Figure 3 introduces the results of the atmospheric correction performance. The optimal AOD_{550}^{**} value presented less uncertainty in the atmospheric correction when compared with the aerosol loadings extracted from the MODIS product. In the eutrophic waters, the difference between the performances of the AOD_{550} values was ~17% (MAPE). In contrast, that difference was ~25% and over 100% in the bright and dark waters, respectively (Figure 3a). The better performance of aerosol loading using the inversion model resulted from the matching of in-situ water signals with the observed TOA signals. Hence, in this optimal condition, the error sources related to the sensor calibration, atmosphere optical complexity, or assumptions adopted for the recovery of aerosol loading, as in the MODIS case [61], were smaller. In general, the AOD_{550} values based on MODIS (average ratio ~3) further underestimated the water reflectance throughout the wavelengths when compared to AOD_{550}^{**} (average ratio ~2) because their aerosol loadings were smaller than the AOD_{550} values extracted from the inversion model. AOD_{550}^{**} was up to two times greater than the aerosol optical depth at 550 nm based on MODIS (Appendix B, Table A1). Both aerosol products caused a significant distortion in the water reflectance at the near-infrared bands for all water types. Water absorption often affects the larger wavelengths, which implies lower reflectance values in this region of the spectrum. That pattern makes the near-infrared bands more sensitive to factors that mask the actual optical behavior of the water, such as the adjacency effect.

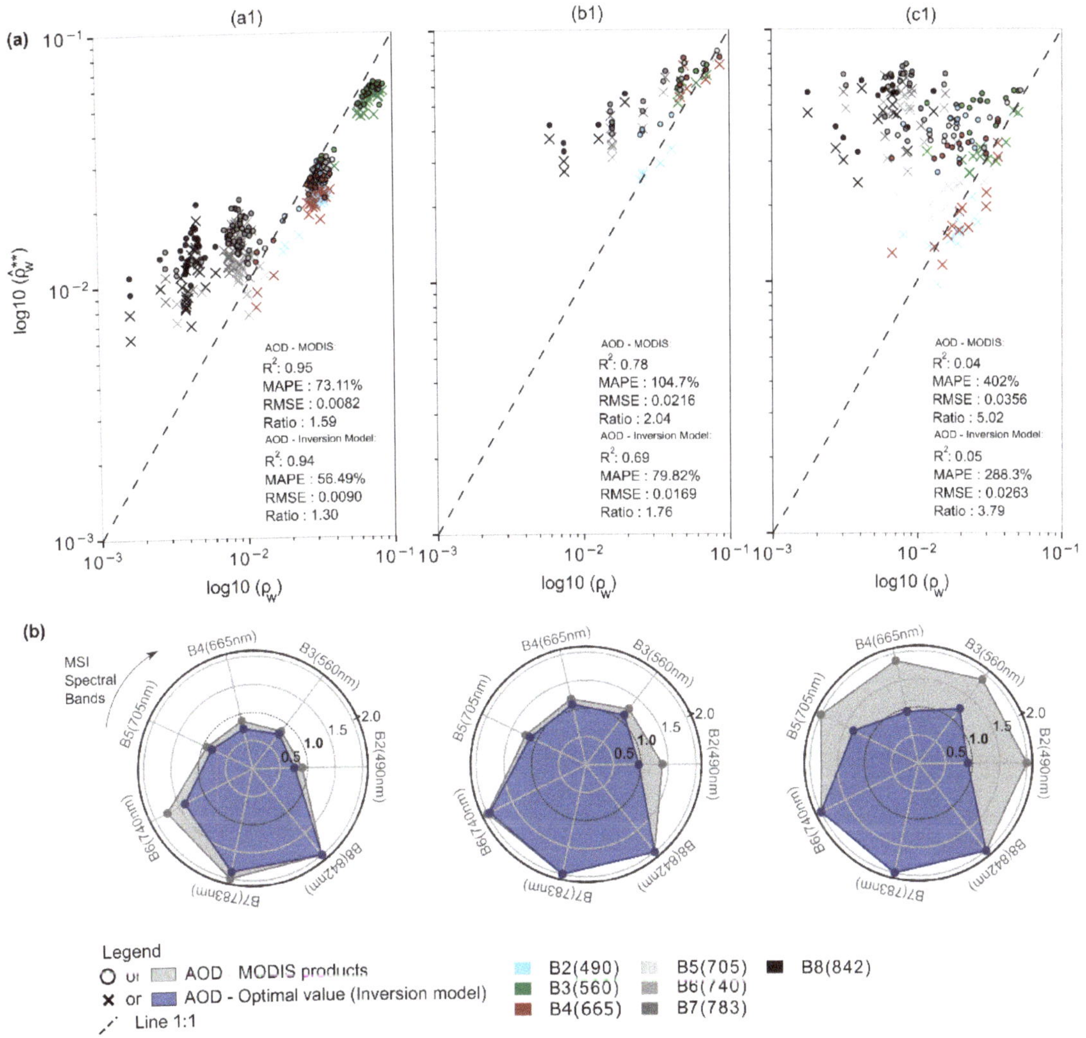

Figure 3. Atmospheric correction performance according to the source of the aerosol loading from the inversion model versus MODIS aerosol optical depth at 550 nm for (**a**) all MSI spectral bands and (**b**) the average ratio per MSI band. Atmospheric correction was performed for the different water optical types: (**a1**) eutrophic (N = 28), (**b1**) bright (N = 5), and (**c1**) dark (N = 13) waters.

In the case of bright and dark waters, the AOD_{550}^{**} produced a more adequate response to AC at shorter wavelengths than the MODIS aerosol loading (Figure 3b). In these environments, the water reflectance was estimated to be two times higher at 740 nm, 783 nm, and 842 nm for both aerosol scattering conditions. Unlike in the eutrophic waters, the MODIS aerosol loading produced slightly better results than those of AOD_{550}^{**} in the range from 443 nm to 705 nm. Little overcorrection of the water reflectance values (average ratio less than 1) occurred in these wavelengths. Differently from the other water bodies, the BIL reservoir (water body inserted in eutrophic water type) presented a low aerosol loading (AOD_{550}^{**} ~0.16 and AOD_{550}^{MD} ~0.10; see Appendix B, Table A1). It illustrated that the estimated aerosol loadings could not correct the water reflectance satisfactorily. This is partly explained by the clean atmosphere condition around the reservoir (small aerosol loading). Overall, the aerosol loadings extracted from MODIS worked better for the AC of inland waters under low AOD_{550} values (smaller than 0.2) than those with high values (larger than 0.3). This was attributed to the challenges of multiple scattering modeling during aerosol retrieval.

3.2. Range of the Adjacency Effect

The methods applied to retrieve the H_{Adj} generated different results (Figure 4). The adaptive methods, SIMEC and AWP-Inland Water, showed substantial differences located mainly close to the land cover. In general, the differences of H_{Adj} around the land–water boundary (or waterline) (distance smaller than 100 m) were up to three times higher than that of the regions far from the waterline (distance larger than 100 m). This difference was even more outstanding in the BRA, MUT, and CON lakes. The SIMEC approach, for instance, resulted in an H_{Adj} several times greater than those estimated by the AWP (average window ~400 m × 400 m) and fixed window (~350 m × 350 m) methods, including infinite H_{Adj} values (larger than 2000 m × 2000 m). SIMEC did not show any relationship with the distance to the waterline. In contrast, the range of the adjacency effect obtained from the AWP algorithm increased with increasing distance value. This relationship (H_{Adj} versus distance to the waterline) was highlighted in the larger (e.g., BIL) and wider (e.g., PIR, BRA, and MUT) water bodies.

Each water body showed an adequate H_{Adj} or proportion of non-water targets within the window for the fixed window and AWP methods, respectively. Despite their empirical nature, the results indicated that the aerosol loading (required for AC) provided information about the proportion of non-water targets needed for the AWP algorithm. Note that, for water bodies under heavy aerosol loadings (larger than 0.3), the desirable proportion of targets was 40–50%, while for lower aerosol loadings (smaller than 0.3), the proportion of non-water targets within the window was 20–30%. Overall, the range of the adjacency effect was higher for lower aerosol loadings. However, PIR presented high H_{Adj} values, despite its high aerosol loading (AOD_{550} ~0.34). In addition to the aerosol scattering effect, the water type also seemed to influence the size of the H_{Adj}. Comparing the PIR and MAM lakes, which were under similar atmospheric characteristics and aerosol scattering effects (Appendix B, Table A1), different H_{Adj} sizes were obtained, with the dark-water lake demanding a smaller H_{Adj} than the bright-water lake. This is explained by the decreased contrast between the water and the targets around the water body. As bright waters have higher reflectance values, more homogeneous surfaces and, consequently, lower adjacency contributions are expected. Conversely, the atmosphere scattering surpassed the reflectance of the adjacent target under high aerosol loadings, generating more significant adjacency effects. In these cases, larger H_{Adj} were needed to compute the existing adjacency effect values.

The MSI-derived water reflectance associated with the H_{Adj} estimation is displayed in Figures 5 and 6 for four MSI near-infrared bands (705, 740, 783, and 842 nm) and three optical water types. It is evident that the H_{Adj} difference effect produced from the three methods increased with increasing wavelength because the larger wavelengths had a sharp reflectance contrast between the water and various land targets (e.g., soil and forest). The AWP algorithm underestimated, significantly, the reflectance values of dark waters. At the same time, the SIMEC and the fixed window approaches in this water type caused an expressive overcorrection of the water reflectance beyond 740 nm.

In the case of dark waters, both the SIMEC and fixed window approaches showed a frequent negative retrieval of water reflectance (or invalid value) at the near-infrared wavelengths (average frequency ~56% and ~25%, respectively). At 842 nm, SIMEC produced up to twenty times more invalid values than AWP-Inland Water. Similarly, the fixed window method produced about ten times more negative results when compared to AWP. In another way, the eutrophic and bright waters presented a satisfactory agreement between the methods, especially between SIMEC and AWP. In these optical scenarios, the number of invalid results was small (~5%, on average). Again, negative values were high at 842 nm compared to other bands. For that band, the SIMEC and fixed window methods generated more significant inconsistencies in estimating water reflectance for eutrophic (~13%) and bright (~17%) waters, respectively.

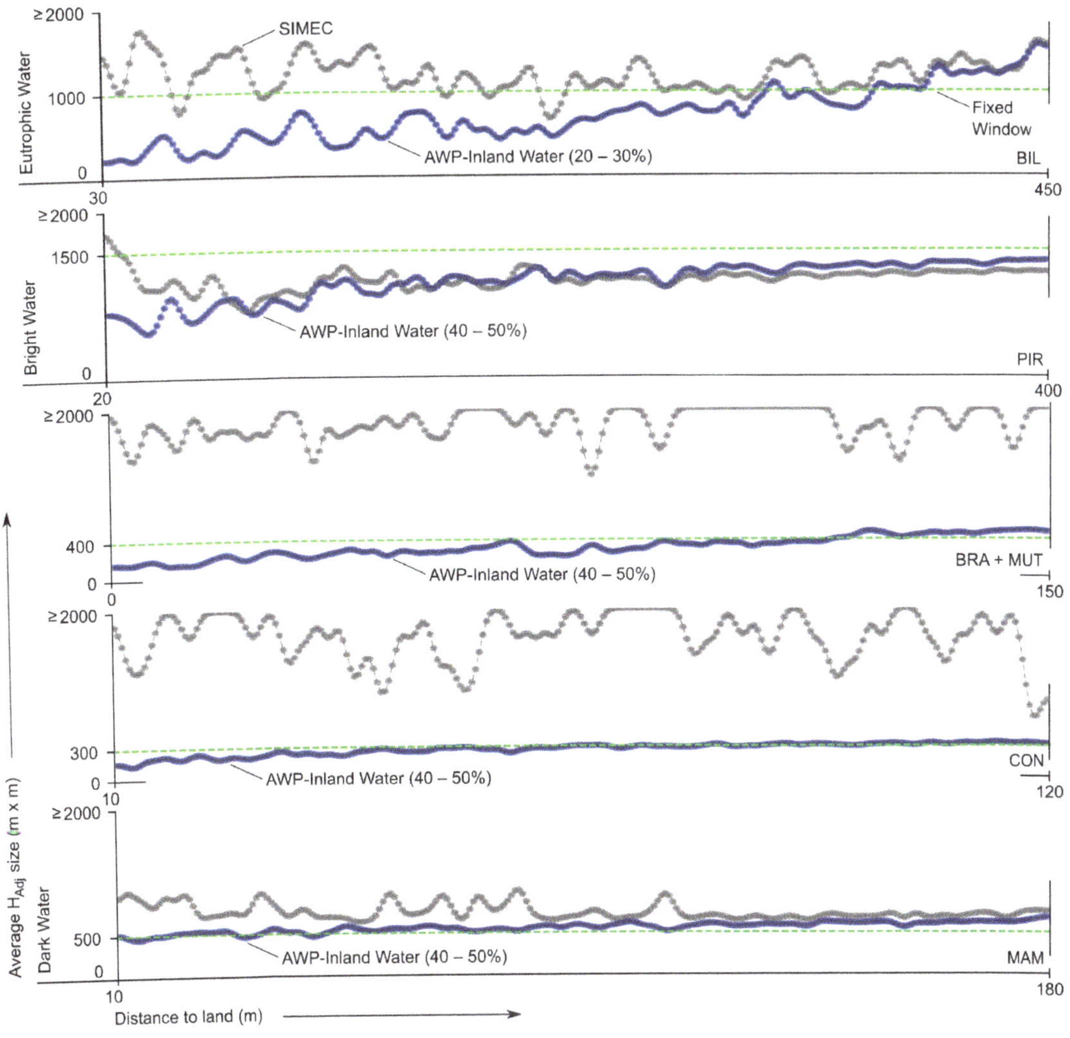

Figure 4. Differences in sizes of H_{Adj} obtained from the fixed and adaptive window approaches for the three selected water types. The analysis was performed with N = 250 pixels per water body. The pixels were randomly selected on the water bodies considering their linear distance from the waterline. The dashed green line refers to H_{Adj} based on the fixed window method, while the gray and blue markers refer to use of the SIMEC and AWP-Inland Water methods, respectively.

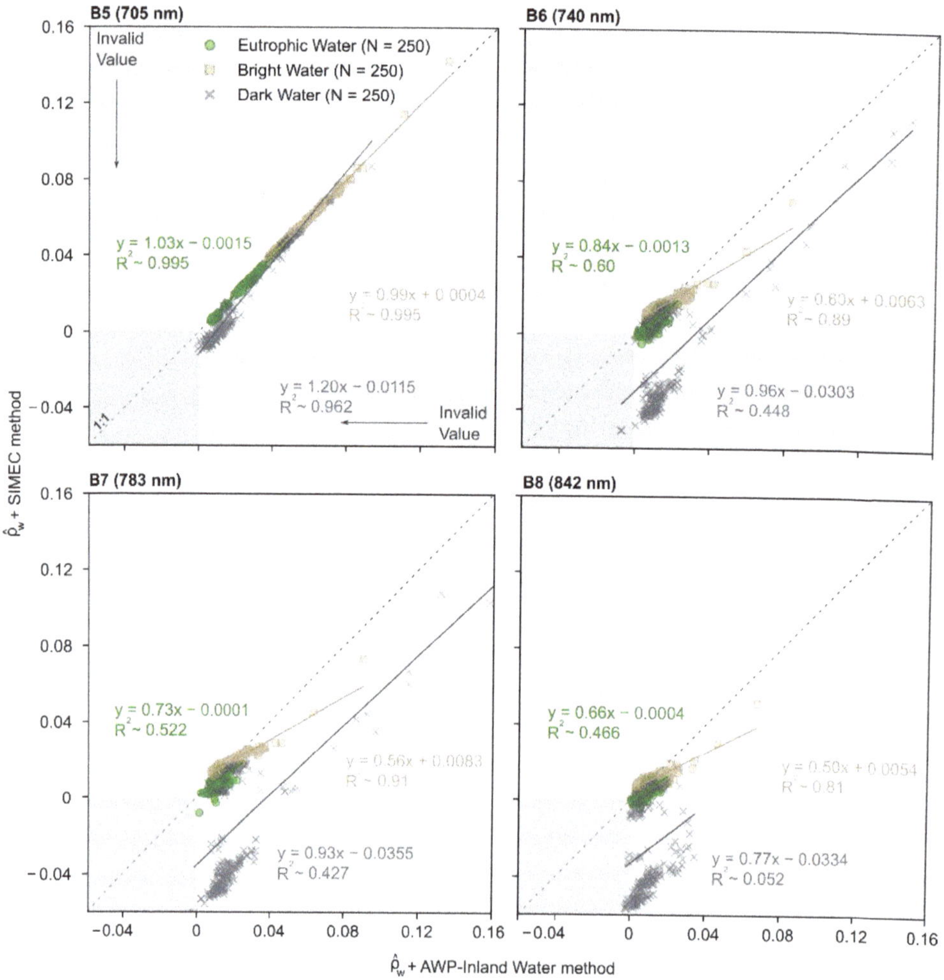

Figure 5. Water reflectance estimated from the MSI considering the range of the adjacency effect generated from SIMEC versus AWP-Inland Water. Analysis was performed for N = 250 pixels per water type. The shaded area (light gray) indicates the invalid value zone (negative values) of water reflectance resulting from AC, including the adjacency effect (Equation (3)).

3.3. *Adjacency Effect Correction*

In general, accurate observations of the satellite-derived water reflectance were obtained after adjacency effect correction for all water types (Figure 7). AWP-Inland Water, as well as SIMEC, demonstrated a good agreement between MSI and in-situ measurements of water reflectance for the eutrophic and bright waters (MAPE smaller than ~28%) (Figure 7a). In these environments, the difference between the performance of these two methods was minimal (~3%). In contrast, only AWP reached a more accurate water reflectance estimate in dark water environments (MAPE ~53%). The number of invalid results (i.e., negative water reflectance values) in the dark waters was remarkably high using SIMEC. In fact, the overcorrection for adjacency effects in the near-infrared wavelengths can leads to negative values due to the low water reflectance. The bright waters showed smaller uncertainties from adjacency effect correction when using the fixed window approach (MAPE ~16.55%).

However, as indicated in Section 3.2, this method often produced invalid results for these water types. This fact suggests that the small number of samples used to validate the adjacency effect correction may have limited a representative statistic in bright waters.

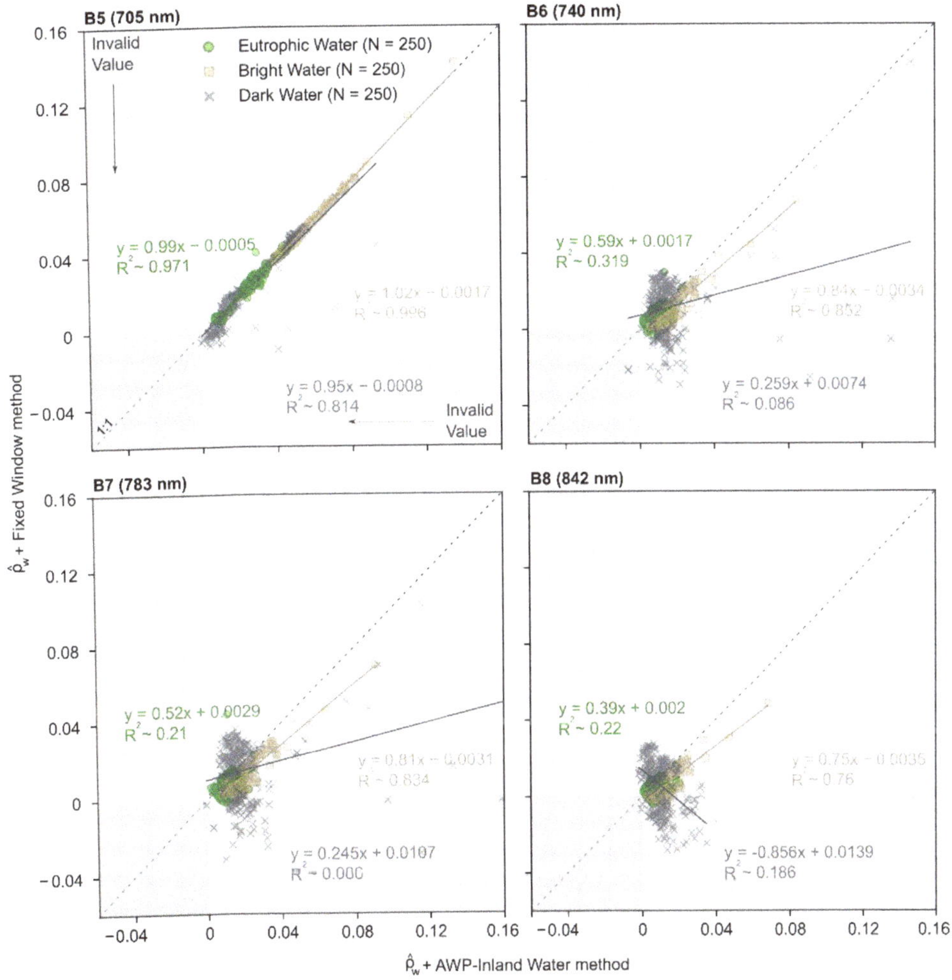

Figure 6. Water reflectance estimated from MSI considering the range of the adjacency effect generated from the fixed window versus AWP-Inland Water methods.

In eutrophic and dark waters, the fixed window approach applied to correct the adjacency effects caused underestimation of the reflectance values at 783 nm and 842 nm (average ratio more than ~1.3 and ~1.6, respectively) (Figure 7b). In the case of bright waters, its performance was superior to that of the other methods at 842 nm considering only the dataset available for validation. AWP slightly underestimated the water reflectance at 842 nm when compared to SIMEC in bright and eutrophic waters. The dark waters were more sensitive to the differences in the H_{Adj} size applied to correct the adjacency effects. SIMEC in all near-infrared wavelengths (from 705 nm to 842 nm) had a poor performance for correcting the target effects around these small lakes. On the other hand, recognizing the challenges of estimating the satellite-derived water reflectance of water bodies with very low reflectance under atmospheric complexity, AWP-Inland Water showed acceptable results, despite the poor results at 783 nm.

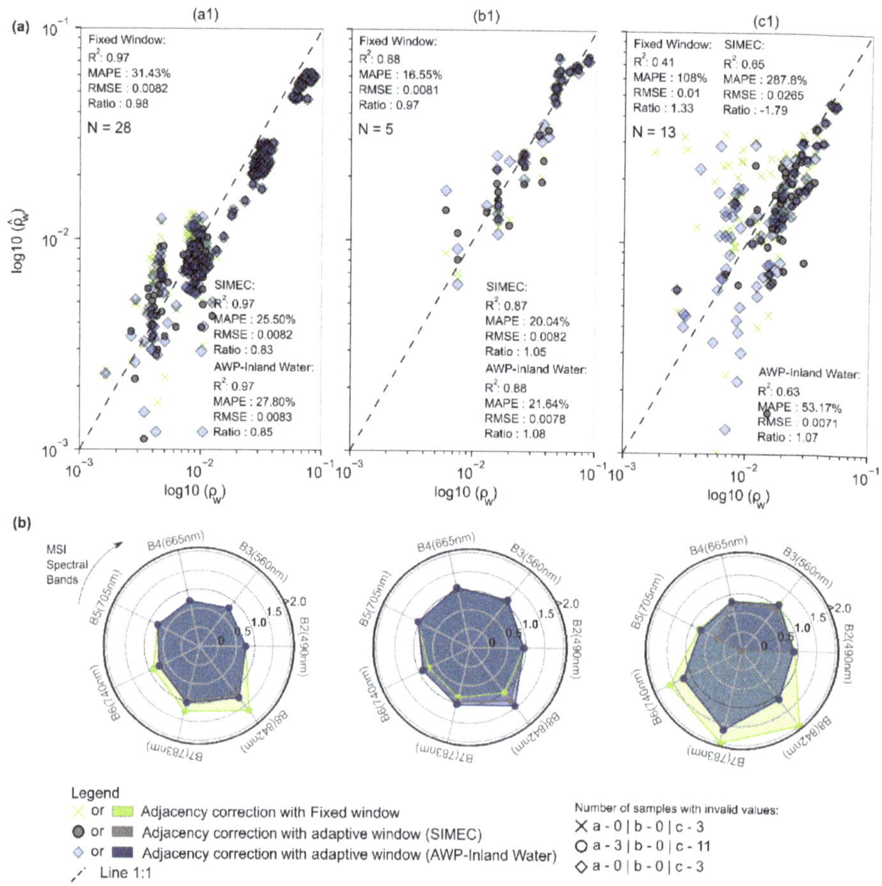

Figure 7. Performance assessment of adjacency effect correction: (**a**) all MSI spectral bands and (**b**) average ratio per MSI band. Adjacency effect correction was performed for the different water types: (**a1**) eutrophic, (**b1**) bright, and (**c1**) dark waters. We used three methods for determining H_{Adj}. Invalid results are not shown in (**a**). However, the nonpositive counts are highlighted in the legend.

Figure 8 shows an overall (i.e., average water spectrum plus its standard deviation) water spectrum shape comparison between MSI and in-situ measurements after adjacency effect correction performed using the fixed window, SIMEC, and AWP-Inland Water approaches to estimate the size of the H_{Adj} according to the water types. At first glance, the adjacency effect correction significantly improved the water spectrum shape across all four near-infrared bands of MSI. The adjacency effect mainly was reduced in this spectral domain and was practically nonexistent in the visible wavelengths. At the water surface, the near-infrared wavelengths usually are more impacted by targets neighboring the water body, as previously discussed. However, errors in estimating the range of the adjacency effect can also produce overcorrection in the visible spectral domain. The inaccurate performance of SIMEC in dark waters changed the water spectrum shape at 560 nm and 665 nm. Evidently, the AWP method showed a good agreement with the field spectra of the different water optical compositions. The physical method employed to correct the adjacency effect, associated with a good computation of H_{Adj}, allowed the correct estimation of the water reflectance under high aerosol loadings (in the case of small Amazon lakes) and lower aerosol loadings (in the case of BIL). If the adjacency effect is not corrected, the water reflectance can heavily impact the retrieval of OACs, especially for bio-optical algorithms that use larger wavelengths to infer about the water composition.

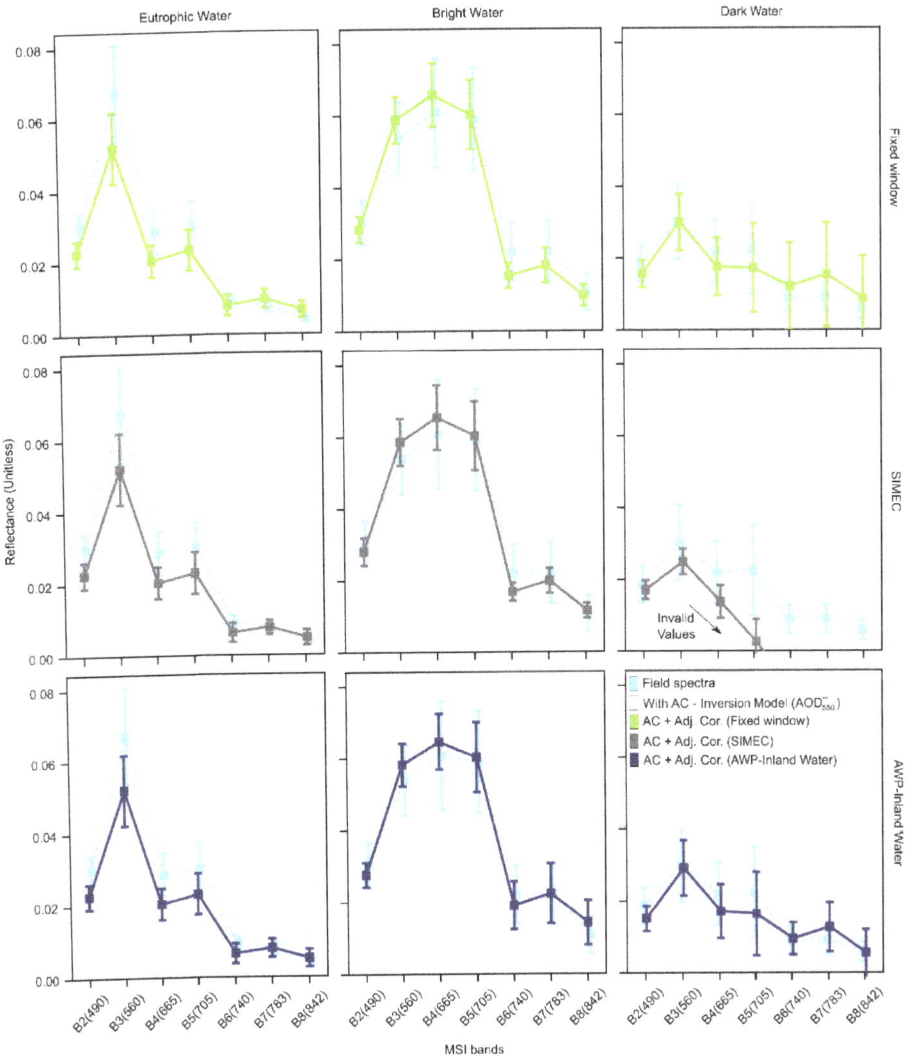

Figure 8. Differences of the water spectrum shape corrected for adjacency effects versus the field spectrum according to H_{Adj} methods and water types.

3.4. Adjacency Effect Influence on Water Bodies

The adjacency effect is highly dependent on the atmosphere scattering processes. Notably, its magnitude increased with increasing AOD_{550} parameters. For imagery acquired under lower aerosol loadings, as in the case of eutrophic waters, the adjacency effect contribution at the top-of-atmosphere was smaller than ~30% throughout the wavelengths (Figure 9). In comparison, in the bright and dark waters observed under high aerosol loadings, this contribution was ~20% at visible wavelengths, reaching up to ~80% in the near-infrared domain. The spectral differences of the adjacency effect depended on the target type surrounding the water body and the optical characteristics of the water itself. For example, at 560 nm, all water bodies presented a peak in adjacency contribution related to the vegetation cover around the water bodies. However, this phenomenon occurred in BIL due to vegetation around it, as well as algae present in the water column. Note that BIL's TOA contribution was approximately ~60% in this band. In the 740 nm–842 nm range, the adjacency contribution was up to ten times higher than that of the dark waters, justifying why this water type was drastically affected by the adjacency effects in our observations. On the other hand, this difference dropped substantially for other water types (~1.2 times and ~2.3 times more than the eutrophic and bright waters, respectively).

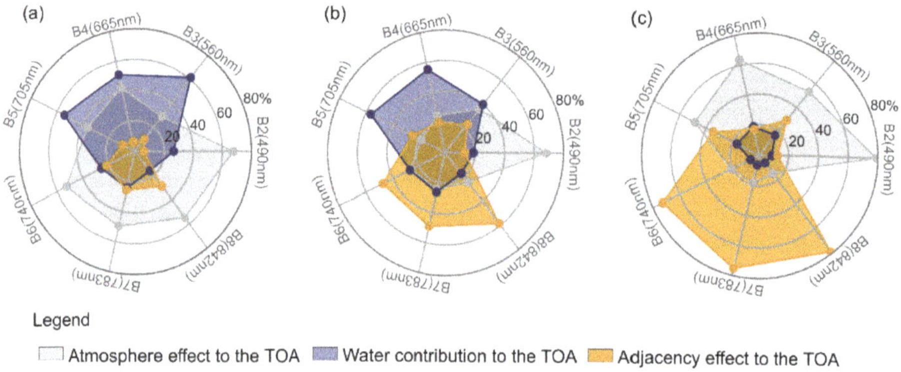

Figure 9. Contribution to the TOA of the reflectance values of water bodies, atmosphere, and adjacency effects in (**a**) eutrophic, (**b**) bright, and (**c**) dark waters. The TOA signal was modeled with 6SV using input data from Appendix B (Table A1). For the water contribution, we used in-situ measurements. Adjacency effect was estimated assuming the H_{Adj} described in Figure 4 (AWP-Inland Water algorithm).

By exploring each water body, the absolute adjacency effect reflectance (ρ_{adj}) estimated in BIL (maximum ρ_{adj} ~3%) was often smaller when compared to the PIR and MAM lakes (~6%) or Juruá River floodplain lakes (~12%) when using the MSI band at 842 nm as a reference (Figure 10a). Overall, the ρ_{adj} was more significant for locations close to the edges and narrow areas, and it was smaller at the center of the water bodies. However, in BIL, the maximum ρ_{adj} values occurred in areas with intense algal bloom. These events of algal bloom produced a significant increase of water reflectance at 842 nm, generating large contrast between the regions with and without algal blooms (Figure 10b). Along the $A \rightarrow A'$ transects, an inverse combination between the amount of ρ_{adj} and the difference of corrected and uncorrected reflectance in eutrophic and bright waters was shown. This also occurred due to heterogeneous surfaces generated from the variability in water composition. On the contrary, in the case of dark water lakes, these two factors were directly related, i.e., the reflectance differences increased with the increase of ρ_{adj}.

The correction of the adjacency effect varied spatially in response to these reflectance changes in the H_{Adj}. The transects showed that the smaller the water reflectance values, the more affected the water signal by the adjacency correction due to higher contrast with the neighborhood signal. For example, for waters with algal bloom or sediment dominance, the shorter spectral reflectance difference from the water and land targets can reduce the effect of surrounding targets on the water body. The water bodies experienced a reflectance decrease of up to −170% after the correction of the adjacency effect at 842 nm. The adjacency effect correction was larger in the small Amazon lakes (average value around −80%). For BIL, these results were even smaller (average value approximately −35%). In this system, the Pedra Branca, Rio Grande, and Rio Pequeno arms were more impacted compared to the other BIL's sections because they are very narrow and are less affected by algal bloom. Interestingly, the adjacency correction induced a slight increase in the water reflectance over the areas with intensive algal bloom (water reflectance difference less than ~6%), which is partially explained by higher target pixel reflectance values than the adjacent pixels.

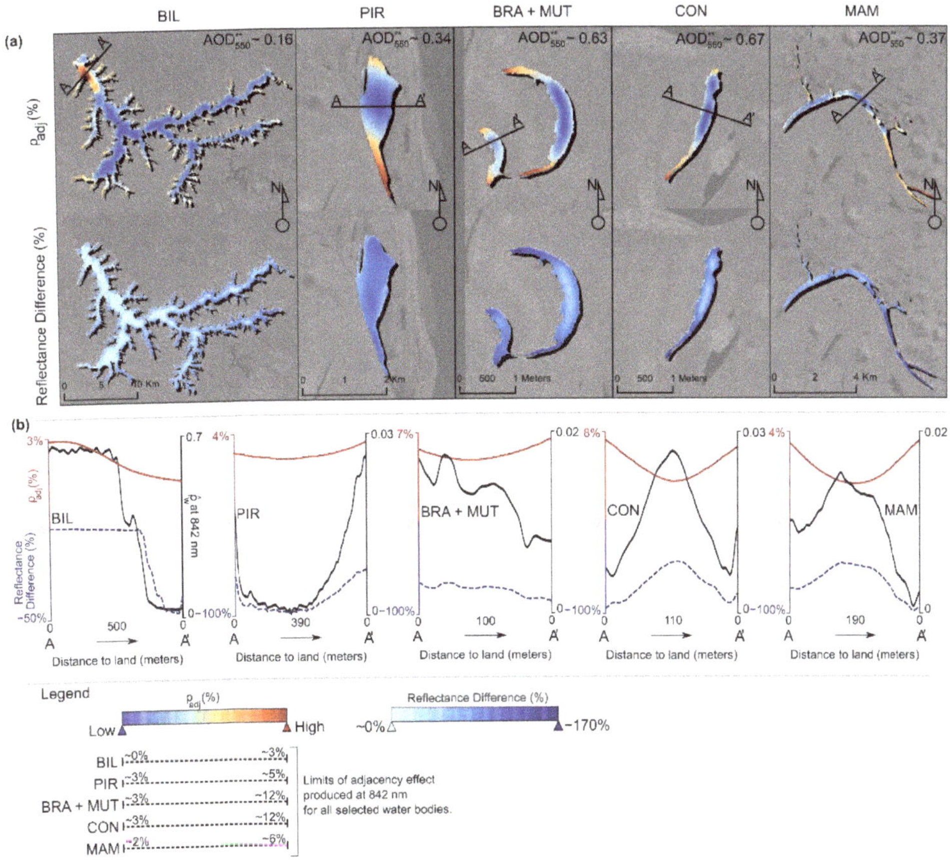

Figure 10. (**a**) Spatial distribution of the ρ_{adj} (in %) and MSI-derived water reflectance difference (in %) before and after adjacency correction at 842 nm. (**b**) The transect of ρ_{adj} (%), reflectance difference (%), and water reflectance in the $A \rightarrow A'$ setting along the water body. In (**a**), the ρ_{adj} was obtained by multiplying the average reflectance of the environment and the ratio between the diffuse atmosphere transmission by the total (upward) ($\rho_{adj}(B_i) = \rho_{env} \times t^{\uparrow}_{dif}/T^{\uparrow}$) (see Equation (4) in [62]). B_i refers to the MSI band at 842 nm. Here, the H_{Adj} was defined using the fixed window approach (Figure 4). The reflectance difference was negative when the corrected water reflectance value was smaller than the uncorrected water reflectance value.

3.5. Sensitivity of Adjacency Effect at the TOA

Here, the sensitivity of the adjacency effect to the proportion of non-water targets within the H_{Adj}, land-cover type around the water body, water optical type, aerosol loading, and aerosol model is investigated. For that, we simulated the adjacency effect contribution at the TOA, assuming fixed atmosphere characteristics to run the 6SV model (Appendix B, Table A2). The sensitivity assessment to the aerosol model was based on the standard models embedded in 6SV (e.g., continental and biomass burning smoke). All runs considered only a water pixel distant 0 m from the waterline. The results in terms of the proportion of non-water targets within the window showed an increase in the adjacency effect magnitude in response to the proportion of non-water targets (Figure 11). For example, at 842 nm, the difference between the adjacency effects of the 40–50% (~33%) and 0–10% (~2%) proportion was ~31%, under a clear atmosphere (AOD_{550} = 0.1). Already in a condition with heavy aerosol loadings (AOD_{550} = 0.5), this difference was even greater at ~53%. The water was more affected by the adjacency effect under high aerosol loadings. Note that, for higher AOD_{550} values, the adjacency effect was several times higher than the water signal at the TOA (up to ~5 times larger for the proportion of 40–50% at 740–842 nm wavelengths). The difference between the adjacency effect and water contribution was tiny (up to ~1.1 times larger) under lower aerosol loadings. The cover type around the water body showed greater differences in the adjacency contributions at 560–705 nm wavelengths due to changes in the spectral response between the bare soil and vegetation targets. Specifically, when varying the land-cover type for bare soil and vegetation, the water body was less impacted by the adjacency effect changes generated from these surrounding targets at larger wavelengths.

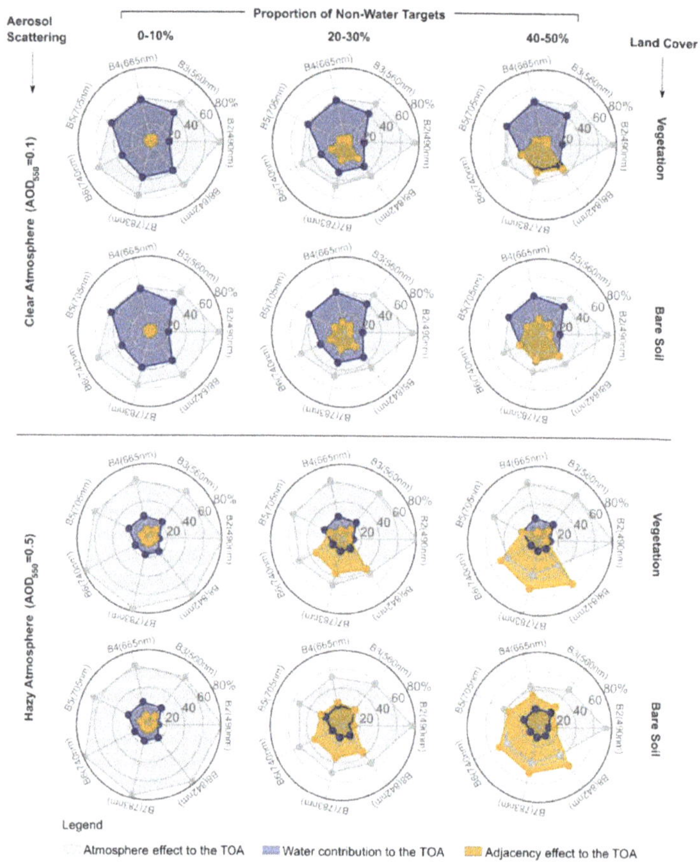

Figure 11. Values of adjacency effect contribution at the TOA (ρ_{adj}/ρ_{TOA} in %) along visible and near-infrared wavelengths considering different proportions of non-water targets within the window, aerosol loadings, and land cover. In the simulations, in-situ measurements of dark water reflectance (see Figure 2) were used as reference. To compose the atmospheric characteristics, we assumed the continental aerosol model available from 6SV. Finally, the adjacency effect was estimated by the AWP-Inland Water approach.

The aerosol models caused different adjacency effect contributions at the TOA (Figure 12). In general, the biomass burning smoke aerosol model produced a smaller adjacency effect when compared to the continental model. The average difference between the adjacency effects generated by the two models was about 10% in the 705–842 nm spectral range. This is explained by the increased radiation absorption from the type of aerosol particle inserted in the biomass burning smoke model. The adjacency effects showed a slight sensitivity to water type at the shorter wavelengths. In this case, the adjacency contribution increased with the increasing water reflectance. The difference in the adjacency effects produced from the variation of dark and bright waters was ~3%.

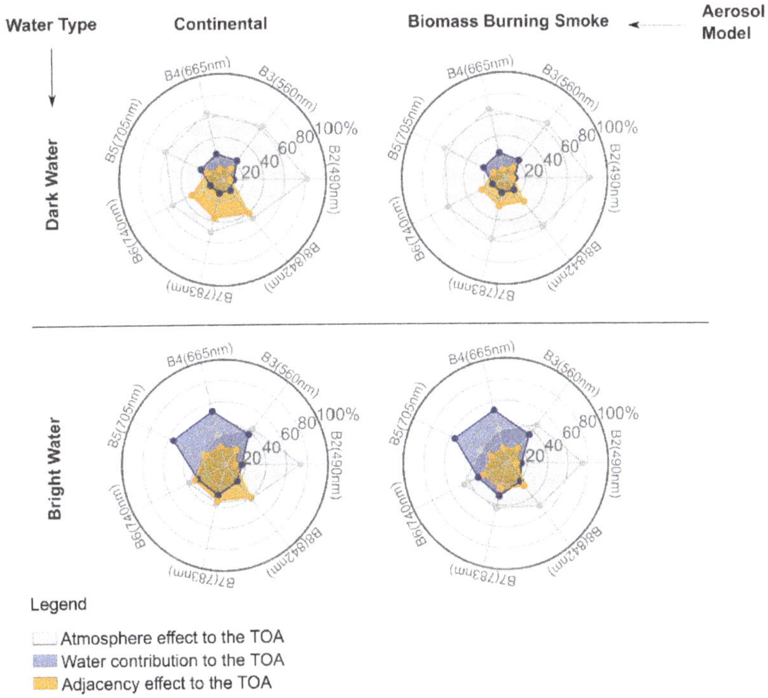

Figure 12. Values of adjacency effect contribution at the TOA (ρ_{adj}/ρ_{TOA} in %) along visible and near-infrared wavelengths considering two aerosol models (continental and biomass burning smoke) and two water types. Here, we used the proportion of non-water targets of 40–50% to define the size of the H_{Adj} when modeling the adjacency effect. The simulation adopted a hazy atmosphere (AOD_{550} = 0.5) and vegetation as the surrounding target.

4. Discussion

4.1. Aerosol and Atmospheric Correction

The recovery of aerosol loading over inland waters has a crucial role in accurately correcting the atmospheric scattering effect on satellite images, and errors in its properties affect the water reflectance measurement. The aerosol loadings extracted from MODIS did not perform well for high aerosol loadings (more than 0.3) compared to the AOD_{550}^{**} based on the inversion model (see Figure 3). Some limitations of the MODIS aerosol product are related to sensor characteristics, surface assumptions, and the aerosol model [61,63,64]. In general, the errors observed for a clear atmosphere occur due to the assumptions adopted for the surface, and for a hazy atmosphere, the errors respond to the aerosol model assumptions [61]. Thus, even though the forest areas surrounding the Amazon floodplain lakes are well-correlated with the MODIS aerosol product (R ~0.88) [65], the atmospheric aerosol complexity surpasses the surface influence due to the high aerosol loadings [66].

Furthermore, the type of aerosol model implemented is another uncertainty source in atmospheric correction over inland waters. Therefore, model selection is crucial to represent the aerosol characteristics in the water body region, and when this is incorrect, errors are introduced in satellite-derived water reflectance under high aerosol loadings [28]. All these inconsistencies associated with AOD_{550} recovery did not stand out in the inversion model, potentially because it estimated the aerosol values in response to the best observation of water reflectance (i.e., field data). Though the initial results are promising, the aerosol loading inversion method requires in-situ water reflectance, and when the method assumptions are not completely fulfilled (e.g., TOA signal free from glint or adjacency contamination), it does not work.

The Amazon region has several aerosol sources, such as biogenic from the rainforest (e.g., fungal spores and volatile organic compounds), black carbon particles from biomass burning, and urban aerosol [67–69]. Those sources have distinct characteristics in wet and dry seasons [67,70], altering aerosol properties [71]. Consequently, the variability of aerosol effects (composition + aerosol load) in this region may not be fully represented by the default aerosol models available for use [72], limiting the application of standard procedures for aerosol recovery. For example, Flores Júnior et al. [73] showed a fair amount of error in OLCI blue spectral bands (more than 100% at 400 nm–490 nm) during AC with the MODIS aerosol product, impacting the use of image-derived reflectance for the retrieval of inherent optical properties. In addition to problems related to the aerosol loading recovery for inland waters, the different water reflectance values displayed over the larger wavelengths can introduce further uncertainty into OACs estimation [19,74]. These wavelengths are strongly affected by gas absorption (e.g., water vapor and oxygen at 700 nm–800 nm) [62] and by the photons of energy reflected from surfaces around the water body (so-called adjacency effect) [17,33]. The adjacency effect depends on the atmosphere scattering conditions and, therefore, is well-correlated with the aerosol optical depth [24]. Adjacency effect correction is typically needed in small water bodies and high aerosol loadings [12], limiting, when neglected, the use of remote sensing data at near-infrared wavelengths for water quality application [10]. Although significant efforts have been made to improve the performance of the atmospheric correction algorithms in aquatic ecosystems over time [75], our results indicated that there are still unresolved challenges for inland waters. In complex atmospheric aerosol conditions, careful estimation of aerosol optical depth in these environments using the inversion model is particularly pertinent to achieve the required accuracy of water reflectance for water quality mapping.

4.2. Estimation of the H_{Adj} over Inland Waters

Three strategies to recover the size of the H_{Adj} were assessed in this study. The estimation of H_{Adj} is essential for a more accurate correction of adjacency effects on inland waters. However, the complex dependence of the range size on multiple factors (e.g., atmosphere scattering, viewing and illumination geometry, spectral characteristics of the surrounding target, and characteristics of satellite sensors) seems to limit a feasible estimative of H_{Adj} for correcting adjacency effects. There is no precise formula available to define the size of the H_{Adj}. Previous studies have shown that the adjacency effect influences in the order of tens of kilometers from the coastline over coastal waters [17,34]. In contrast, values in good agreement with the aerosol scale height in the atmosphere (between 0.5 and 1.0 km) have generally been used for land and water applications [25,27,28]. This study computed an H_{Adj} between 0.1 and 2 km, considering all the methods employed to recover the range of the adjacency effect. The three strategies investigated showed a vast difference concerning the estimation of H_{Adj}. For example, the average difference between SIMEC and AWP-Inland Water in dark-water lakes was ~0.8 km. In contrast, this difference was smaller for the fixed window and AWP methods (average difference ~0.06 km). These results strongly affected the MSI-derived water reflectance, especially at the near-infrared wavelengths. As previously discussed, these wavelengths refer to the spectral domain with the more significant adjacency contribution on inland waters. They are commonly related to an increase in land reflectance and a decrease in water reflectance. The resulting estimates of H_{Adj} demonstrated a linear relationship with the aerosol optical depth at 550 nm, where the size of the H_{Adj} decreased with increasing aerosol loading observed on water bodies. A similar pattern was reported in [27]. The adjacency effect receives a greater contribution from distant targets under relatively small aerosol loading (less than 0.2). On the other hand, if the aerosol loading is high (more than 0.3), the adjacency effect is more affected by the water and its neighbors due to decreasing atmosphere transmittance and the increase in forward scattering (related to aerosol particles) [24]. Differently, for higher aerosol loading (AOD_{550} of 0.3 to 0.5; see Figure 5 in [76]) combined with a more elevated surface reflectance, the surface reflectance exercises an effect on the size of the H_{Adj}, modifying the relationship between the H_{Adj} and aerosol loading, as we observed in the case of bright water.

Generally, our approach (AWP-Inland Water) had the best performance in determining the water reflectance, particularly in dark water environments (see Figure 7). Over inland waters, especially small lakes and reservoirs, the water body geometry (e.g., shape and size of water body) is often variable. Thus, a dynamic and challenging relationship between the water pixels and their surrounding targets is established. This configuration did not benefit the use of the fixed window method for inland waters, since it tended to over- and under-correct the reflectance of water pixels near and far from the waterline, respectively. The improved correlation between MSI and in-situ measurements of water reflectance shown by AWP-Inland Water indicated that the iterative changes in the window sizes explain the best performance of the method. The AWP-Inland Water method fulfills the essential relationship between window size and distance from the land (see Figure 4) by considering what is inside the window (water or non-water targets) to compose the adjacency effect magnitude. However, the proportion of non-water targets within the window can be variable for each water body in response to the adjacency effect. The relationship between the proportion of targets versus aerosol loading determined in this study can guide its definition in future AWP applications. In comparison to the fixed window and AWP methods, SIMEC demonstrated a good performance in the eutrophic and bright waters but a deficient performance in the dark waters. SIMEC often generated infinite H_{Adj} values (larger than 2 km $\times$ 2 km) and, consequently, invalid results (see Figure 5). The dark waters were the most affected by the production of invalid results. As previously discussed in Section 2.5.1., in conditions where the NIR similarity spectrum assumptions are not satisfied, the SIMEC approach may not work [18,33]. Considering that the similarity spectrum was designed for turbid and coastal waters [59], this may prevent its use for inland waters, particularly in dark waters.

4.3. Influence of Adjacency Effect on Water Reflectance Data

The relationship between the magnitude of the adjacency effect versus aerosol scattering is well-discussed [24]. In our case, it is evident that high aerosol loadings favored the occurrence of the adjacency effect. In the eutrophic waters observed under lower aerosol loadings, the water reflectance surpassed the adjacency effect at the top-of-atmosphere for the whole spectral range (see Figure 9). On the other hand, in the small lakes, the adjacency effect was more remarkable. In agreement with these results, a previous study indicated the sensitivity of the small Amazon floodplain lakes to the adjacency effects [12]. These environments maximize adjacency effects due to atmospheric aerosol complexity (Section 4.1), lake size and shape, canopy stand, the vigor of the surrounding forest, and water optical type. Note that the proportional contribution of the effect was more significant for the dark-water lakes than for the bright-water lake. This is explained by the lower reflectance of dark waters due to water absorption and colored dissolved organic matter. The impact of scattered radiation from surrounding targets was significantly larger closer to waterline. However, in conditions of algal blooms or high sediment loadings, the water body pixels close to these events were impacted by adjacency correction due to the contrast between spectral signatures caused by the different concentrations of water optical components (see Figure 10b). The algal blooms and high sediment loadings could significantly increase water reflectance in the near-infrared domain, making the water reflectance similar to that of land targets. This explains why the pixels with higher water reflectance values were less affected by adjacency correction, as depicted in Figure 10b. When the water reflectance is much higher (e.g., in the case of an intense algal bloom in the near-infrared region) than that of its neighboring pixels, the adjacency effect leads to a decrease in the water reflectance value. Note that the adjacency effect is a flux of photons produced by atmospheric scattering directed from bright to dark targets [77]. Then, on no-uniform surfaces, this effect causes a decrease in photons on high-reflectivity surfaces (e.g., bloom areas) as a function of the low-reflectivity surfaces (e.g., areas around the blooms). In this way, adjacency effect correction compensates for the photons lost by the bright targets, i.e., it removes the incremented photons on dark surfaces and returns them to bright surfaces. Hence, small increases may occur in water reflectance when corrected for adjacency effects, giving positive differences between the corrected and uncorrected MSI-image reflectance, as reported here. This behavior is frequently observed in the adjacency correction of surfaces with a high contrast in coverage [27,28,30].

4.4. Sensitivity and Challenges of Adjacency Effect

The adjacency effect magnitude has been little investigated in applications with inland waters. Based on this study, the main factors influencing the increase or decrease of adjacency effects over inland waters were the aerosol loading, aerosol model, land-cover type, and H_{Adj}. Indeed, heavy aerosol loadings and higher atmospheric scattering conditions (as in the continental aerosol model) maximized the adjacency effect, since it originates from the atmospheric scattering [24]. Thus, the results indicated that it may not be necessary to implement adjacency effect correction for cases of very low AOD_{550} (smaller than 0.1) (see Figure 11). Furthermore, the land-cover type around water bodies also influences the adjacency contributions. As the reflectance of the surrounding targets increases, so did the adjacency effect [17,24]. The effect related to bare soil cover was more significant in the visible domain compared to vegetation cover. However, these two covers produced a minimal difference in the adjacency contributions at larger wavelengths due to their reflectance similarity. Regarding water type (Figure 12), the adjacency effect was less sensitive to the variation of water optical composition (~3% at the visible domain). A previous study has shown that this difference depends on the type of land cover around water bodies [17]. The changes in the adjacency contribution at the TOA caused by water type are more significant when the land targets had low reflectance due to the importance given to water reflectance in modeling the adjacency effect. In addition, other factors that may influence the adjacency effect have been reported in the literature, such as viewing and illumination geometry, satellite spatial resolution, and glint contributions [34,58,78].

The physical approach [15,21] performed well in correcting the adjacency effects over inland waters. However, the strong sensitivity of the method to the size of H_{Adj}, as remarked in our study, may limit an accurate quantification of the adjacency effect in real applications. Additionally, the APSF weights defined from the molecular and aerosol scattering effects (see Equation (9)) also seem to influence the adjacency contribution. Sei [22] showed that the increase of APSF weight produced much larger ranges of adjacency effect because the targets around the water pixel received greater importance compared to lower APSF weights. Comparing the weight approach with a methodology based on three-dimensional radiative transfer simulations applied over coastal waters, Bulgarelli and Zibordi [17] reported an increase of adjacency effects close to land (~70%) from the use of APSF weights to compose the effect magnitude. Even though adjacency effect determination requires highly complex approaches and the physical method has its limitations, our positive experience with this method and the AWP method in estimating H_{Adj} opened a feasible way to correct adjacency effects over inland waters.

5. Conclusions

This paper assessed the feasibility of a physical approach based on APSF to correct the adjacency effect in medium spatial resolution satellite imagery on small lakes surrounded by dense forest cover and a large urban water reservoir considering variable aerosol loadings. This study showed that an iterative inversion model minimized the limitations related to MODIS-derived aerosol loadings and surpassed other sources of uncertainty (e.g., the aerosol model). This careful estimation of aerosol loading is proper in regions of complex atmospheric aerosol conditions (e.g., hazy atmospheric conditions), where the uncertainty of MODIS aerosol estimates is higher, or when there are no aerosol ground-based stations (e.g., AEORONET) around the interest area, as in the case of our study areas.

Adjacency effects impact the reflectance of inland water, and the application of physical methods can remove them from satellite imagery with various degrees of success. In general, none of the empirical approaches used to determine the H_{Adj} presented an outstanding performance for all the selected water types. The three approaches indicated a better recovery of water reflectance in eutrophic and bright waters. Regarding dark waters, only AWP-Inland Water (MAPE ~53%) exhibited improvements, partly because: (i) this water type was outside of the range suggested by SIMEC (~289%) and (ii) the water body geometry changes (e.g., shape and size of water bodies) decreased the performance of the fixed window approach (~108%). In addition, the combination of low water reflectance (e.g., dark waters) and higher adjacency contribution made it challenging to estimate water reflectance accurately. As shown in Figure 9, the adjacency contribution at the TOA exceeded the dark water contribution in such a way that the sensitivity of these waters to the adjacency effect was high. Thus, defining a proper H_{Adj} is crucial for the best adjacency correction from the physical approach and is very challenging due to the factors that can impact it (e.g., land-cover type around the water body and the distance of water pixels from land cover). Even though the AWP-Inland Water method exhibited acceptable results, there are limitations for operational applications, such as defining the optimal proportion of non-water targets, which may vary according to the magnitude of adjacency effect and the increase in running time due to its pixel-by-pixel iteration.

Over inland waters, the adjacency problem was maximized for small water bodies, higher aerosol loadings (more than 0.1), and dark waters (water reflectance smaller than 4%). On the contrary, our results showed that under lower aerosol loadings (smaller than 0.1) the adjacency effects were very small. The adjacency effect contribution at the TOA (see Figure 11), taking into account a critical scenario (i.e., pixels close to land, dark waters, and a high proportion of non-water targets within the window) was very close to water contribution. In this sense, under clear atmospheric conditions associated with larger water bodies is possible neglecting the adjacency effect in medium spatial resolution MSI imagery. Finally, this paper encourages the application and validation of physical methods for the correction of adjacency effects on inland waters, as well as the development and validation of approaches to more adequately determine H_{Adj} size.

Author Contributions: Conceptualization, R.S.P., V.S.M. and E.M.L.M.N.; methodology and application, R.S.P. and V.S.M.; field data validation, R.S.P., F.N.B. and C.C.F.B.; writing—original draft preparation, R.S.P.; writing—review and editing, V.S.M., E.M.L.M.N., C.C.F.B., L.A.S.d.C. and F.N.B. All authors have read and agreed to the published version of the manuscript.

Funding: This research was funded through the 2017–2018 Belmont Forum and BiodivERsA joint call for research proposals under the BiodivScen ERA-Net COFUND program, and by funding from the following organizations: the French National Research Agency (ANR), the São Paulo Research Foundation (FAPESP), the National Science Foundation (NSF), the Research Council of Norway, the German Federal Ministry of Education and Research (BMBF), and the Foundation BNP Paribas through SABERES Project. The field campaigns were sponsored by the MAS-BNDES (project code 1022114003005); the National Council for Scientific and Technological Development (CNPq) (project code 461469/2014-6); the FAPESP projects (2014/23903-9, 2018/12083-1, 2020/14613-8); and the Coordenação de Aperfeiçoamento de Pessoal de Nível Superior—Brasil (CAPES) (Finance Code 001).

Data Availability Statement: In-situ water reflectance data are available from the authors upon reasonable request. MSI Sentinel-2 (A and B) images (Level 1C) are freely available from the Copernicus Open Access Hub website (https://scihub.copernicus.eu/ (accessed on 10 December 2021)).

Acknowledgments: The authors would like to thank the Instrumentation Laboratory for Aquatic Systems (LabISA)/INPE for optical instrument support used for the in-situ measurements. We also thank ESA for the available Sentinel-2 data.

Conflicts of Interest: The authors declare no conflict of interest.

Appendix A

Accurate estimation of the AOD_{550} is important to model and remove the atmospheric scattering and adjacency contribution from the images. The potential use of the AOD_{550} recovered from the image-based approach in atmospheric correction has been demonstrated in several studies [7,57,79]. Briefly, this approach explores the difference between the surface reflectance (no aerosol) and the TOA reflectance (aerosol) of a target contained in the image and with known spectral responses at specific wavelengths to estimate the remote sensing aerosol [79]. Applications using the vegetation as reference targets for obtaining the AOD_{550} have indicated poor performances in the atmospheric correction of aquatic environments, generally overestimating the water reflectance [11,12]. To accommodate these inconsistences, we used MSI reflectance simulated from in-situ data as reference. The idea was to obtain an optimal AOD_{550} value that allowed the matching of the water signal observed in-situ with that at the top-of-atmosphere based on the inversion of a radiative transfer equation. Three assumptions were needed to apply the method: (i) the in-situ water reflectance in the deep-blue waveband (at 443 nm) is known; (ii) the difference between the water reflectance and the TOA is due to the atmosphere content, i.e., the TOA signal is free from other factors, such as sun and sky glint, bottom, and adjacency effect; and (iii) for overwater, the retrieved AOD_{550} value is assumed spatially invariant.

In this paper, an optimal AOD_{550} value was estimated using an iterative process based on a bisection method. Commonly, this method is applied to solve root-finding problems of mathematical equations. A bisection method uses an initial interval, where the equation's root is contained, that is iteratively divided into subintervals equally spaced in such a way that there is a better approximation for the value of interest [80]. The iterative process used for the AOD_{550} retrieval can be written as:

$$f\left(AOD_{550}^{(k)}, B_i\right) = \left|\hat{\rho}_w^{**}\left(AOD_{550}^{(k)}, B_i\right) - \rho_w^{*}(B_i)\right| \quad (A1)$$

where the function f expresses the difference between the corrected reflectance of atmospheric effect ($\hat{\rho}_w^{**}$) (Equation (7)) and in-situ observed reflectance ρ_w^{*} (Equation (2)) for a given $AOD_{550}^{(k)}$ value and spectral band B_i. k refers to the iteration number. In the zero-order approximation ($k = 0$), the $AOD_{550}^{(0)}$ value was calculated using the midpoint of the range $[AOD_{550}^{(min)}, AOD_{550}^{(max)}]$. This range included the optimal AOD_{550} value, and its limits were defined at 0 and 2:

$$AOD_{550}^{(k)} = \frac{AOD_{550}^{(min)(k)} + AOD_{550}^{(max)(k)}}{2} \quad (A2)$$

In the first iteration, the interval initial was divided into two halves, such as $[AOD_{550}^{(min)}, AOD_{550}^{(0)}]$ and $[AOD_{550}^{(0)}, AOD_{550}^{(max)}]$. To know which half of the interval in which the optimal AOD_{550} value is contained, a simple observation of function sign f at the midpoint is performed. If $f\left(AOD_{550}^{(min)}\right) \cdot f\left(AOD_{550}^{(0)}\right) < 0$, the optimal AOD_{550} value is contained in the range of $[AOD_{550}^{(min)}, AOD_{550}^{(0)}]$. On the contrary, if $f\left(AOD_{550}^{(0)}\right) \cdot f\left(AOD_{550}^{(max)}\right) < 0$, it is inserted in the range of $[AOD_{550}^{(0)}, AOD_{550}^{(max)}]$. This process was repeated up to the k-order approximation. The convergence condition of the iterative process is $f\left(AOD_{550}^{(k)}, B_i\right) \leq 0.0001$. These iterative steps contain only simple mathematical operations. Therefore, the estimative of an optimal AOD_{550} value for atmospheric correction of the images is not very time-consuming. The total time spent to recover the AOD_{550} parameters using a single field sample and a spectral band was around 600 s. Some strategies can be adopted to reduce the total computation time, such as parallel operations, reducing the size of the initial interval, and increasing the tolerance value of the convergence condition.

All in-situ samples were used in the inversion model of AOD_{550}, except for BIL, where only the samples collected far from the reservoir border were applied. In each water body, the optimal AOD_{550} value was recovered using the MSI deep blue waveband and the average of the aerosol loading values observed for each in-situ sample. The estimative of the optimal AOD_{550} value over water bodies was derived from the following equation:

$$AOD^{**}_{550}(w, B_{db}) = \frac{1}{N}\sum_{j=1}^{N} AOD^{j}_{550}(w,\ B_{db}) \tag{A3}$$

where AOD^{**}_{550} is the average value of the aerosol optical depth for a given water body w, AOD^{j}_{550} is the aerosol recovered for a single in-situ sample, B_{db} is the MSI deep blue waveband, and N is the total number of in-situ samples. Note that we did not have in-situ measurements of aerosol optical depth data next to the water bodies. Therefore, it was not possible to validate the optimal AOD_{550} value obtained by the inversion method directly. Thus, its validation was performed indirectly through the AC.

Appendix B

Table A1. Overview of input data required for the AC for water bodies included in this study.

Input Data	BIL	CON	BRA + MUT	MAM	PIR
Solar Zenith Angle	48.22°	29.52°	29.52°	27.78°	27.78°
Solar Azimuth Angle	33.37°	53.65°	53.65°	61.70°	61.70°
View Zenith Angle	3.74°	2.83°	2.83°	9.44°	9.44°
View Azimuth Angle	111.67°	194.68°	194.68°	101.95°	101.95°
Ozone (cm-atm)	0.282	0.262	0.262	0.271	0.271
Water Vapor (g/cm^3)	1.482	3.418	3.562	4.407	4.247
Altitude (km)	0.716	0.071	0.072	0.043	0.041
Aerosol Model			Continental		
AOD at 550 nm *	0.100	0.331	0.272	0.164	0.170
AOD at 550 nm **	0.162	0.656	0.633	0.369	0.342

* AOD_{550} recovered with the MODIS products (MCD19A2 Collection 6). ** AOD_{550} recovered with the inversion model using the deep blue waveband.

Table A2. Atmospheric parameters and geometric conditions (viewing and illumination) used in the theoretical simulations.

Solar Zenith Angle	View Zenith Angle	Solar Azimuth Angle	View Azimuth Angle	Target Altitude	Aerosol Model	Atmospheric Profile	Band Range
33°	6°	53°	141°	0.189 (km)	*	Tropical (default)	443–842 (nm)

* Continental or biomass burning smoke model.

References

1. Vörösmarty, C.J.; McIntyre, P.B.; Gessner, M.O.; Dudgeon, D.; Prusevich, A.; Green, P.; Glidden, S.; Bunn, S.E.; Sullivan, C.A.; Reidy Liermann, C.; et al. Global threats to human water security and river biodiversity. *Nature* **2010**, *467*, 555–562. [CrossRef] [PubMed]
2. Boretti, A.; Rosa, L. Reassessing the projections of the World Water Development Report. *Nature* **2019**, *15*, 15. [CrossRef]
3. UNESCO, UN-Water. United Nations World Development Report 2020: Water and Climate Change. Paris: UNESCO. 2020. Available online: https://www.unwater.org/publications/world-water-development-report-2020/ (accessed on 20 January 2022).

4. Pahlevan, N.; John, R.S.; Franz, B.A.; Zibordi, G.; Markham, B.; Bailey, S.; Schaaf, C.B.; Ondrusek, M.; Greb, S.; Strait, C.M. Landsat 8 remote sensing reflectance (Rrs) products: Evaluations, intercomparisons, and enhancements. *Remote Sens. Environ.* **2017**, *190*, 289–301. [CrossRef]
5. Pahlevan, N.; Sarkar, S.; Franz, B.A.; Balasubramanian, S.V.; He, J. Sentinel-2 MultiSpectral Instrument (MSI) data processing for aquatic science applications: Demonstrations and validations. *Remote Sens. Environ.* **2017**, *201*, 47–56. [CrossRef]
6. Vanhellemont, Q.; Ruddick, K. Advantages of high quality SWIR bands for ocean colour processing: Examples from Landsat-8. *Remote Sens. Environ.* **2015**, *161*, 89–106. [CrossRef]
7. Vermote, E.; Justice, C.; Claverie, M.; Franch, B. Preliminary analysis of the performance of the Landsat 8/OLI land surface reflectance product. *Remote Sens. Environ.* **2016**, *185*, 46–56. [CrossRef]
8. Cairo, C.; Barbosa, C.; Lobo, F.; Novo, E.; Carlos, F.; Maciel, D.; Flores Júnior, R.; Silva, E.; Curtarelli, V. Hybrid chlorophyll-a algorithm for assessing trophic states of a tropical brazilian reservoir based on MSI/Sentinel-2 data. *Remote Sens.* **2020**, *12*, 40. [CrossRef]
9. Kutser, T.; Paavel, B.; Verpoorter, C.; Ligi, M.; Soomets, T.; Toming, K.; Casal, G. Sensing of Black Lakes and Using 810 nm Reflectance Peak for Retrieving Water Quality Parameters of Optically Complex Waters. *Remote Sens.* **2016**, *8*, 497. [CrossRef]
10. Maciel, D.A.; Barbosa, C.C.F.; Novo, E.M.L.M.; Flores Júnior, R.; Begliomini, F.N. Water Clarity in Brazilian Water Assessed Using Sentinel-2 and Machine Learning Methods. *ISPRS J. Photogramm. Remote Sens.* **2021**, *182*, 134–152. [CrossRef]
11. Toming, K.; Kutser, T.; Laas, A.; Sepp, M.; Paavel, B.; Nõges, T. First experiences in mapping lake water quality parameters with sentinel-2 MSI imagery. *Remote Sens.* **2016**, *8*, 640. [CrossRef]
12. Martins, V.S.; Barbosa, C.C.F.; de Carvalho, L.A.S.; Jorge, D.S.F.; Lobo, F.L.; Novo, E.M.L.M. Assessment of atmospheric correction methods for sentinel-2 MSI images applied to Amazon floodplain lakes. *Remote Sens.* **2017**, *9*, 322. [CrossRef]
13. Otterman, J.; Fraser, R.S. Adjacency effects on imaging by surface reflection and atmospheric scattering: Cross radiance to Zenith. *Appl. Opt.* **1979**, *197*, 2852–2860. [CrossRef] [PubMed]
14. Richter, R.; Bachmann, M.; Dorigo, W.; Müller, A. Influence of the Adjacency Effect on Ground Reflectance Measurements. *IEEE Geosci. Remote Sens. Lett.* **2006**, *3*, 565–569. [CrossRef]
15. Tanré, D.; Herman, M.; Deschamps, Y. Influence of the background contribution upon space measurements of ground reflectance. *Appl. Opt.* **1981**, *20*, 3676–3683. [CrossRef]
16. Bulgarelli, B.; Kiselev, V.; Zibordi, G. Simulation and analysis of adjacency effects in coastal waters: A case study. *Appl. Opt.* **2014**, *53*, 1523–1545. [CrossRef] [PubMed]
17. Bulgarelli, B.; Zibordi, G. On the detectability of adjacency effects in ocean color remote sensing of mid- latitude coastal environments by SeaWiFS, MODIS-A, MERIS, OLCI, OLI and MSI. *Remote Sens. Environ.* **2018**, *209*, 423–438. [CrossRef]
18. Sterckx, S.; Knaeps, E.; Ruddick, K. Detection and correction of adjacency effects in hyperspectral airborne data of coastal and inland waters: The use of the near infrared similarity spectrum. *Int. J. Remote Sens.* **2011**, *32*, 6479–6505. [CrossRef]
19. Warren, M.A.; Simis, S.G.H.; Selmes, N. Complementary water quality observations from high and medium resolution Sentinel sensors by aligning chlorophyll-a and turbidity algorithms. *Remote Sens. Environ.* **2021**, *265*, 112651. [CrossRef]
20. Sander, L.C.; Schott, J.R.; Raqueño, R. A VNIR/SWIR atmospheric correction algorithm for hyperspectral imagery with adjacency effect. *Remote Sens. Environ.* **2001**, *71*, 252–263. [CrossRef]
21. Vermote, E.F.; Tanré, D.; Deuzé, J.L.; Herman, M.; Morcrette, J.J. Second Simulation of the Satellite Signal in the Solar Spectrum, 6S: An Overview. *IEEE Trans. Geosci. Remote Sens.* **1997**, *35*, 675–686. [CrossRef]
22. Sei, A. Analysis of adjacency effects for two Lambertian half-spaces. *Int. J. Rem. Sens.* **2007**, *28*, 1873–1890. [CrossRef]
23. Vermote, E.F.; Tanré, D.; Deuzé, J.L.; Herman, M.; Morcrette, J.J. *Second Simulation of the Satellite Signal in the Solar Spectrum (6S)*; 6S User Guide Version 3.0.; NASA-GSFC: Greenbelt, MD, USA, 2006.
24. Minomura, M.; Kuze, H.; Takeuchi, N. Adjacency effect in the atmospheric correction of satellite remote sensing data: Evaluation of the influence of aerosol extinction profiles. *Opt. Rev.* **2001**, *8*, 133–141. [CrossRef]
25. Martins, V.S.; Kaleita, A.; Barbosa, C.C.F.; Fassoni-Andrade, A.C.; Lobo, F.L.; Novo, E.M.L.M. Remote sensing of large reservoir in the drought years: Implications on surface water change and turbidity variability of Sobradinho reservoir (Northeast Brazil). *Remote Sens. Appl. Soc. Environ.* **2018**, *13*, 275–288. [CrossRef]
26. Wang, T.; Du, L.; Yi, W.; Hong, J.; Zhang, L.; Zheng, J.; Li, C.; Ma, X.; Zhang, D.; Fang, W.; et al. An adaptive atmospheric correction algorithm for the effective adjacency effect correction of submeter-scale spatial resolution optical satellite images: Application to a WorldView-3 panchromatic image. *Remote Sens. Environ.* **2021**, *259*, 112412. [CrossRef]
27. Houborg, R.; McCabe, M.F. Adapting a regularized canopy reflectance model (REGFLEC) for the retrieval challenges of dryland agricultural systems. *Remote Sens. Environ.* **2016**, *186*, 105–120. [CrossRef]
28. Houborg, R.; McCabe, M.F. Impacts of dust aerosol and adjacency effects on the accuracy of Landsat 8 and RapidEye surface reflectance. *Remote Sens. Environ.* **2017**, *194*, 127–145. [CrossRef]
29. Keukelaere, L.; Sterckx, S.; Adriaensen, S.; Knaep, E.; Reusen, I.; Giardino, C.; Bresciani, M.; Hunter, P.; Neil, C.; Van der Zande, D.; et al. Atmospheric correction of Landsat-8/OLI and Sentinel-2/MSI data using iCOR algorithm: Validation for coastal and inland waters. *Eur. J. Remote Sens.* **2018**, *51*, 525–542. [CrossRef]
30. Kiselev, V.; Bulgarelli, B.; Heege, T. Sensor independent adjacency correction algorithm for coastal and inland water systems. *Remote Sens. Environ.* **2015**, *157*, 85–95. [CrossRef]

31. Pahlevan, N.; Mangin, A.; Balasubramanian, S.V.; Smith, B.; Alikas, K.; Arai, K.; Barbosa, C.; Bélanger, S.; Binding, C.; Bresciani, M.; et al. ACIX-Aqua: A global assessment of atmospheric correction methods for Landsat-8 and Sentinel-2 over lakes, rivers, and coastal waters. *Remote Sens. Environ.* **2021**, *258*, 112366. [CrossRef]
32. Pereira-Sandoval, M.; Ruescas, A.; Urrego, P.; Ruiz-Verdú, A.; Delegido, J.; Tenjo, C.; Soria-Perpinyà, X.; Vicente, E.; Soria, J.; Moreno, J. Evaluation of Atmospheric Correction Algorithms over Spanish Inland Waters for Sentinel-2 Multi Spectral Imagery Data. *Remote Sens.* **2019**, *11*, 1469. [CrossRef]
33. Sterckx, S.; Knaeps, E.; Kratzer, S.; Ruddick, K. SIMilarity Environment Correction (SIMEC) applied to MERIS data over inland and coastal waters. *Remote Sens. Environ.* **2015**, *157*, 96–110. [CrossRef]
34. Bulgarelli, B.; Zibordi, G. Adjacency radiance around a small island: Implications for system vicarious calibrations. *Appl. Opt.* **2020**, *59*, 63–69. [CrossRef] [PubMed]
35. Ribeiro, M.S.F.; Tucci, A.; Matarazzo, M.P.; Viana-Niero, C.; Nordi, C.S.D. Detection of Cyanotoxin-Producing Genes in a Eutrophic Reservoir (Billings Reservoir, São Paulo, Brazil). *Water* **2020**, *12*, 903. [CrossRef]
36. Wengrat, S.; Bicudo, D.C. Spatial evaluation of water quality in an urban reservoir (Billings Complex, southeastern Brazil). *Acta Limnol. Bras.* **2011**, *23*, 200–216. [CrossRef]
37. Alcantara, E.; Coimbra, K.; Ogashawara, I.; Rodrigues, T.; Mantovani, J.; Rotta, H.R.; Park, E.; Cunha, D.G.F. A satellite-based investigation into the algae bloom variability in large water supply urban reservoirs during COVID-19 lockdown. *Remote Sens. Appl. Soc. Environ.* **2021**, *23*, 100555. [CrossRef]
38. Leme, E.; Silva, E.P.; Rodrigues, P.S.; Silva, I.R.; Martins, M.F.M.; Bondan, E.F.; Bernardi, M.M.; Kirsten, T.B. Billings reservoir water used for human consumption presents microbiological contaminants and induces both behavior impairments and astrogliosis in zebrafish. *Ecotoxicol. Environ. Saf.* **2018**, *151*, 364–373. [CrossRef]
39. Lobo, F.L.; Nagel, G.W.; Maciel, D.A.; de Carvalho, L.A.S.; Martins, V.S.; Barbosa, C.C.F.; Novo, E.M.L.M. AlgaeMAp: Algae Bloom Monitoring Application for Inland Waters in Latin America. *Remote Sens.* **2021**, *13*, 2874. [CrossRef]
40. Affonso, A.G.; Queiroz, H.L.; Novo, E.M.L.M. Abiotic variability among different aquatic systems of the central Amazon floodplain during drought and flood events. *Braz. J. Biol.* **2015**, *75*, 60–69. [CrossRef]
41. Silva, M.P.; de Carvalho, L.A.S.; Novo, E.; Jorge, D.S.F.; Barbosa, C.C.F. Use of optical absorption indices to assess seasonal variability of dissolved organic matter in Amazon floodplain lakes. *Biogeosciences* **2020**, *17*, 5355–5364. [CrossRef]
42. Jorge, D.S.F.; Barbosa, C.C.F.; de Carvalho, L.A.S.; Affonso, A.G.; Lobo, F.L.; Novo, E.M.L.M. SNR (signal-to-noise ratio) impact on water constituent retrieval from simulated images of optically complex Amazon lakes. *Remote Sens.* **2017**, *9*, 644. [CrossRef]
43. Maciel, D.A.; Barbosa, C.C.F.; Novo, E.M.L.M.; Cherukuru, N.; Martins, V.S.; Flores Júnior, R.; Jorge, D.S.; de Carvalho, L.A.S.; Carlos, F.M. Mapping of diffuse attenuation coefficient in optically complex waters of amazon floodplain lakes. *ISPRS J. Photogramm. Remote Sens.* **2020**, *170*, 72–87. [CrossRef]
44. ESA, European Space Agency. Mission Search. Available online: https://directory.eoportal.org (accessed on 10 December 2021).
45. Ciancia, E.; Campanelli, A.; Lacava, T.; Palombo, A.; Pascucci, S.; Pergola, N.; Pignatti, S.; Satriano, V.; Tramutoli, V. Modeling and Multi-Temporal Characterization of Total Suspended Matter by the Combined Use of Sentinel 2-MSI and Landsat 8-OLI Data: The Pertusillo Lake Case Study (Italy). *Remote Sens.* **2020**, *12*, 2147. [CrossRef]
46. Hestir, E.L.; Brando, V.E.; Bresciani, M.; Giardino, C.; Matta, E.; Villa, P.; Dekker, A.G. Measuring freshwater aquatic ecosystems: The need for a hyperspectral global mapping satellite mission. *Remote Sens. Environ.* **2015**, *167*, 181–195. [CrossRef]
47. ESA, European Space Agency. User Guides. Available online: https://sentinel.esa.int/web/sentinel/user-guides/sentinel-2-msi/product-types/level-1c (accessed on 10 December 2021).
48. Barbosa, C.C.F.; Novo, E.M.L.M.; Martınez, J.M. Remote sensing of the water properties of the Amazon floodplain lakes: The time delay effects between in-situ and satellite data acquisition on model accuracy. In Proceedings of the International Symposium on Remote Sensing of Environment: Sustaining the Millennium Development Goals, Stresa, Italy, 4–9 May 2009.
49. Marinho, R.R.; Harmel, T.; Martinez, J.; Filizola Junior, N.P. Spatiotemporal Dynamics of Suspended Sediments in the Negro River, Amazon Basin, from In Situ and Sentinel-2 Remote Sensing Data. *ISPRS Int. J. Geo-Inf.* **2021**, *10*, 86. [CrossRef]
50. Warren, M.A.; Simis, S.G.H.; Martinez-Vicente, V.; Poser, K.; Bresciani, M.; Alikas, K.; Spyrakos, E.; Giardino, C.; Ansper, A. Assessment of atmospheric correction algorithms for the Sentinel-2A MultiSpectral Imager over coastal and inland waters. *Remote Sens. Environ.* **2019**, *225*, 267–289. [CrossRef]
51. Capobianco, J.P.R.; Whately, M. *Billings 2000: Ameaças e Perspectivas Para o Maior Reservatório de Água da Região Metropolitana de São Paulo. Relatório do Diagnóstico Socioambiental Participativo da Bacia Hidrográfica da Billings no Período 1989–99*; Instituto Socioambiental: São Paulo, Brazil, 2002.
52. Affonso, A.G.; Queiroz, H.L.; Novo, E.M.L.M. Limnological characterization of floodplain lakes in Mamirauá Sustainable Development Reserve, Central Amazon (Amazonas State, Brazil). *Acta Limnol. Bras.* **2011**, *23*, 95–108. [CrossRef]
53. Barbosa, C.C.F. Sensoriamento Remoto da Dinâmica da Circulação da Água do Sistema Planície de Curuai/Rio Amazonas. Ph.D. Thesis, National Institute for Space Research (INPE), São José dos Campos, Brazil, 2005.
54. Nagel, G.W.; Novo, E.M.L.M.; Martins, V.S.; Campos-Silva, J.V.; Barbosa, C.C.F.; Bonnet, M.P. Impacts of meander migration on the Amazon riverine communities using Landsat time series and cloud computing. *Sci. Total Environ.* **2022**, *806*, 150449. [CrossRef]
55. Mobley, C.D. Estimation of the Remote-Sensing Reflectance from Above-Surface Measurements. *Appl. Opt.* **1999**, *38*, 7442–7455. [CrossRef]

56. Mobley, C.D. Polarized reflectance and transmittance properties of windblown sea surfaces. *Appl. Opt.* **2015**, *54*, 4828–4849. [CrossRef]
57. Lobo, F.L.; Costa, M.P.F.; Novo, E.M.L.M. Time-series analysis of Landsat- MSS/TM/OLI images over Amazonian waters impacted by gold mining activities. *Remote Sens. Environ.* **2014**, *157*, 170–184. [CrossRef]
58. Santer, R.; Schmechtig, C. Adjacency effects on water surfaces: Primary scattering approximation and sensitivity study. *Appl. Opt.* **2000**, *39*, 361–375. [CrossRef] [PubMed]
59. Ruddick, K.G.; Cauwer, V.; Park, Y.J.; Moore, G. Seaborne measurements of near infrared water- leaving reflectance: The similarity spectrum for turbid waters. *Limnol. Oceanogr.* **2006**, *51*, 1167–1179. [CrossRef]
60. Xu, H. Modification of normalised difference water index (NDWI) to enhance open water features in remotely sensed imagery. *Int. J. Remote Sens.* **2006**, *27*, 3025–3033. [CrossRef]
61. Levy, R.C.; Remer, L.A.; Kleidman, R.G.; Mattoo, S.; Ichoku, C.; Kahn, R.; Eck, T.F. Global evaluation of the Collection 5 MODIS dark-target aerosol products over land. *Atmos. Chem. Phys.* **2010**, *10*, 10399–10420. [CrossRef]
62. Vermote, E.; El Saleous, N.; Justice, C.O.; Kaufman, Y.J.; Privette, J.L.; Remer, L.; Roger, C.; Tanré, D. Atmospheric correction of visible to middle-infrared EOS-MODIS data over land surfaces: Background, operational algorithm and validation. *J. Geophys. Res. Atmos.* **1997**, *102*, 17131–17141. [CrossRef]
63. Levy, R.C.; Mattoo, S.; Munchak, L.A.; Remer, L.A.; Sayer, A.M.; Patadia, F.; Hsu, N.C. The Collection 6 MODIS aerosol products over land and ocean. *Atmos. Meas. Tech.* **2013**, *6*, 2989–3034. [CrossRef]
64. Lyapustin, A.; Wang, Y.; Korkin, S.; Huang, D. MODIS Collection 6 MAIAC algorithm. *Atmos. Meas. Tech.* **2018**, *11*, 5741–5765. [CrossRef]
65. Martins, V.S.; Lyapustin, A.; de Carvalho, L.A.S.; Barbosa, C.C.F.; Novo, E.M.L.M. Validation of high-resolution MAIAC aerosol product over South America. *J. Geophys. Res.* **2017**, *122*, 7537–7559. [CrossRef]
66. Seidel, F.C.; Popp, C. Critical surface albedo and its implications to aerosol remote-sensing. *Atmos. Meas. Tech. Discuss.* **2012**, *4*, 7725–7750. [CrossRef]
67. Artaxo, P.; Rizzo, L.V.; Brito, J.F.; Barbosa, H.M.J.; Arana, A.; Sena, E.T.; Cirino, G.G.; Bastos, W.; Martin, S.T.; Andreae, M.O. Atmospheric aerosols in Amazonia and land use change: From natural biogenic to biomass burning conditions. *Faraday Discuss.* **2013**, *165*, 203–236. [CrossRef]
68. Löbs, N.; Barbosa, C.G.G.; Brill, S.; Walter, D.; Ditas, F.; Sá, M.O.; Araújo, A.C.; Oliveira, L.R.; Godoi, R.H.M.; Wolf, S.; et al. Aerosol measurement methods to qualify spore emissions from fungi and cryptogamic covers in the Amazon. *Atmos. Meas. Tech.* **2020**, *13*, 153–164. [CrossRef]
69. Shrivastava, M.; Andreae, M.O.; Artaxo, P.; Barbosa, H.M.J.; Berg, L.K.; Brito, J.; Ching, J.; Easter, R.C.; Fan, J.; Fast, J.D. Urban pollution greatly enhances formation of natural aerosols over the Amazon rainforest. *Nat. Commun.* **2019**, *10*, 1046. [CrossRef] [PubMed]
70. Artaxo, P.; Martins, J.V.; Yamasoe, M.A.; Procópio, A.S.; Pauliquevis, T.M.; Andreae, M.O.; Guyon, P.; Gatti, L.V.; Leal, A.M.C. Physical and chemical properties of aerosols in the wet and dry seasons in Rondônia, Amazonia. *J. Geophys. Res.* **2002**, *107*, 8081. [CrossRef]
71. Fan, J.; Zhang, D.R.; Giangrande, S.E.; Li, Z.; Machado, L.A.; Martin, S.T.; Yang, Y.; Wang, J.; Artaxo, P.; Barbosa, H.M.J.; et al. Substantial convection and precipitation enhancements by ultrafine aerosol particles. *Science* **2018**, *359*, 411–418. [CrossRef]
72. Taylor, M.; Kazadzis, S.; Amiridis, V.; Kahn, R.A. Global aerosol mixtures and their multiyear and seasonal characteristics. *Atmos. Environ.* **2015**, *116*, 112–129. [CrossRef]
73. Flores Júnior, R.; Flores Júnior, R.; Barbosa, C.C.F.; Maciel, D.A.; Novo, E.M.L.M.; Martins, V.S.; Lobo, F.L.; de Carvalho, L.A.S.; Carlos, F.M. Hybrid Semi Analytical Algorithm for estimating chlrophyll-a concentration in Lower Amazon Floodplain waters. *Front. Remote Sens.* **2022**, *3*, 834576. [CrossRef]
74. Feng, L.; Hu, C. Land adjacency effects on MODIS Aqua top-of-atmosphere radiance in the shortwave infrared: Statistical assessment and correction. *J. Geophys. Res. Ocean.* **2017**, *122*, 4802–4818. [CrossRef]
75. Gordon, H.R. Evolution of Ocean Color Atmospheric Correction: 1970–2005. *Remote Sens.* **2021**, *13*, 5051. [CrossRef]
76. Kaufman, Y.J.; Joseph, J.H. Determination of surface albedos and aerosol extinction characteristics from satellite imagery. *J. Geophys. Res.* **1982**, *87*, 1287–1299. [CrossRef]
77. Lyapustin, A.; Kaufman, Y.J. Role of adjacency effect in the remote sensing of aerosol. *J. Geophys. Res.* **2001**, *106*, 11909–11916. [CrossRef]
78. Duan, S.; Li, Z.; Gao, C.; Zhao, W.; Wu, H.; Qian, Y.; Leng, P.; Gao, M. Influence of adjacency effect on high-spatial-resolution thermal infrared imagery: Implication for radiative transfer simulation and land surface temperature retrieval. *Remote Sens. Environ.* **2020**, *245*, 111852. [CrossRef]
79. Kaufman, Y.J.; Wald, A.E.; Remer, L.A.; Gao, B.; Li, R.; Flynn, L. The MODIS 2.1-μm Channel—Correlation with Visible Reflectance for Use in Remote Sensing of Aerosol. *IEEE Trans. Geosci. Rem. Sens.* **1997**, *35*, 1286–1298. [CrossRef]
80. Ruggiero, M.A.G.; Lopes, V.L.R. Zero reais de funções reais. In *Cálculo Numérico: Aspectos Teóricos e Computacionais*, 2nd ed.; Pearson: São Paulo, Brazil, 1996; pp. 27–104.

remote sensing

MDPI

Article

Vertical Ground Displacements and Its Impact on Erosion along the Karachi Coastline, Pakistan

Shamsa Kanwal [1,*], Xiaoli Ding [1,2], Songbo Wu [1] and Muhammad Sajjad [3]

1 Department of Land Surveying and Geo-Informatics, The Hong Kong Polytechnic University, Hong Kong SAR 999077, China; xl.ding@polyu.edu.hk (X.D.); song.bo.wu@connect.polyu.hk (S.W.)
2 Research Institute for Land and Space, The Hong Kong Polytechnic University, Hong Kong SAR 999077, China
3 Centre for Geo-Computation Studies and Department of Geography, Hong Kong Baptist University, Hong Kong SAR 999077, China; msajjad@hkbu.edu.hk
* Correspondence: shamsa.kanwal@connect.polyu.hk

Abstract: This study employed remote sensing (optical and synthetic aperture radar) and data analysis techniques to quantify vertical ground displacements and assess their contribution to coastline erosion. To provide evidence from Pakistan, we selected the coast of Karachi—a mega-city located along the dynamic coastline of the Indus River Delta—which has been experiencing severe coastal erosion during the last few decades. Observations from the C-band Envisat/ASAR and Sentinel-1A sensors over the 2004–2010 and 2014–2016 periods, respectively, enabled us to study vertical ground displacements in the study area, providing a long-term assessment during 2004–2016. Results suggest that some areas along the Karachi coastline are subsiding at comparable rates to or even much higher than the relative sea-level rise (SLR, ~1.9 mm/yr), which may amplify the rates of relative SLR in coming years, along with accelerating coastal erosion. Various parts of the study area along the coast are unstable and undergoing displacement. Landsat images from 1989 to 2018 (10-year temporal resolution) were further used to examine the state of coastline erosion using three statistical approaches (i.e., End Point Rate (EPR), Linear Regression Rate (LRR), and Least Median of Squares (LMS)). While the erosion underlaid the majority of the eastern sections of the study area, the ground displacements were spatially heterogeneous across the study area and along the coastline. Erosion rates of ~2.4 m/yr spatially corresponded with ground displacement rates of up to ~−1.4 cm/yr, but not all the coastline segments with high annual mean erosion rates were associated with local mean subsidence. The causes of ground displacements and coastline erosion were analyzed, and results were interpreted by integrating spatial ancillary information. Results indicate that rapid urbanization, construction on reclaimed land, coastline erosion favoring seawater intrusion, failed drainage/sewerage networks, and soil liquefaction are contributing to the site specific variations in the land displacement in Karachi.

Keywords: remote sensing; coastal dynamics; coastline erosion; vertical ground displacement; InSAR

Citation: Kanwal, S.; Ding, X.; Wu, S.; Sajjad, M. Vertical Ground Displacements and Its Impact on Erosion along the Karachi Coastline, Pakistan. *Remote Sens.* **2022**, *14*, 2054. https://doi.org/10.3390/rs14092054

Academic Editors: Paweł Terefenko, Jacek Lubczonek, Marta Wlodarczyk-Sielicka and Katarzyna Bradtke

Received: 9 March 2022
Accepted: 21 April 2022
Published: 25 April 2022

1. Introduction

Various large-scale and long-term environmental processes, such as seismic events, construction activities, soil liquefaction, and groundwater abstraction along with climate change in a particular low-lying coastal area, may manifest in the form of ground instabilities (e.g., Long Beach in the United States, Venice in Italy, Tokyo in Japan, Bangkok in Thailand, or Shanghai in China), making these problems more severe [1–3]. The damage is not always known until it is too late. Many of the river deltas are experiencing accelerated subsidence and coastline erosion. As a result, these vulnerable deltaic coasts and communities are at increased risk of permanent flooding and sea-level rise (SLR) caused by reduced sediment supplies due to upstream dam constructions and anthropogenic intrusions of the estuaries themselves [4]. It is noted that parts of coastlines around the world are undergoing tectonic subsidence (e.g., the south-eastern coast of England, Atlantic coast of the US, south-western Denmark, the Netherlands, northern Germany, and the Campania Region of Italy coast), while in other areas an increase in natural resource exploitation (e.g., parts of the Sindh coast, Pakistan) has induced subsidence [5,6]. In comparison, a myriad of factors contributing to the erosion of coastlines along the subsiding Indus River Delta (our area of interest, located along the Arabian Sea) have been identified; this erosion is associated with a drastic reduction in sediment influx, mangrove depletion, seawater intrusion, and sea-level rise [7]. While ground instabilities negatively impact the environment, urban infrastructure, and human life, the scientific evidence on ground subsidence in the study area is inadequate and its role in driving coastal erosion is not quite clear [1,8–10].

Geographically, being a constituent part of the Indus Deltaic Region (IDR), the NW-SE-oriented Karachi coastline spans more than 70 km. This coastal area is exposed to monsoon-induced waves and tides. Being influenced by coastal and riverine processes of the Arabian Sea (such as sediment discharge, storm surges, and currents), the study area's coastline has been enduring dynamic changes [7]. As a result, several segments along the IDR have undergone subsequent severe erosion, particularly along the coastline of East Karachi. The Karachi coastline has experienced an average landward retreat rate of ~2.43 m/yr, with the highest erosion rate being along the banks of the creeks where the Indus River Delta begins. More recently, several studies have emerged that postulate causative factors aggravating erosion along the Karachi coast.

SLR and coastal erosion are mutually interconnected. Khan, et al. [11] carried out a relative SLR analysis through calculations with Karachi station's ten years of hourly tide-gauge (TG) data. In their study, authors estimated ~3.6 mm/yr of mean SLR from 2007–2016. SLR of ~1.9 mm/yr was reported for the Karachi coastline by Kanwal, et al. [7] from 1916 to 2015 using the Permanent Service for Mean Sea Level (PSMSL) data archive. Considering the evidence from these studies, it seems that the coastline erosion along the Karachi coast is linked with rising sea levels.

Another viewpoint is that abrupt reduction in sediment load and water reaching the Arabian Sea has amplified the effects of waves' actions and tidal currents, thus impeding mangrove growth and resulting in accelerated seawater intrusion (SWI) and coastline erosion along the entire Sindh coastline. Kidwai, et al. [12] reported a significant drop in the sediment load reaching the IDR coastal plains, mainly due to dams, barrage construction, and channeling in upstream areas. This reduction in the sediment supply has escalated the erosional effects of tides and waves, resulting in seawater intrusion (SWI) into coastal plains. The results from these studies agreed well with those of Kanwal, et al. [13], who investigated the evolution of the barrier islands (BIs) along Sindh coastal regions in Pakistan and concluded that ~65% of these BIs underwent high erosion (i.e., >2 m/yr), during 1989–2018. Their study implies that the ~15% of these BIs experiencing moderate erosion (i.e., <2 m/yr), are likely to be subjected to high erosion (i.e., >2 m/yr) in the wake of SLR and reduced sediment influx received by the IDR plains in the near future, without proper measures.

Erosional effects could be severe and damaging in low-lying coastal regions if this phenomenon is coupled with other hazards, such as ground displacement, storm surges, and SLR. Although ground displacements have been observed in different parts of Karachi and reported by various media outlets, the issue has not been frequently discussed in published literature until recently [3,14–17]. For instance, Amin, et al. [18] and Kanwal, et al. [19] utilized synthetic aperture radar images to identify ground instabilities in the coastal megacity of Karachi. Comprehensive historical information on the characteristics of ground subsidence is neither readily available nor applied in the planning of mitigation efforts. Such integration is being hindered by the unavailability of chronological data. However, earlier research has not fully established the relationship between the subsidence at the coast and the ongoing erosion. The lack of published data highlights the gravity of the issue. It further suggests that the serious implications of the induced subsidence on accelerating the coastline deterioration have been underestimated. In this context, the present study aims to assist evidence-based decision-making by: (i) leveraging the synthetic aperture radar interferometry (InSAR) techniques to determine the spatial distribution and rates of vertical ground displacement in Karachi and (ii) unraveling the link, if any, between vertical ground displacement and coastline erosion derived from Landsat images along low-lying regions of the developing coast of Karachi.

The capability of remote sensing satellites carrying Synthetic Aperture Radar (SAR) sensors to achieve regular and frequent data acquisition has been proven to have significant advantages in terms of wide-area coverage, cost, and equipment installation for improved ground displacement detection, mapping, and monitoring. In addition, the collected data have facilitated the recording of longitudinal information over large geographical areas. The prodigious InSAR technique has been utilized worldwide in various fields for the extraction of ground displacement ascribed to landslides [20], deltaic subsidence [21], and coast deformation [22,23] among other related factors. InSAR measures ground displacements along a line of sight (LOS) by estimating the phase difference between two SAR images. In rapidly changing landscapes, some landmarks (scatterers) may be only visible in a subset of the SAR acquisition but not for the entire observation period. In rapidly developing areas, these partially/temporarily as well as persistently available scatters, collectively termed Temporarily Coherent Points (TCPs), can allow us to estimate the ground displacement in such changing landscapes.

In such urban developing areas, particularly those characterized by the complex geographic and geologic environments found near low-lying coasts, the InSAR technique can help monitor ground instabilities [24]. Integrating multi-scale standpoints of ground displacements that could cause widespread instability with its quantitative assessment provides essential information to study the relative sea-level rise scenarios in coastal lands. This information could be useful for effective and efficient resource planning and management to implement conservation practices in a particular area of interest to avoid damages and minimize risks, such as frequent flooding, and increase in relative sea level.

2. Study Area

The metropolitan city of Karachi is located on the southern-most border of Pakistan along the coast of the Arabian Sea (Figure 1), with its geographical extent marked by longitudes 66°37′E–67°37′E and latitudes 24°45′N–25°15′N. At present, Karachi appears to be in the unfortunate position of having precarious tectonic framework, as it sits next to the Arabian, Indian, and Eurasian plates' triple junction [14,25,26]. Neotectonics of this area show various NE—SW-trending fold structures crossed by numerous active NW-SE-trending active faults of local and regional extents and transcurrent or transverse nature in and around the city, endangering about 23 million lives inhabiting an area of about 3527 km^2. Comprehensive information about these faults, along with the structural elements and coastal geology of Karachi, can be found in [25]. The known faults and their movements are expected to provide reasonably good information for assessing the potential of vertical ground displacements in the Karachi area.

In addition to the larger tectonic framework, shown in Figure 1, the city faces multiple environmental and geological issues. Karachi is a constituent part of the indented coastline of the Indus River Delta, which is highly vulnerable to erosion and inundation, mainly due to anthropogenic activities, such as groundwater extraction, and development, in conjunction with hazards, such as strong tides, wave action, and increases in relative sea level. In most parts of the city, slopes are between 0–4°, particularly in areas near the coast with flat topography. Such a minimal decrease in ground elevation relative to mean sea level can impede the city's existence due to the intensification of coastal erosion and acceleration of other environmental issues, such as seawater intrusion, sea-level rise, and changes in the natural lagoon environment nearby to the Indus Deltaic system. To address this problem, this study aims to map ground displacements through interferometric processing of the available historical archives of C-band SAR datasets and study his displacement's influence on coastline erosion identified through Landsat images.

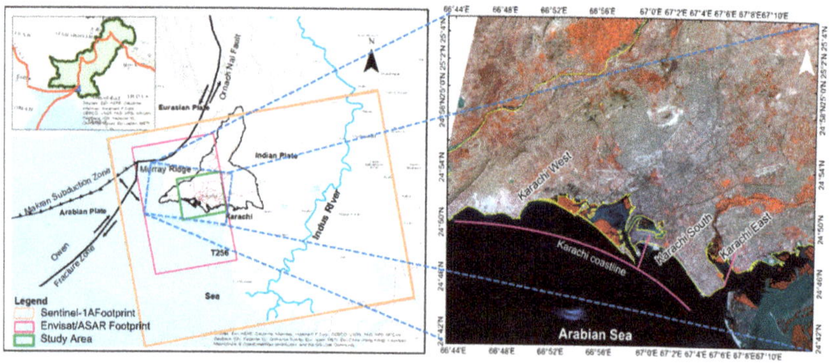

Figure 1. Inset (left) maps the location of the coastal mega-city of Karachi relative to the Arabian-Indian-Eurasian plates' tectonic triple junction. The green box marks the coverage of the study area, for which 33 scenes from ENVISAT/ASAR (2004–2010), 23 images from Sentinel-1A (2014–2016), and 4 images of Landsat (1989–2018) are processed. Each administrative unit along the Karachi coastline is represented on the right, overlaid upon a Landsat Operational Land Imager (OLI) false color image (2018).

3. Materials and Methods

To conduct this study, a two-fold methodology is adopted: (i) assess the coastline's spatial changes using the Digital Shoreline Analysis System (DSAS) tool from the United States Geological Survey (USGS) for 1989–2018 and (ii) quantify the vertical ground displacements using InSAR analysis of images acquired for the 2004–2010 and 2014–2016 periods. The results are demonstrated and contrasted to corroborate the relationship between coastline erosion and vertical ground displacements in the study area.

3.1. Vertical Ground Displacement Detection Using TCPInSAR

To quantitatively extract the ground displacement in the changing landscapes of the low-lying coastal mega-city of Karachi, where ground motion is likely to affect the rate of relative sea-level rise, the multi-temporal InSAR method is used. For this purpose, this study mines the historical archives of the European Space Agency (ESA) SAR C-band mission Envisat/ASAR (available at http://eo-virtual-archive4.esa.int/) and extracts time-series acquisitions suitable for ground displacement monitoring through interferometric processing to demonstrate the capabilities of SAR datasets and sensors. Table S1 (given in separate Supplementary Materials File) lists all the 33-ascending single-look complex images (Track 256, IM mode, and HH polarization) used in this study, acquired during the 2004–2010 period by the ESA. This approach allows us to visualize the ground displacements in Karachi from 2004–2010.

As part of the new European Copernicus Programme, the C-band Sentinel-1A satellite has been in orbit since April 2014. The presented analysis is further extended over the same area using a set of 23 images from Sentinel-1A (IW mode, path 42, and VV polarization) retrieved from the ASF DAAC and processed by ESA (see Supplementary Table S2). Careful selection of the SLC images enables us to study ground displacements over the same geographical extent by Envisat/ASAR from Sentinel-1A. Despite some data gaps, this study allows us to construct a longitudinal analysis of the Karachi coast from 2004 to 2016. The results are interpreted, and the causes of the ground displacements are examined by integrating multidisciplinary geospatial data layers (i.e., land use, faults, lithology, and topography) [26].

The Temporary Coherent Point InSAR (TCPInSAR) approach (see Figure 2), which is designed to monitor displacement in the changing landscapes of metropolitan areas [27] as persistent scatterers are more abundant in such environments [28], is applied to investigate the ground displacements in the study area. However, this approach can also investigate ground displacements in rural areas and wetlands [29,30]. A temporary coherent point

(TCP) preserves its coherence value through multiple SAR acquisitions [31]. The procedures for identifying TCPs have been reported in detail by Zhang [32]. All the identified TCPs are used to construct a Local Delaunay Triangulation Network [31]. This is useful for preventing the island effect due to the removal of considered arcs (generated from a pair of TCPs) containing phase ambiguities as part of the data processing chain. The TCPInSAR algorithm separates orbital artifacts and ground displacement signals through a joint model mechanism [33]. This model simultaneously analyzes the difference of interferometric phase at arcs, the rate of ground displacements, the topographic residuals, and the polynomial coefficients indicative of orbital errors at TCPs, and reflects their relationship with each other.

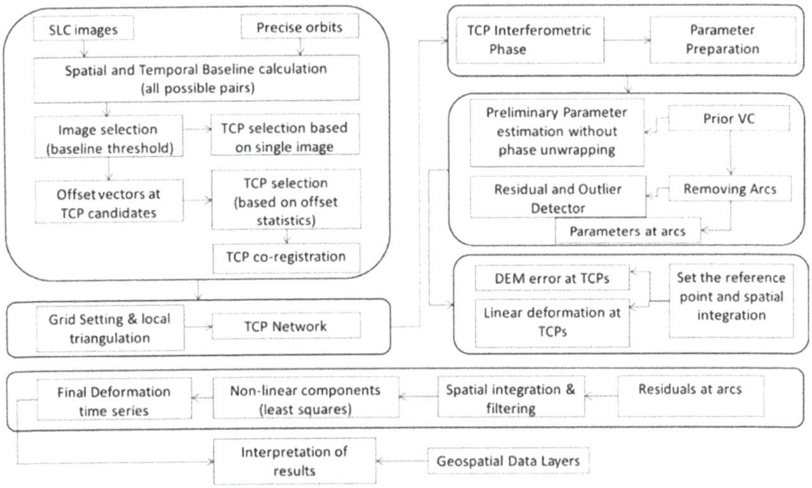

Figure 2. Flow chart of the methodology adopted in this study.

All the raw data are processed into readable SLC images. Every second SLC image is carefully chosen, co-registered, and interferograms (IFG) are created. An IFG in its raw form is potentially affected by various sources of errors, such as atmospheric contamination, topographic artifacts, and decorrelation noise. To find the optimal phase value for ground displacement rate estimation, it is required to eliminate as many of these aforementioned errors as possible from the IFG phase value [34]. For subsequent interferometric processing, an Envisat/ASAR image taken on 2 September 2005 is selected as the reference. In order to reduce the decorrelation noise, 28 ENVISAT/ASAR images are selected to generate IFGs per their perpendicular baseline and temporal baseline criteria (220 m Bp and 420 days Bt), as shown in Figure S1 of the Supplementary Materials. The temporal and perpendicular baseline criteria are chosen to ensure that all of the involved images produce a good quality IFG. Similarly, for Sentinel-1A datasets, only selected interferometric pairs are utilized to generate the IFGs, per the perpendicular and temporal baseline criteria.

To remove the topographic artifacts, a 1-Arc-Second digital elevation model (DEM) from Shuttle Radar Topography Mission (SRTM) is processed via a standard two-pass differential InSAR (DInSAR) method [20,35]. The coherence method is used to choose the TCPs. The optimum coherence threshold value is selected through the process of trial and error. The coherence value of the IFGs is further enhanced with a multi-looking step (down-sampling with one look in the azimuth and five looks in the range direction), and de-noised using a non-local framework for SAR images known as NL-SAR [36]. It should be noted that the technique of TCPInSAR can find the observations represented by arcs (pairs of TCPs) without ambiguities by analyzing the least-squares (LS) residuals [25], and without requiring phase unwrapping.

In this study, approximately 2.1 million TCPs are identified and used to construct 410,529 arcs by processing 28 interferometric pairs from Envisat/ASAR with Bp less than 220 m. To estimate the rate of ground displacements and various noises, such as topographic residuals, and orbital errors, a large design matrix (14,368,515 × 2,830,907) is constituted representing the number of observations (i.e., 14,368,515) and the number of unknowns (i.e., 2,830,907) representing topographic residuals and ground displacements rates at those identified TCPs with five polynomial coefficients of polynomial denoting the orbital artifacts in the observations, keeping one image as reference image and one TCP as the reference coherent point. These parameters are resolved by employing the LS method. The arcs of TCPs containing phase ambiguities are removed by employing a phase ambiguity detector [37], which leaves fewer resultant TCPs at the end (i.e., ~2 million). The examination of topographic residuals depicts the difference between the observed topography and 1-arc SRTM DEM from −10 m to 20 m, with 4.28 m standard deviation (SD). One point location is selected as a reference point for retrieving the ground displacements in the study area.

For Sentinel data processing, the process mentioned above (see Figure 2) is repeated with a reference image dated 28 August 2015. Only 34 Sentinel-1A image pairs are utilized to produce the IFGs, meeting the perpendicular (Bp 100 m) and temporal baseline (Bt 100 days) criteria, as shown in Figure S1 of the Supplementary Materials. This is done to maintain optimal coherence value and lessen the probability of the phase wrapping step. Then, the TCPInSAR procedure is applied burst-wise over the good quality IFGs. Table S2 given in the Supplementary Materials, describes the details of the relevant bursts that cover the same geographical area as Envisat/ASAR to generate mosaic SLC images for interferometric analysis.

Processing of Sentinel-1A images requires good care, especially if they are from different paths and frames. It is imperative to keep the number of bursts in each observation image the same; otherwise, it might cause various errors/noise, e.g., co-registration errors. However, the set number of bursts for each frame can vary. The Sentinel-1A images are comprised, in total, of three bursts for sub-swaths 1 and 2 of the Sentinel-1A images that show the broad area coverage offered by Sentinel-1A. This ensures an overview of the ground displacement status over a large area at a good spatial resolution. TCPInSAR (coherence method) is applied to the Sentinel-1A images over the Karachi area using burst information, and details are given in Table S2 and provided in the Supplementary Materials. By processing Sentinel-1A IFGs, ~6.3 million TCPs are identified and processed to construct ~810,119 arcs. Similarly, a sizeable sparse design matrix (27,544,046 × 3,674,889) is constructed to estimate the rate of ground displacements, topographic residuals, and orbital errors. After applying the method of LS for resolving the parameters, arcs containing phase ambiguities are removed. It is worth noting that LOS ground displacements combine both vertical and horizontal displacements. Vertical ground displacements are separated from LOS displacements in the final step for further analysis. Even though it would be ideal to validate the displacements using ground-survey-based GPS data, we cannot incorporate them at this stage due to the unavailability of any data, which is a major issue in developing countries. However, as the approach utilized here has been validated with GPS data in many parts of the world [38,39], it is reasonable to proceed and make best use of the available data to provide important insights regarding the subject matter.

Lastly, vertical displacement velocities from both datasets were computed and overlaid on the land-use map of Karachi to examine the role of different land-use activities in causing these displacements. The land-use/land-cover dataset was developed by a research group from the University of Karachi, Pakistan in 2010. About 85% of land-use data were developed using High-Resolution Imagery from QuickBird (available at https://earth.esa.int/eogateway/catalog/quickbird-full-archive). The remaining 15% were developed through field surveys and ground truthing conducted by a group of students from the Department of Geography at the University of Karachi. Since the core land-use map of the city has not changed significantly since, and the year of mapping lies within the time period of this study, we have obtained and use this data for the comparison and analysis of ground displacement results. Furthermore, the vertical displacement rates in each administrative zone along the coast and their coastline erosion rates are compared and analyzed to examine the influence of vertical ground displacement in deriving coastline changes. Details of coastline change estimation are given in the following section, Section 3.2.

3.2. Coastline Change Estimation Using DSAS

In the present study, coastline positions are derived from Landsat data using the DSAS tool [40,41] developed by the USGS as an extension of the ArcMap software. We select high-quality images from Landsat climate data records (CDR) captured in the pre-monsoon (December–April) non-flooding season to delineate the coastline positions. Table S3 (see Supplementary Materials) details the acquisition details of the images used in this study. The astronomical tidal height prediction records for the tide gauge station of Port Muhammad Bin Qasim (24.7833°N, 67.3500°E) from Tides 4 Fishing (https://tides4fishing.com/) shows that the tidal level range is ~0.20–2.80 m during the Landsat pass-overs for the 1989–2018 period. Due to the constraints of tidal level (i.e., ≤2.8 m) and availability of images of the study area, four atmospherically corrected and cloud-free Landsat images from 1989, 1999, 2009, and 2018 (path-152 and row-043) at 30 m spatial resolution are considered for extraction of the coastline positions and are pre-processed for change analysis.

The land-water boundary is delineated by computing the Normalized Difference Water Index (NDWI) using the near-infrared (NIR) band and short-wave infrared (SWIR) band. To extract the coastline change rate from NDWI images, the following steps are carried out: (i) delineating coastline; (ii) generating baseline; (iii) casting transects; (iv) computing distances between the baseline and the coastline at each transect; and (v) estimating the rate of coastline change at each transect [4]. To compute the rate of coastline change, 1500 transects with 500 m spacing along the baseline for the study area are analyzed, employing three statistical methods (i.e., End Point Rate (EPR), Least Median of Squares (LMS), and Linear Regression Rate (LRR)). The first model, EPR, estimates the coastline change between coastline positions separated by a certain number of years using:

$$EPR = \frac{dist_1 - dist_2}{y_1 - y_2} \tag{1}$$

where $dist_1$ and $dist_2$ represent the distance of coastline positions from the baseline, and y_1 and y_2 are the dates the coastline positions are observed.

The second method, Linear Regression Rate (LRR) of coastline change, determines the least-squares regression of all the coastline points for each casted transect. The line of regression takes the following form:

$$D = a + mt \tag{2}$$

where D represents the distance separating the observed coastline and the baseline, a represents the y-intercept, t represents time in years [41], and m denotes the slope of the linear regression line (i.e., coastline change rate). The value of linear regression (i.e., R-squared, R^2), is computed for each transect:

$$R^2 = 1 - \sqrt{\frac{\sum_{i=1}^{n}(dist - distp)^2}{\sum_{i=1}^{n}\left(dist - \overline{dist}\right)^2}} \tag{3}$$

where $dist$ represents the known distance separating the observed coastline and the baseline; $distp$ represents the predicted value of $dist$ estimated using the regression equation; $\overline{dist}$ represents the mean position of the observed coastlines; and n represents the number of the observed coastlines. Following Nassar et al. (2019) $R^2 > 0.87$ is considered as the limit of certainty in this study [41]. The standard error of the regression line slope representing an uncertainty of the calculated rate of coastline change is noted as Confidence Interval of regression line (LCI). In this study, the LCI is considered at a 95% confidence interval ($p = 0.05$, LCI95).

In the third method based on the Least Mean Square (LMS) model, the best-fit equation for the regression slope line is determined using the median of the squared residuals. The LRR model is based on the mean offset of the sample to compute the slope line equation by contrast.

We compare the affinity between results from these various methods using R^2 measures. Compared to the affinity between EPR–LMS ($R^2 = 0.80$) and LMS–LRR ($R^2 = 0.91$), the obtained value of R^2 is high for EPR-LRR (0.94). For this reason, and with LRR being the most widely used approach for coastline change and analysis, the LRR method is applied for further analysis of the coastline changes and erosion rate estimation. The quality of the statistical results from LRR model is validated through correlation analysis between the model-predicted coastline and image-based delineated coastline for the year 2018. The correlation between the model-predicted and the image-derived coastlines was found to be 90%. Since the LRR model-predicted coastline values and the image-based coastline values were in close agreement (~90%), the LRR model coastline change estimates are considered acceptable for further analysis.

4. Results

4.1. Displacement along the Karachi Coast (2004–2016)

The TCPInSAR algorithm is applied to the selected IFGs to estimate the vertical ground displacement rate of the Karachi area, as shown in Figure 3. The maps obtained in SAR image coordinates are initially geocoded into Universal Transverse Mercator (UTM) coordinates. This is done purposely to integrate the results with ancillary geospatial information (i.e., land use, geological features, and coastline erosion rates). The vertical ground displacement results are color-coded red to blue. The green shades represent ground displacement values around zero; the tones of blue correspond to uplifts, and the tones of red correspond to subsidence.

The local ground displacements are evident in Envisat/ASAR results, with a maximum vertical ground displacement velocity greater than ±15 mm/yr. The magnitude of the vertical ground displacement rate is spatially dynamic in the study area (Figure 3). The sites experiencing the most conspicuous uplift are located in the town of Baldia and some neighboring areas from the town of Kemari. This uplift trend is related to the water level variations in underground reservoirs, as this location is surrounded by the Liyari river to the east and the Hub River to the west. An unusual positive ground displacement rate is also observed in the Port Muhammad Bin Qasim area (see Figure 4), although the reason for uplifting requires further investigation. It is acknowledged that these ground displacements are estimated relative to a reference location deemed stable during the period of this study. It is likely that the chosen reference point might also have experienced some instability and movements during the period selected for this study. Despite the unavailability of in situ data for result validation, the SAR-based analysis in this study and observations from field visits suggest that the site-specific ground displacements in the coastal mega-city of Karachi could be attributed to multiple factors, mainly excessive extraction of groundwater for domestic and industrial usage, or SWI engendered by geomorphological variations in the sub-soils (more details provided in the Discussion section). However, additional evaluations might be a prerequisite for confirming this—presenting an interesting research question for future studies. In addition, the underlying geology should also be considered to understand the drivers triggering the ground displacements in the study region, as highlighted in the results [26]. Therefore, in this case study, a revised and updated geological map of Karachi (Map Series No. 3) prepared by the Geological Survey of Pakistan is combined with ancillary geospatial information to understand the role of geology and lithology in causing displacements in the study area. See Section 5.1.1 for more details.

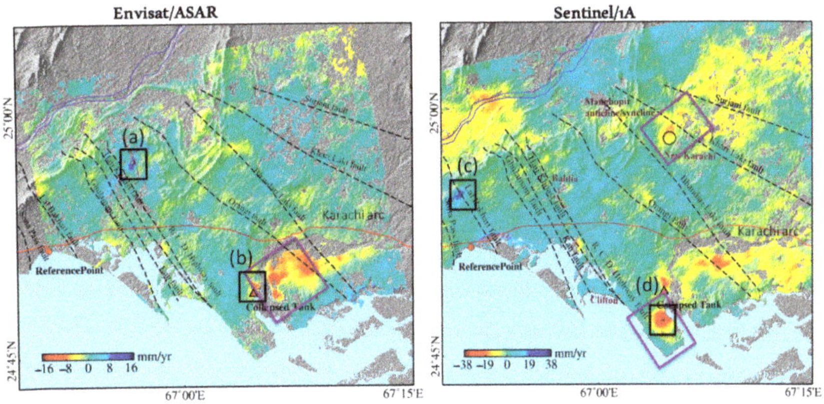

Figure 3. Vertical ground displacements in Karachi mapped through ENVISAT/SAR images over 2004–2010 (**left**) and Sentinel-1A images over 2014–2016 (**right**). Geologic features are overlaid, and major locations undergoing ground displacements are labeled (a–d) and highlighted. Positive values represent uplift, negative values represent ground subsidence.

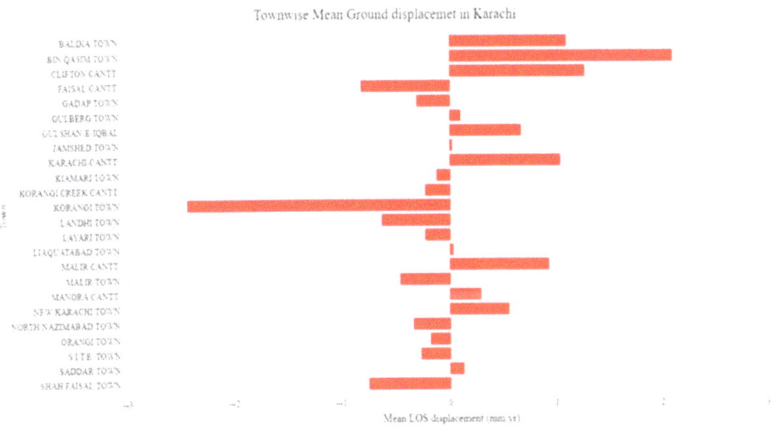

Figure 4. Town-wise mean ground displacements in Karachi.

The highest negative vertical ground displacement (subsidence) rates are found in the town of Clifton Cantonment, in the Defense Housing Authority (DHA) and Korangi Industrial Area (KIA) located on the downside of the seasonal river of Malir, both with a mean rate of vertical ground displacement greater than ~−5 mm/yr (see Figures 3 and 4). The town of Korangi houses about 3000 facilities for various industries, including steel, textile, automobile, pharmaceutical, chemical, flour, and engineering mills. This area is topped with irregular and loose deposits of semi-consolidated gravel in and around Karachi, such as the Malir River area, which are excessively used for construction purposes. The Malir Riverbed is formed by the recent/sub-recent stream bed deposits (Qsd) comprising loose fine-to-coarse sand, gravels, pebble, and boulders. Some of these materials, such as gravel, are of economic importance. Therefore, the subsidence in Korangi is most likely associated with this area's industrial and construction activities.

Uplift is also evident in some areas in the western end of the area of study, mainly those neighbouring the Mangopir Anticline (MA). The MA, and the adjoining Lalji syncline, are major large fault-induced Kirthar fold structures of the region that are cut across by the numerous northwest-trending sinistral fault lines (such as the Sona Pass fault), as shown in Figure 3. Displacement and movement of these wrench faults are antithetic to the southward-creeping Karachi Arc [42], which supports and reflects the ongoing uplift in the areas highlighted in the map (Figure 3). Sarwar and Alizai [43] also report that the triple junction tectonics [44] and the active fold belt geologic activity in the coastal region [43] reflect on the ground displacements in this area.

It is evident from the TCPInSAR-derived vertical map from Sentinel-1A acquisitions (Figure 3) that quite a large part of the area remains unstable after 2010. The map shows a spread of higher ground-displacement values to the new areas seen by Envisat/ASAR, particularly in the industrial zones of the city and reclaimed lands of the DHA Phase-VIII. Figure 5 summarizes the minimum, maximum, and mean regional change in vertical ground displacement in Karachi during 2004–2016. Subsidence in Karachi's east zone (with the highest subsidence rate reaching ~16 mm/yr in 2004–2010) has increased up to 38 mm/yr during 2014–2016 in Karachi's southern zone, including the neighboring Bundal and Buddo Islands (BI-5) (See Table 1).

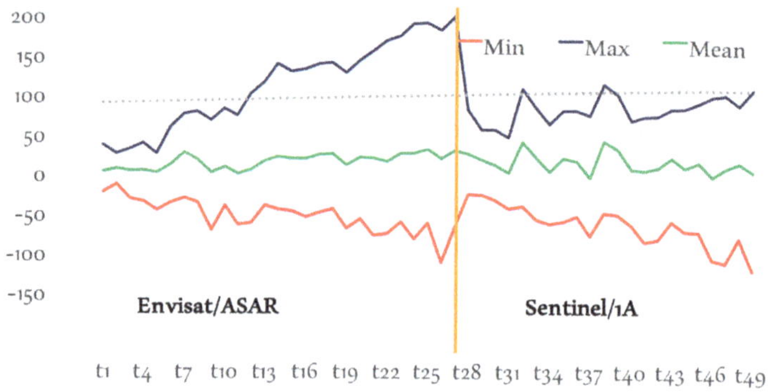

Figure 5. Minimum, maximum, and mean regional change in vertical ground displacement in Karachi during 2004–2016. The vertical orange line marks the data gap between two datasets (i.e., 2011–2013).

Similarly, in a rapidly urbanizing world, land-use and land-cover changes are also important factors to account for. Figure S2 in the Supplementary Materials reveals that most of the ground subsidence is identified in built-up and residential areas—particularly in Karachi's eastern region in the study area. Both of these are well-known factors of ground subsidence. On the contrary, the uplift is notable in Karachi's southern region, particularly in areas of commercial land use, followed by areas of transportation and build-up.

Table 1. Administrative-zone-wise statistical summary and trend of vertical ground displacements in Karachi (2004–2016).

Zone	Sites	Envisat/ASAR			Sentinel/1A			Trend
	(Towns)	Min	Max	Mean	Min	Max	Mean	
West Karachi	Baldia, S.I.T.E, Orangi	−10.81	24.04	−0.21	−18.93	33.082	0.58	↓
South Karachi	Kiamari, Saddar, Layari, Clifton, DHA, Manora	−9.92	5.96	0.797	−37.96	12.792	0.46	↑
East Karachi	Shah Faisal, Korangi, Landhi, Korangi Creek Cantonment, Gulshan e Iqbal, Jamshed	−15.62	6.57	−1.04	−22.44	10.929	−1.83	↑
Malir	Bin Qasim, Malir Cantonment, Gadap	−18.49	10.90	1.061	−16.03	11.78	−3.35	↑
Central Karachi	North Nazimabad, New Karachi, Gulberg, Liaqatabad	−5.28	5.86	0.134	−16.23	12.683	−0.97	↑

4.1.1. Vertical Ground Displacement Time Series

To examine the trends of ground displacement over time, the vertical ground displacement time series of four sites are represented in Figure 6; see Figure 3 for locations. Figure 6a illustrates the time series of an area of highest uplift bounded by the RCD Highway fault and the Orangi faults. Figure 6b profiles the time series of the area around the location of the collapsed tank. Figure 6c shows the time series of an area with the relatively lowest and most-stable subsidence over time, whereas Figure 6d shows profiles of the area of highest subsidence, which is increasing over time in the lands of the DHA reclaimed from the Arabian Sea.

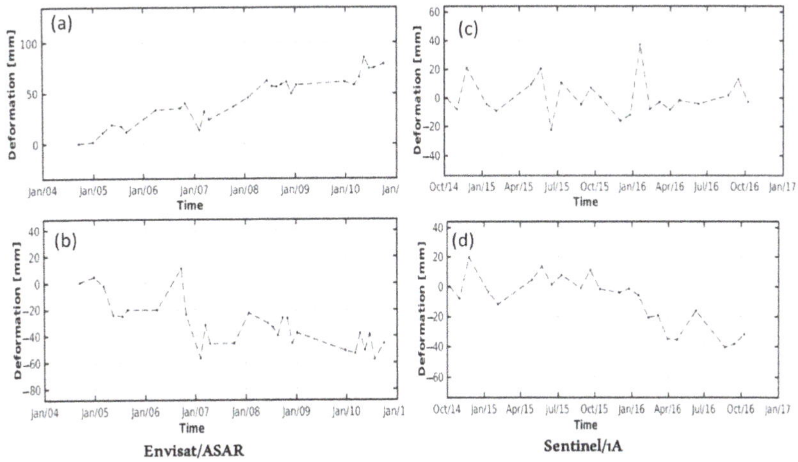

Figure 6. Time series of vertical ground displacement for the selected representative points (**a**–**d**) mapped in Figure 3. (**a**) Represents the location undergoing uplift near the Mangopir Hills in West Karachi; (**b**,**d**) mark the subsiding location in the area where water tanker collapsed in the DHA in South Karachi; whereas (**c**) represents a relatively stable location between the Haji Khoast Fault and Sona Pass fault in West Karachi.

4.2. Erosion Rate along the Karachi Coast (1989–2018)

The most developed part of the Sindh coast is the NW-SE-oriented coastal stretch of Karachi, characterized by raised beaches, marine terraces, sand dunes, shallow lagoons, and sea cliffs. A closer inspection of the coastline's change shows that the coastline of the study area oscillates between accretion and erosion (Figure 7). Annual coastal erosion is increasing along the west-east axis (same as the littoral drift), reaching its maximum of ~2.4 m/yr in East Karachi, located next to the Buddo and Bundal barrier islands, which have been eroding at an even higher rate (i.e., ~14.33 m/yr) [13]. In contrast, the displacement patterns are variable across the city over the 2004–2016 timeframe and along the coastline, precluding the probability of a relationship between coastline erosion and ground subsidence in general. The spatial distribution map of erosion along the coastline shows large distinctiveness for South Karachi relative to the other two sub-zones (East Karachi and West Karachi). An accretion rate up to ~8.34 m/yr is noted at the south side of the entrance to Korangi creek and the southern segments of the study area's coastline (i.e., the South Karachi sub-zone) (Figure 7). These noticeable changes in accretion observed along South Karachi's coastline are apparent in Figure 7, and they surface notably in the corresponding images from Landsat in different years. This finding is consistent with that of Luijendijk, et al. [45], who also found South Karachi to be experiencing very high accretion during 1984–2016. It is worth mentioning that the accretion measured in this sub-zone is artificial and predominantly attributed to the lands reclaimed from the Arabian Sea for the development of coastal infrastructure [46]. A comparison of the satellite images before and during the study period (more details are given in the following sub-section, Discussion) corroborates the coastline assessment's findings in this sub-zone (South Karachi).

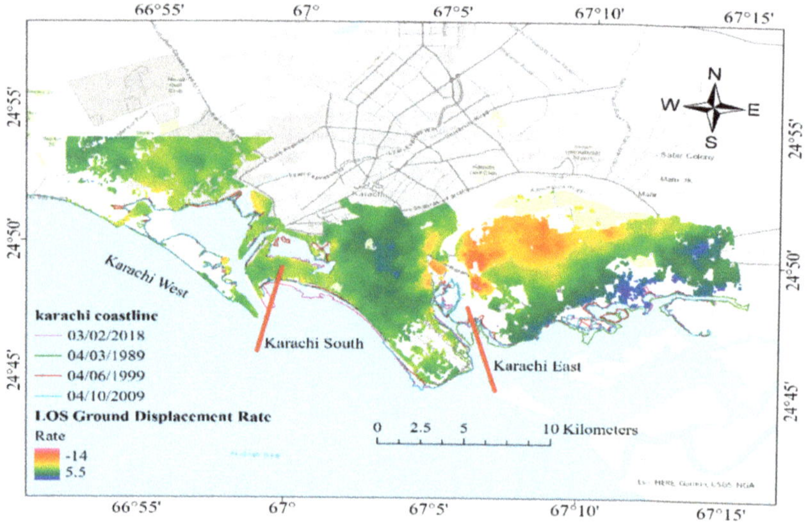

Figure 7. Evolution of Karachi's coastline from 1989–2018, obtained from Landsat along with LOS ground displacement (2004–2010).

Topographic slope controls vulnerability to coastal erosion, seawater intrusion, and permanent flooding. In comparison, the West Karachi sub-zone in the far west of the study area has a steep gradient (i.e., ~0–15.5%), whereas the other two sub-zones (i.e., South Karachi and East Karachi) have ~0–3.5% slope gradients. Most of the inland areas along the coastline are already at the same elevation level of the Arabian Sea. The slope of the land along Karachi's coastline is decreasing eastward while coastline erosion is increasing eastward [7], endangering the eastern segments of the study area, where Port Muhammad Bin Qasim, a major deep seaport of the country, is located towards Korangi creek. The development of a navigational channel to facilitate the passing of cargo ships via Phitti Creek to the port has caused erosion of up to ~−2.43 m/yr. In 1970, when this seaport started functioning, natural barrier islands located next to Port Muhammad Bin Qasim formed the first line of shielding against coastal flooding and storms. Erosion in the East Karachi sub-zone suggests that these natural barriers gaurding against the erosional effects coming from the Arabian Sea have already been badly impacted. This finding is broadly in accord with the earlier work of Kanwal, et al. [13], which also found that the islands were undergoing very high erosion (~8.9 m/yr). However, more assessment in this regard is required—this is left for future work.

4.3. Influence of Ground Displacement on Coastal Erosion in Karachi

The coastline changes and vertical ground displacement findings are not contrasting. While erosion processes underlie most of the eastern parts of the study area, the rate of vertical ground displacement is spatially variable along the coastline across the study area (See Figure 8). In other words, not all coastline segments with high annual mean erosion rates are linked with mean subsidence (Table 2). This implies that subsidence can exacerbate ongoing erosion and its impacts. Regions that exhibit coastal erosion show a variable ground subsidence signal along the coastline, with larger values in Karachi's southern and eastern zones, particularly in the areas located next to the Indus Delta during 2004–2010 [17]. A similar interaction is observed for the coastline segment where the rate of coastline erosion up to ~2.4 m/yr corresponds spatially with a rate of vertical ground displacement up to ~−1.4 cm/yr. These observations illustrate the regional variability of these complex drivers responsible for coastal changes of the low-lying deltaic coasts.

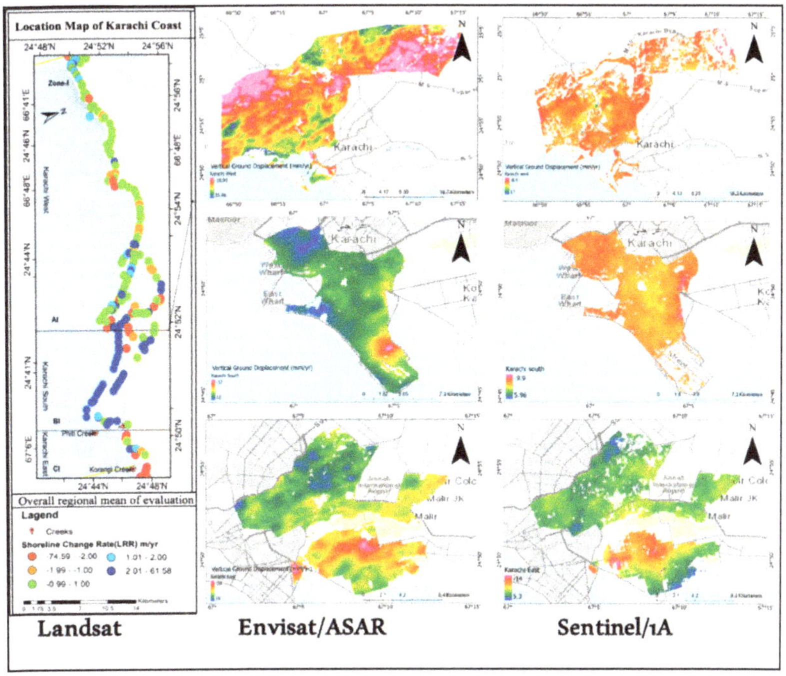

Figure 8. Karachi coastline change rates estimated by the LRR method using USGS DSAS for the period 1989–2018 (**left**). The zone-wise map shows the regional distribution of mean vertical ground displacement in the coastal zones (**middle** and **right**). These distribution maps show an expansion in extent between subsidence and uplifting over the time in the study area from 2004 to 2016 with substantial subsidence in the DHA, Korangi, and Landhi areas.

It is noted that the present trend of coastline erosion increasing (west-east) will continue to cause coastal changes in the eastern segments of Karachi (including Korangi Creek and the outlying natural barrier islands), mainly due to both marine processes and human intervention, if timely and appropriate actions are not taken. In Karachi's eastern zone, the subsidence rate reached its highest at 16 mm/yr during 2004–2010, but this has increased up to 38 mm/yr during 2014–2016 in Karachi's southern zone and the neighboring Bundal and Buddo barrier islands. The maximum subsidence rate quantified in this study is greater than the tide gauge-measured rate of SLR (i.e., 1.9 mm/yr) measured by Kanwal, et al. [7] and 3.6 mm/yr Khan, et al. [11] (see Table 2). This indicates that although subsidence alone does not account for the high rates of coastline erosion, it can intensify the erosive processes by inviting more seawater inland and is associated with more frequent over-washes and lagoon breaching [47].

Table 2. Coastal evolution and possible contribution of erosion, ground displacement, and mean sea levels along different section of Karachi's coast.

Section	Coastline Change Rate (m/yr)—Landsat			Coastal Zone Displacement Rate (m/yr)—TCPInSAR			Mean Sea Level (mm/yr)
	Min	Max	Mean	Min	Max	Mean	
West Karachi	−0.43	−4.112	2.168	−0.33	−7.14	8.60	1.9
South Karachi	8.34	−74.59	62.04	0.22	−5.8	2.92	
East Karachi	−2.43	−10.66	2.5	2.1	−13.95	7.7	

5. Discussion

This study provides a detailed assessment of ground displacements and their possible association with coastal erosion in the largest city of Pakistan—Karachi—via leveraging integrated remote sensing (RS) and geographic information systems (GIS) practices. The results presented here are of particular interest to coastal/urban planners, decision-makers, and relevant stakeholders. For instance, the areas with the largest ground subsidence (i.e., Korangi and reclaimed areas) should be prioritized when evaluating the possible causes and designing measures to prevent them. This could progressively help avoid impacts to society and reduce the likelihood of financial losses. In addition, this research provides a roadmap to establish a baseline inventory of ground displacement information at localized levels (i.e., Figure 3). This assessment can further be extended to higher-resolution administrative units to provide reference to local bodies for designing appropriate action plans. For example, Figure S3 in the Supplementary Materials provides a comparison of different union councils (the smallest admin-level for resource allocation in Pakistan). The LOS ground displacement in ~55% of union councils (29 out of 52) should be a point of immediate concern for the relevant authorities.

5.1. Potential Factors Inducing Ground Displacement

5.1.1. Role of Geology and Lithology

As mentioned earlier, the entirety of Karachi lies on the junction where two tectonic plates meet, with several active faults passing through the city and causing geological activity in the study area, as shown in Figure 9 [48]. Although these geological movements may cause ground displacements in this area of interest, the amount and distribution of ground displacements are not fully understood. Sarwar and Alizai [43] suspect that epicenters of a group of earthquakes located between the Malir River and Hyderabad motorway might explain the current shallow ground displacements in this area (Figure 9). Areas along and south of the suspected Malir-Lyari Rivers fault zone, from where earthquake reports regularly come, are the same places where higher ground displacement rates are observed. This could be a sign that the Malir River fault is active. To understand the role of currently mapped geologic faults in the study area and their contribution to cause ground displacements estimated through interferometric processing, ancillary geospatial information related to the current faults in the study area is incorporated. However, no significant association between the currently mapped faults and ongoing vertical ground displacements is observed in the study area. Sarwar and Alizai [43] anticipate geological structures in the city that are currently unmapped, as marked by the deflection in the riverine flow. Another potential factor causing ground displacements could be flash flooding in the Malir Riverbed, which might, in turn, increase the moisture level of the soils, weaken the bond strength between the soil particles, and thus causing the ground to subside. The Malir River Basin used to have an extensive network of streams that is now being lost from the city's geomorphology due to unplanned urban encroachments. Our results put emphasis on paying due care and consideration to the environmental aspects of aforesaid unplanned urban and geological development activities and encroachments.

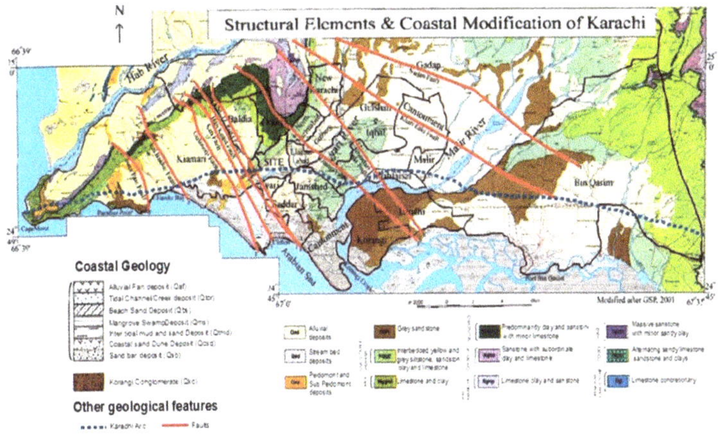

Figure 9. Fault lines (red lines) and Karachi Arc (thick dotted blue line) overlaid on the Geological Map of Karachi. Karachi Arc appears to be creeping eastward, which is directly related to active subsidence of the Hyderabad graben that underlies it (Source: [27,43]).

5.1.2. Unregulated and Excessive Groundwater Extraction

Another factor related to the site-specific variability in the displacement rate is the uneven distribution and unregulated installation of private groundwater extraction wells. In line with other studies [18,49,50], it is noticed during field visits for this study that acute water shortages in the Karachi city have forced residents to install private groundwater extraction wells on a large scale to meet their domestic water needs. The lack of water supply has given way to excessive groundwater extraction in the Clifton and DHA coastal areas, which is likely to cause subsidence, a phenomenon identified along the National Highway [51]. Amin, et al. [18] argued that uncontrolled groundwater withdrawal has caused ground subsidence in the Karachi city, based on the recorded data of the depleting groundwater levels in Qayyumabad, investigated by Khan, et al. [52]. However, no data on groundwater table for areas along the coast and sewerage network plan for the city are available. Consequently, it cannot be included in the analyses. If available, integration of this information could facilitate simulation of aquifer compaction and scenario analysis. Assessment of vertical ground displacements (subsidence) can provide more information with respect to different factors, such as interbed storage, aquifer compaction due to groundwater extraction, and overuse of the land. For example, Jafari, et al. [49] has undertaken similar work for one of the cities in Iran. Similarly, Jafari, et al. [50]) forecasted subsidence due to the over-extraction of water using a MODFLOW model and InSAR integration.

5.1.3. Structural Damage Due to Failed Sewerage System

Vertical ground displacements in the study area are also linked to land-use activities, such as mining, shallow compaction or structural instabilities, and sewerage network breakage. During the collection of photographic evidence (Figure 10), it was learned from the local people and petroleum experts that another probable source of site-specific/localized variations in the ground displacements, observed in different parts of the study area (accelerating SWI), could be broken sewerage and water supply lines (a potential source of soil liquefaction in the study area) [53]. A drain collapse could cause the ground around it to subside over time via water leaking from the pipe and into the soil. The structural damage caused by failed sanitation and drainage networks explains the site-specific variability in the ground displacement rates.

Figure 10. Signs of ground displacements in the study area. (**a**) Cracks in the building walls due to differential ground displacement; (**b**) elevation of a residential house in the University of Karachi's Professors colony has dropped to ~2 m below its base-level due to soil liquefaction caused by seawater intrusion. During the field visit, signs of soil salinity were observed in different parts of the University of Karachi.

Lodi, et al. [54] also posited immense leakage and overflows from damaged water supply and sewerage lines throughout the Karachi, particularly in the coastal areas, which is likely to wash away the underlying soil and cause soil liquefaction and settlement. Furthermore, JICA [53] reported that sewer line waste had been mixed with water supply in some areas, causing unhygienic conditions and diseases. It is unfortunate that no comprehensive GIS information about the water supply system and sewerage network is currently available or accessible to researchers, hindering the ground displacement evaluation along the broken sewer lines and the health of this mega-city's sewerage as well as its water supply system. From reports from various media sources, the National Institute of Oceanography (NIO), and other institutions, it is evident that Karachi is indeed facing the dual threat of ground displacements and sea-level rise. However, it is unclear whether ground displacement gives rise to sea levels, favoring coastline erosion, or vice versa.

5.1.4. Land Reclamation

Many high-rise buildings along Clifton beach, DHA Phase-VII, and several housing projects have been developed on offshore reclaimed lands, as shown in Figure 11. Observations from different stakeholders, reported in the final Environmental Impact Assessment (EIA) reports of various housing projects, highlight that signs of subsidence are apparent in the coastal area and any further stress is likely to produce a catastrophic situation [55,56].

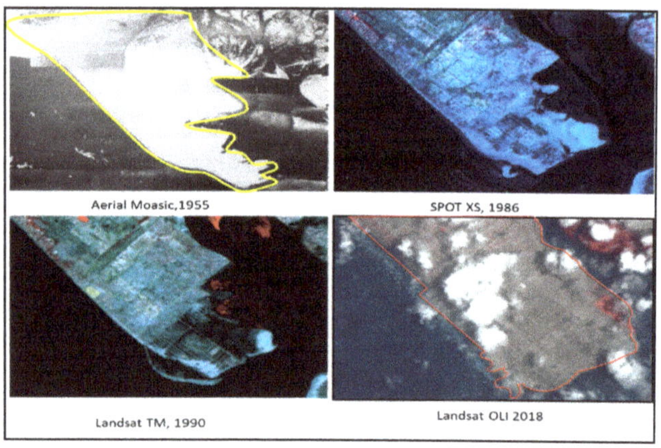

Figure 11. Expansion of Defense Phase VIII over land reclaimed from the Arabian Sea during the 1955–2018 period.

5.1.5. Liquefaction

The coastal region of Karachi is highly prone to soil liquefaction in the case of any severe seismic events of magnitude seven or above [54]. This problem was realized after scientific investigations of the water storage tank collapse in 2006 in the DHA. Mahmud and Sheikh [57] reported that an 80-foot high water tank in the DHA collapsed and was vertically submerged 70 feet into the ground in 8 min due to an earthquake event that induced liquefaction. The remaining 10 feet of the tank stayed above the surface. Soils in this part of the study area are particularly susceptible to liquefaction threats, being rich in water and increased salinity with poor cementation of soil grains [25]. This study found a higher average subsidence rate (>15 mm/yr) near the location of the water tank collapse, as shown in Figure 11. Different areas in the DHA and Clifton regions are unstable and are thus undergoing ground deformation, as large parts of these areas are comprised of land reclaimed from the sea. Another potential reason for these localized ground displacements is damaged sewerage and water lines, which might increase the possibility of soil liquefaction. Recently, various commercial projects have been announced for the coast, which raises many concerns [58] due to the land's potential for liquefaction and subsidence. Along the study area's coastline, the analysis of six specific sites for comparison of their coastline erosion and InSAR derived coastal displacement (Figure 12) shows the potential of InSAR to examine and explain localized changes in coastal areas. However, as the specific sites selected for the time series analysis are not all located close to the coastline, their vertical ground displacement is not directly affected by the coastline's erosive regime. Therefore, it can be considered to reflect subsidence only [47].

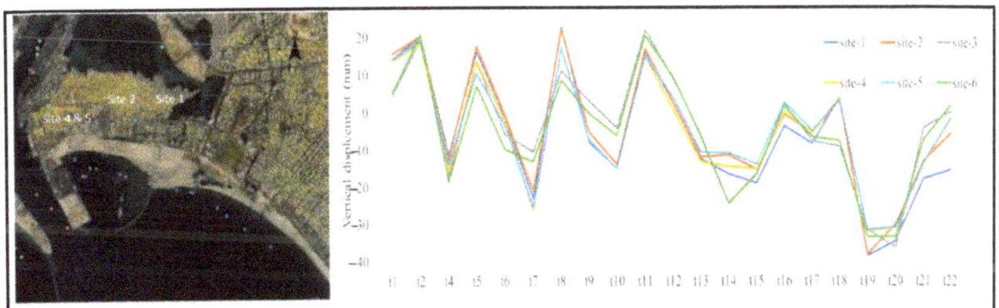

Site	Town	Land use	Erosion (m/yr)	Mean vertical displacement (mm)
site-1	Karachi south	Residential, and Residential with Commercial Services	-1	-6.14
site-2	Karachi south	Built-up Open/Barren Land,	-9.63	-4.186
site-3	Bin Qasim	Built-up Open/Barren Land	-0.25	-4.204
site-4	Karachi south	Industrial and Manufacturing	-9.97	-5.50
site-5	Karachi south	Built-up Open/Barren Land	51.54	-4.71
site-6	Korangi Cantonment	Military Area (built-up)	2.52	-5.39

Figure 12. Location of six sites selected for coastline erosion and mean vertical displacement (**left**) spatial analysis. Mean vertical ground displacement time series from Envisat/ASAR 2014–2016 (**top right**) and land use, coastline change rate, and displacement values at each site in different coastal towns of the study area (**bottom right**).

6. Limitation and Way Forward

There are some limitations of this ground displacement study that restrict the generalization of the results at this stage. For instance, because GPS records were not available in the Karachi city and neighboring Indus River Delta, subsidence could only be detected relative to a reference point that was assumed to be stable. Calibration with in-situ GPS devices allows precise measurements of absolute ground displacement rates and allows InSAR to resolve more processes. Wherever possible, in situ GPS records should be exploited to ensure that InSAR can provide absolute subsidence measurements. However,

GPS specialists measuring absolute subsidence might collaborate with oceanographers or wetland botanists who understand local oceans and flora in order to evaluate the true stability of low-lying lands in the coastal regions. Furthermore, validation of the ground displacements observed using InSAR data was mainly inhibited by the unavailability of ground surveys. Therefore, the present study primarily focused on analyzing the results from Envisat and extending the analysis using Sentinel-1A for further investigation and validation. We believe that monitoring the progressive ground displacement with SAR interferometry can offer valuable evidence for assessing the impacts of ground displacement and its influence on coastal erosion and relative sea levels, and provide important references for adaptation strategies [59]. The results presented in this study will potentially influence the monitoring of risk to infrastructure on a regular basis, the collection of ground surveys for validation of future data, and outcomes based on the location of the ground displacements and geological patterns.

Similarly, with the launch of new SAR satellites and sophisticated processing of the altimetry products, unprecedented near real-time monitoring of vulnerable coastal/delta cities will become possible. SAR data must be made freely and easily accessible to researchers by space agencies for all of these applications. A priority for future work may also be extending ground displacement work to the other two zones from the IDR (i.e., the East and West Indus River zones).

7. Conclusions

It is well documented that the location of the coastline in sinking basins depends on the rates of ground displacement and sediment influx. For the Indus River Delta and its surroundings, including the Karachi coastline, these two factors have varied together with relative sea-level rise, causing coastline regression and transgression over time. Therefore, we believe that the SLR-based assessments of inundation risk due to flooding have underestimated the actual risk in Karachi's coastal regions. The analysis of datasets shows visible displacement in various sections of the city and erosion along the low-lying Karachi coastline. To gauge the risk to infrastructure based on the spatial distribution and pattern of ground displacements, available high-resolution remote sensing satellite data and geological information should be acquired and incorporated. In this context, the primary purpose of this study was to examine the interplay between ground displacements (2004–2016) and coastline changes (1989–2018) in Karachi (the most populous city in Pakistan) via combining the Landsat images and InSAR data.

It was observed that the annual mean coastline erosion increased from west to east. In contrast, the ground displacement patterns were highly variable across the city and along the coastline, precluding a link between coastal erosion rates and subsidence in general. Using InSAR in the proximity of the coastline was quite complex due to the ever-changing condition of the dynamic coastal landscape. Within a distance of a few hundred meters from the coastline, the ground displacement rates obtained by InSAR could be affected by sediment deposition. The vertical ground displacements obtained through interferometric processing of both the Envisat/ASAR and Sentinel-1A images showed that quite a large part of the study area was not stable. This is unsurprising since the Indus River delta, which is about 79 miles from Karachi, is already subsiding at the rate of 1.3 mm/yr, as confirmed by Syvitski, et al. [3]. Taken together, the findings of this study suggest that some areas along the Karachi coast are subsiding at rates comparable to the SLR, which may amplify the rates of the relative SLR in years to come. The miscellaneous data suggest that the coastal areas face the hazard of seismo-tectonic incidents leading to settling the land in some parts and uplifts in other areas caused by mining activities, shallow compaction, structural instabilities, excessive groundwater extraction, sewerage network breakage, and/or liquefaction.

There is a lack of research in the study area that jointly discusses the localized dynamics of coastal erosion and ground displacements. To the best of our understanding, the present study is among the pioneer attempts, particularly in developing countries, to highlight unseen factors (i.e., sewer lines breaking, overuse of land, and coastline erosion) which might be responsible for ground displacement. In addition, we examined displacement's influence along the Karachi coastline by utilizing Landsat images in combination with InSAR. Kanwal, et al. [7] discussed that the coastline retreat was caused by sediment supply to the Indus River and SLR. Amin et al. (2021) discussed the ongoing subsidence in small segments of Karachi. This study demonstrates that ground displacements and their influence on the coastline erosion in the study region are important drivers for coastal changes and should also be considered along with other factors in further research for coastal hazard assessment at the regional and local scale. This work will serve as a baseline and be used as a blueprint to track coastline changes and ground deformations once longer records of InSAR and in situ data become available. In addition to contributing to the explanation of the longitudinal erosion in the study area, the outcomes of this project are relevant for real estate developers who are reclaiming land from the Arabian Sea to meet the housing needs of the city. This study will potentially influence the monitoring of risk to infrastructure on a regular basis, the collection of ground surveys for validation of future data, and outcomes based on the location of ground displacements and geological patterns.

Supplementary Materials: The following supporting information can be downloaded at: https://www.mdpi.com/article/10.3390/rs14092054/s1, Figure S1: Perpendicular and temporal baselines of the selected interferometric pairs (blue line) (a) ENVISAT ASAR, and (b) Sentinel-1A IW SLC (green dots); Figure S2: Ground displacement in different land use land cover areas in the study area; Figure S3: Union council wise localized assessment on ground displacement; Table S1: Coastal Evolution and possible contribution at each section of Karachi coast; Table S2: Sentinel-1A SLC images used for ground displacements monitoring during 2014–2016 period over Karachi (*Master image); Table S3: Details of Landsat images used in the study for coastline delineation and erosion estimation.

Author Contributions: Conceptualization and methodology, S.K.; formal analysis, S.K. and S.W.; investigation and data curation, S.K.; resources, X.D.; and writing—original draft preparation, S.K.; writing—review and editing, S.K., M.S., S.W. and X.D.; visualization, S.K., S.W. and M.S.; supervision, project administration, and funding acquisition, X.D. All authors have read and agreed to the published version of the manuscript.

Funding: The research was jointly supported by the Research Grants Council (RGC) of the Hong Kong Special Administrative Region (PolyU 152164/18E and PolyU 152233/19E), the Research Institute for Sustainable Urban Development (RISUD), The Hong Kong Polytechnic University, and the Innovative Technology Fund (ITP/019/20LP). The first author would like to thank the RGC for the Hong Kong PhD Fellowship provided.

Data Availability Statement: The data presented in this study are openly available in the repositories, acknowledged and mentioned in the data source within this article.

Acknowledgments: The authors would like to thank three anonymous reviewers for their insightful and valuable comments which greatly improved the quality of this article.

Conflicts of Interest: The authors declare no conflict of interest.

References

1. Kontgis, C.; Schneider, A.; Ozdogan, M.; Kucharik, C.; Tri, V.P.D.; Duc, N.H.; Schatz, J. Climate Change Impacts on Rice Productivity in the Mekong River Delta. *Appl. Geogr.* **2019**, *102*, 71–83. [CrossRef]
2. Syvitski, J.P.; Kettner, A.J.; Overeem, I.; Giosan, L.; Brakenridge, G.R.; Hannon, M.; Bilham, R. Anthropocene Metamorphosis of the Indus Delta and Lower Floodplain. *Anthropocene* **2013**, *3*, 24–35. [CrossRef]
3. Syvitski, J.P.M.; Kettner, A.J.; Overeem, I.; Hutton, E.W.H.; Hannon, M.T.; Brakenridge, G.R.; Day, J.; Vörösmarty, C.; Saito, Y.; Giosan, L.; et al. Sinking Deltas Due to Human Activities. *Nat. Geosci.* **2009**, *2*, 681–686. [CrossRef]
4. Dai, Z.; Mei, X.; Darby, S.; Lou, Y.; Li, W. Fluvial Sediment Transfer in the Changjiang (Yangtze) River-Estuary Depositional System. *J. Hydrol.* **2018**, *566*, 719–734. [CrossRef]
5. Aucelli, P.P.C.; Di Paola, G.; Incontri, P.; Rizzo, A.; Vilardo, G.; Benassai, G.; Buonocore, B.; Pappone, G. Coastal Inundation Risk Assessment Due to Subsidence and Sea Level Rise in a Mediterranean Alluvial Plain (Volturno Coastal Plain—Southern Italy). *Estuar. Coast. Shelf Sci.* **2017**, *198*, 597–609. [CrossRef]

6. Milliman, J.D.; Haq, B.U. *Sea-Level Rise and Coastal Subsidence: Causes, Consequences, and Strategies*; Springer Science & Business Media: Berlin/Heidelberg, Germany, 2013; Volume 2.
7. Kanwal, S.; Ding, X.; Sajjad, M.; Abbas, S. Three Decades of Coastal Changes in Sindh, Pakistan (1989–2018): A Geospatial Assessment. *Remote Sens.* **2020**, *12*, 8. [CrossRef]
8. Do, A.T.; de Vries, S.; Stive, M.J.F. Beach Evolution Adjacent to a Seasonally Varying Tidal Inlet in Central Vietnam. *J. Coast. Res.* **2018**, *34*, 6–25. [CrossRef]
9. Duc, D.M.; Nhuan, M.T.; Van Ngoi, C. An Analysis of Coastal Erosion in the Tropical Rapid Accretion Delta of the Red River, Vietnam. *J. Southeast Asian Earth Sci.* **2012**, *43*, 98–109. [CrossRef]
10. Duc, D.M.; Yasuhara, K.; Hieu, N.M.; Lan, N.C. Climate Change Impacts on a Large-Scale Erosion Coast of Hai Hau District, Vietnam and the Adaptation. *J. Coast. Conserv.* **2017**, *21*, 47–62. [CrossRef]
11. Khan, F.A.; Khan, T.M.A.; Ahmed, A.N.; Afan, H.A.; Sherif, M.; Sefelnasr, A.; El-Shafie, A. Complex Extreme Sea Levels Prediction Analysis: Karachi Coast Case Study. *Entropy* **2020**, *22*, 549. [CrossRef]
12. Kidwai, S.; Ahmed, W.; Tabrez, S.M.; Zhang, J.; Giosan, L.; Clift, P.; Inam, A. The Indus Delta—Catchment, River, Coast, and People. *Coasts Estuaries* **2019**, *2019*, 213–232.
13. Kanwal, S.; Ding, X.; Sajjad, M.; Nazeer, M.; Zia, I. Remote Sensing of Narrowing Barrier Islands along the Coast of Pakistan over Past 30 Years. *J. Mar. Sci. Eng.* **2021**, *9*, 295. [CrossRef]
14. Bilham, R.; Lodi, S.; Hough, S.; Bukhary, S.; Khan, A.M.; Rafeeqi, S.F.A. Seismic Hazard in Karachi, Pakistan: Uncertain Past, Uncertain Future. *Seismol. Res. Lett.* **2007**, *78*, 601–613. [CrossRef]
15. Fattahi, H.; Amelung, F.; Chaussard, E.; Wdowinski, S. Coseismic and Postseismic Deformation Due to the 2007 M5.5 Ghazaband Fault Earthquake, Balochistan, Pakistan. *Geophys. Res. Lett.* **2015**, *42*, 3305–3312. [CrossRef]
16. Panhwar, M.H. *Seepage of Water of the River Indus and Occurrence of Fresh Ground Water in Sindh. The Indus River: Biodiversity, Resources, Humankind*; Oxford University Press: Delhi, India, 1999; pp. 180–197.
17. Rahman, A.-U.; Khan, A.N.; Shaw, R. *Disaster Risk Reduction Approaches in Pakistan*; Springer: Tokyo, Japan, 2015.
18. Amin, G.; Shahzad, M.I.; Jaweria, S.; Zia, I. Measuring Land Deformation in a Mega City Karachi-Pakistan with Sentinel SAR Interferometry. *Geocarto Int.* **2021**, 1–15. [CrossRef]
19. Kanwal, S.; Ding, X.; Zhang, L. Measurement of Vertical Deformation in Karachi Using Multi-Temporal Insar. In Proceedings of the IEEE International Geoscience and Remote Sensing Symposium (IGARSS 2018), Valencia, Spain, 23–27 July 2018.
20. Shi, X.; Liao, M.; Li, M.; Zhang, L.; Cunningham, C. Wide-Area Landslide Deformation Mapping with Multi-Path ALOS PALSAR Data Stacks: A Case Study of Three Gorges Area, China. *Remote Sens.* **2016**, *8*, 136. [CrossRef]
21. Wang, H.; Wright, T.J.; Yu, Y.; Lin, H.; Jiang, L.; Li, C.; Qiu, G. InSAR Reveals Coastal Subsidence in the Pearl River Delta, China. *Geophys. J. Int.* **2012**, *191*, 1119–1128. [CrossRef]
22. Tomás, R.; Li, Z.; Liu, P.; Singleton, A.; Hoey, T.; Cheng, X. Spatiotemporal Characteristics of the Huangtupo Landslide in the Three Gorges Region (China) Constrained by Radar Interferometry. *Geophys. J. Int.* **2014**, *197*, 213–232. [CrossRef]
23. Tosi, L.; Teatini, P.; Strozzi, T. Natural versus Anthropogenic Subsidence of Venice. *Sci. Rep.* **2013**, *3*, 2710. [CrossRef]
24. Wang, J.; Gao, W.; Xu, S.; Yu, L. Evaluation of the Combined Risk of Sea Level Rise, Land Subsidence, and Storm Surges on the Coastal Areas of Shanghai, China. *Clim. Chang.* **2012**, *115*, 537–558. [CrossRef]
25. Hamid, G. Geotechnical Evaluation of Rocks and Soils of Karachi Area and Their Impact on Urbanization. Ph.D. Thesis, University of Karachi, Karachi, Pakistan, 2011.
26. Gehlot, S.; Ketelaar, V.B.H.; Verbree, E.; Hanssen, R.F. Conceptual Framework for PS-InSAR Deformation Interpretation Assisted by Geo-Information Technology. *High Resolut. Earth Imaging Geospat. Inf.* **2005**, 17–20.
27. Zhang, L.; Lu, Z.; Ding, X.; Jung, H.-S.; Feng, G.; Lee, C.-W. Mapping Ground Surface Deformation Using Temporarily Coherent Point SAR Interferometry: Application to Los Angeles Basin. *Remote Sens. Environ.* **2012**, *117*, 429–439. [CrossRef]
28. Soergel, U. *Radar Remote Sensing of Urban Areas*; Springer Science & Business Media: Berlin/Heidelberg, Germany, 2010; Volume 15.
29. Liu, G.; Jia, H.; Nie, Y.; Li, T.; Zhang, R.; Yu, B.; Li, Z. Detecting Subsidence in Coastal Areas by Ultrashort-Baseline TCPInSAR on the Time Series of High-Resolution TerraSAR-X Images. *IEEE Trans. Geosci. Remote Sens. Lett.* **2013**, *52*, 1911–1923. [CrossRef]
30. Sun, Q.; Zhang, L.; Ding, X.; Hu, J.; Liang, H. Investigation of Slow-Moving Landslides from ALOS/PALSAR Images with TCPInSAR: A Case Study of Oso, USA. *Remote Sens.* **2014**, *7*, 72–88. [CrossRef]
31. Zhang, Y.; Zhang, J.; Wu, H.; Lu, Z.; Guangtong, S. Monitoring of Urban Subsidence with SAR Interferometric Point Target Analysis: A Case Study in Suzhou, China. *Int. J. Appl. Earth Obs. Geoinf.* **2011**, *13*, 812–818. [CrossRef]
32. Zhang, L. *Temporarily Coherent Point SAR Interferometry*; The Hong Kong Polytechnic University: Hong Kong, China, 2012.
33. Zhang, L.; Ding, X.; Lu, Z.; Jung, H.-S.; Hu, J.; Feng, G. A Novel Multitemporal InSAR Model for Joint Estimation of Deformation Rates and Orbital Errors. *IEEE Trans. Geosci. Remote Sens.* **2013**, *52*, 3529–3540. [CrossRef]
34. Ashrafianfar, N.; Busch, W.; Dehghani, M.; Haghighatmehr, P. Differential SAR Interferometric Technique for Land Subsidence Monitoring Due to Groundwater Over-Exploitation in the Hashtgerd. In Proceedings of the 'Fringe 2009 Workshop', Frascati, Italy, 30 November–4 December 2009.
35. Han, J.-Y. *Interferometric Synthetic Aperture Radar Observation of Vertical Land Displacement in the Vicinity of the All-American Canal at the United States and Mexico Border*; ProQuest: Salt Lake City, UT, USA, 2008; Volume 69-03, p. 1535.
36. Deledalle, C.-A.; Denis, L.; Tupin, F.; Reigber, A.; Jager, M. NL-SAR: A Unified Nonlocal Framework for Resolution-Preserving (Pol)(In)SAR Denoising. *IEEE Trans. Geosci. Remote Sens.* **2015**, *53*, 2021–2038. [CrossRef]

37. Zhang, L.; Ding, X.; Lu, Z. Deformation Rate Estimation on Changing Landscapes Using Temporarily Coherent Point InSAR. In Proceedings of the Fringe Conference, Frascati, Italy, 19–23 September 2011.
38. Wang, Q.; Zhao, Q.; Ding, J.; Fedotov, A.A.; Badenko, V.; Liu, M.; Pepe, A. Investigation of the Ground Displacement in Saint Petersburg, Russia, Using Multiple-Track Differential Synthetic Aperture Radar Interferometry. *Int. J. Appl. Earth Obs. Geoinf.* **2020**, *87*, 102050. [CrossRef]
39. Xu, W.; Wu, S.; Materna, K.; Nadeau, R.; Floyd, M.; Funning, G.; Chaussard, E.; Johnson, C.W.; Murray, J.R.; Ding, X.; et al. Interseismic Ground Deformation and Fault Slip Rates in the Greater San Francisco Bay Area from Two Decades of Space Geodetic Data. *J. Geophys. Res. Solid Earth* **2018**, *123*, 8095–8109. [CrossRef]
40. Armenio, E.; De Serio, F.; Mossa, M. Analysis of Data Characterizing Tide and Current Fluxes in Coastal Basins. *Hydrol. Earth Syst. Sci.* **2017**, *21*, 3441–3454. [CrossRef]
41. Nassar, K.; Mahmod, W.E.; Fath, H.; Masria, A.; Nadaoka, K.; Negm, A. Shoreline Change Detection using DSAS Technique: Case of North Sinai Coast, Egypt. *Mar. Georesources Geotechnol.* **2019**, *37*, 81–95. [CrossRef]
42. Niamatullah, M.; Imran, M. Structural Geometry and Tectonics of Southern Part of Karachi Arc—A Case Study of Pirmangho & Lalji Area. In Proceedings of the PAPG/SPE Annual Technical Conference, Islamabad, Pakistan, 17–18 November 2009.
43. Sarwar, G.; Alizai, A. Riding the Mobile Karachi Arc, Pakistan: Understanding Tectonic Threats. *J. Himal. Earth Sci.* **2013**, *46*, 9–24.
44. Sarwar, G. Earthquakes and the Neo-Tectonic Framework of the Kutch-Hyderabad-Karachi Triple Junction Area, Indo-Pakistan. *Pak. J. Hydrocarb. Res.* **2004**, *14*, 35–40.
45. Luijendijk, A.; Hagenaars, G.; Ranasinghe, R.; Baart, F.; Donchyts, G.; Aarninkhof, S. The State of the World's Beaches. *Sci. Rep.* **2018**, *8*, 6641. [CrossRef]
46. Sengupta, D.; Chen, R.; Meadows, M.E. Building beyond Land: An Overview of Coastal Land Reclamation in 16 Global Megacities. *Appl. Geogr.* **2018**, *90*, 229–238. [CrossRef]
47. Gómez, J.; Kwoll, E.; Walker, I.; Shirzaei, M. Vertical Land Motion as a Driver of Coastline Changes on a Deltaic System in the Colombian Caribbean. *Geosciences* **2021**, *11*, 300. [CrossRef]
48. Ellouz-Zimmermann, N.; Lallemant, S.J.; Castilla, R.; Mouchot, N.; Leturmy, P.; Battani, A.; Buret, C.; Cherel, L.; Desaubliaux, G.; Deville, E.; et al. Offshore Frontal Part of the Makran Accretionary Prism: The Chamak Survey (Pakistan). In *Thrust Belts and Foreland Basins*; Springer: Berlin, Germany, 2007; pp. 351–366. [CrossRef]
49. Jafari, F.; Javadi, S.; Golmohammadi, G.; Karimi, N.; Mohammadi, K. Numerical Simulation of Groundwater Flow and Aquifer-System Compaction Using Simulation and InSAR Technique: Saveh Basin, Iran. *Environ. Earth Sci.* **2016**, *75*, 833. [CrossRef]
50. Jafari, F.; Javadi, S.; Karimi, N. Forecasting of Subsidence Due to Groundwater over Exploitation using MODFLOW and Interferometry Technique in Radar Imagery. In Proceedings of the 36th IAHR World Congress, The Hague, The Netherlands, 28 June–3 July 2015.
51. EMC. *Environmental Impact Assessment (EIA) "Ferere Excellency—A High Density Zone of Civil Lines, Karachi" Residential Cum Commercial Project*; EMC Paksitan Pvt. Ltd.: Karachi, Pakistan, 2017.
52. Khan, A.; Raza, S.A.; Fatima, A.; Haider, S.W. Assessment of Groundwater Quality in Coastal Region a Case Study of Qayyumabad, Karachi, Pakistan. *Asian Rev. Environ. Earth Sci.* **2020**, *7*, 9–17. [CrossRef]
53. JICA. *The Study on Water Supply and Sewerage System in Karachi in the Islamic Republic of Pakistan*; Japan International Cooperation Agency: Tokyo, Japan, 2008.
54. Lodi, S.H.; Sultan, W.; Bukhary, S.S.; Rafeeqi, S.F.A. Liquefaction Potential along the Coastal Regions of Karachi. *J. Himal. Earth Sci.* **2015**, *48*, 89–98.
55. KMC; Environmental Impact Assessment (EIA). *Grade Separated Traffic Improvement Plan from Park Tower Intersection to A.T. Naqvi Roundabout*; Environemtnal Management Consultatns: Karachi, Pakistan, 2014.
56. EMC. *Environmental Impact Assessment Report "Royal Marina Tower" Residential Building Project*; Environemtnal Management Consultatns: Karachi, Pakistan, 2017.
57. Mahmud, S.A.; Sheikh, S.A. Earthquake Shocks and Liquefaction Threat to Dha and Clifton Areas, Karachi Pakistan. A Case Study. In *Geotechnical Engineering for Disaster Mitigation and Rehabilitation*; Springer: Berlin/Heidelberg, Germany, 2008.
58. Ilyas, F. Concern over 100 High-Rises under Construction in Karachi. *Dawn*, 28 May 2016.
59. Liu, G.; Jia, H.; Zhang, R.; Li, Z.; Chen, Q.; Luo, X.; Cai, G. Ultrashort-Baseline Persistent Scatterer Radar Interferometry for Subsidence Detection. *ISPRS Ann. Photogramm. Remote Sens. Spat. Inf. Sci.* **2012**, *I-7*, 41–48. [CrossRef]

remote sensing MDPI

Technical Note

Classification of Wetland Vegetation Based on NDVI Time Series from the HLS Dataset

Yang Ju [1] and Gil Bohrer [2,*]

1 Environmental Science Graduate Program, Ohio State University, Columbus, OH 43210, USA; ju.116@buckeyemail.osu.edu
2 Department of Civil, Environmental & Geodetic Engineering, Ohio State University, Columbus, OH 43210, USA
* Correspondence: bohrer.17@osu.edu

Abstract: Natural wetlands are intrinsically heterogeneous and typically composed of a mosaic of ecosystem patches with different vegetation types. Hydrological and biogeochemical processes in wetlands vary strongly among these ecosystem patches. To date, most remote sensing classification approaches for wetland vegetation either rely on coarse images that cannot capture the spatial variability of wetland vegetation or rely on very-high-resolution multi-spectral images that are detailed but very sporadic in time (less than once per year). This study aimed to use NDVI time series, generated from NASA's HLS dataset, to classify vegetation patches. We demonstrate our approach at a temperate, coastal lake, estuarine marsh. To classify vegetation patches, a standard time series library of the four land-cover patch types was built from referencing specific locations that were identified as "pure" pixels. These were identified using a single-time high-resolution image. We calculated the distance between the HLS-NDVI time series at each pixel and the "pure"-pixel standards for each land-cover type. The resulting true-positive classified rate was >73% for all patch types other than water lily. The classification accuracy was higher in pixels of a more uniform composition. A set of vegetation maps was created for the years 2016 to 2020 at our research site to identify the vegetation changes at the site as it is affected by rapid water elevation increases in Lake Erie. Our results reveal how changes in water elevation have changed the patch distribution in significant ways, leading to the local extinction of cattail by 2019 and a continuous increase in the area cover of water lily patches.

Keywords: HLS data; Lake Erie; NDVI; vegetation classification

Citation: Ju, Y.; Bohrer, G. Classification of Wetland Vegetation Based on NDVI Time Series from the HLS Dataset. *Remote Sens.* **2022**, *14*, 2107. https://doi.org/10.3390/rs14092107

Academic Editors: Jacek Lubczonek, Paweł Terefenko, Katarzyna Bradtke and Marta Wlodarczyk-Sielicka

Received: 31 March 2022
Accepted: 25 April 2022
Published: 27 April 2022

1. Introduction

Coastal wetlands provide critical ecosystem services, such as wildlife habitat, fisheries support, carbon sequestration, flood protection, and water quality improvement. Specifically, around the US Great Lakes, as Harmful Algal Bloom (HAB) events are intensifying in extent and frequency, coastal wetlands are becoming increasingly important for mitigating nutrient runoff from agriculturally dominated watersheds to the lakes [1]. Natural wetlands are intrinsically heterogeneous and are typically composed of a mosaic of ecosystem patches with different plant types and hydrological conditions. The adaptation of these plant communities to a water-dominated environment is the basis for their use in improving the water quality in constructed wetlands [2]. Understanding the effects of wetland vegetation on the environmental conditions and ecological function of the wetland is key to determining which plants to grow in a constructed wetland with regards to the specific goals of the wetland-construction project (e.g., maximizing nutrient removal), and to correctly model the global contribution of wetlands to global greenhouse gas emissions and carbon sequestration. Wetland vegetation generates drag on the flow and therefore can influence the water movement through and around a vegetated patch [3]. The plant density and life forms affect the drag and thus control the flow velocity and residence time of the water in different parts of the wetlands, as well as the rate of deposition of suspended solids [4]. Furthermore, emergent plants with high transpiration rates can lower the water level [5]. The rates of nitrogen processing also vary among wetland vegetation communities [6]. Emergent vegetation shows a higher nitrogen retention rate and is more efficient at removing nitrogen than submerged macrophytes [7]. Identifying patch types is particularly important in predicting their contribution to nutrient uptake and sedimentation in coastal wetlands in agricultural watersheds, where the wetlands provide a last "line of defense" to improve runoff water quality. Vegetation uptakes carbon by photosynthesis, and different plants have different characteristic leaf areas and photosynthesis rates. Furthermore, vegetation plays an important role in methane transport, and, thus, the vegetation patch types have a strong impact on carbon sequestration and greenhouse gas flux rates in wetlands [8–11].

Traditionally, mapping the vegetation communities in a wetland requires intensive fieldwork, and some wetland areas are not easy to access. Remote sensing provides an effective, scalable, and economical alternative by which to monitor and map plant communities in wetlands. Many studies have been conducted to discern wetland vegetation communities using remote sensing imagery. Multispectral remote sensing data, for example, Landsat imagery and hyperspectral remote sensing imagery, have been used to discriminate and map wetland vegetation types [12–17]. However, classifying wetland plants comes with additional challenges in comparison to classification of dryland terrestrial plants [18]. Specifically, the boundaries between vegetation patches within a wetland are difficult to identify because of the intrinsically small scale of the wetlands and patches within them, combined with the short ecotones and sharp demarcation between vegetation patches, which is caused by the small scale of the spatial variability of wetland vegetation [17,19,20]. Thus, moderate-resolution remote sensing imagery, for example, MODIS, which has 250 m spatial resolution, is usually not sufficient for differentiating wetland vegetation patches.

The other challenge in classifying wetland vegetation is that the reflectance of different types of wetland canopies is often very similar, and these indistinct reflectance signatures are combined with the reflectance of the underlying soil and water surface [21–23]. To overcome this difficulty, high spatial resolution imagery has been used to perform the classification to reduce the spectral mixing effect of different wetland vegetations and underlying soil or hydrological regimes within single pixel [24–26]. However, high spatial resolution imagery is typically only available for a given wetland (a small location) with a very sparse return time (at best, 2–3 cloud-free images per year, usually less, and, in most years and locations, none). Therefore, it is not guaranteed that the imagery covers the wetland of interest during the growing season when the vegetation is visible. Alternatively, the spectral difference of wetland vegetation at different times of the year has been demonstrated to be effective in discriminating dryland vegetation types globally [27] and, specifically, in arid/semiarid regions' dryland vegetation [28]. This approach has been demonstrated in wetlands for the identification of individual target wetland plant species in salt marshes [28–31]. Similar to our work, Sun et al. [32] used a time series of the Normalized Difference Vegetation Index (NDVI) derived from HJ-1 at a 30 m resolution to classify and map multiple vegetations at a salt marsh in China, but did not continue to monitor vegetation changes over a long (multiple years) period. Sun et al. [32] further indicated the Landsat data are suitable for similar analyses. Not every remote sensing dataset is suitable for constructing time series in terms of wetland vegetation classification. Moderate spatial resolution data have more frequent and regularly repeated images of the same location, which is important for constructing a time series, but its spatial resolution is not high enough to discern the patches of different wetland vegetation types. Moreover, the frequent cloud cover in coastal region results in significant data gaps in remote sensing data [32]. The harmonization of the Landsat and Sentinel-2 (HLS) data by NASA has made it possible to acquire imagery at a moderate-to-high spatial resolution and at a high temporal resolution, which has created the opportunity to build time series of wetland vegetation characteristics at sufficient spatial and temporal resolutions. NDVI is one of the most widely used indices to assess vegetation from remote sensing imagery, and the construction of the NDVI time series has enabled the quantification of the penological dynamics of different wetland vegetations.

The overall goal of this study was to develop a classification approach for wetland vegetation that utilizes vegetation-type characteristic plant phenology, as observed by the NDVI time series from NASA's HLS dataset. The classification approach that we demonstrate and test here could be used to classify wetland vegetation on a large scale.

2. Materials and Methods

2.1. Study Area

The study was focused on Old Woman Creek (OWC) National Estuarine Research Reserve (NERR) (Figure 1a). It is a temperate, mineral soil, estuarine marsh, located at the south-central shore of Lake Erie, longitude: $-82°30'3''$, Latitude: $41°22'39''$. It is an Ohio State nature preserve and is managed as a cooperative partnership between NOAA NERR and the Ohio Department of Natural Resources (ODNR). It experiences rapid fluctuations in hydrology and water levels as the river mouth is periodically blocked by a sand barrier that is deposited at the lake shore and affected by additional long-term (interannual) water level fluctuations due to the recent water elevation increase of Lake Erie (~1 m over the last decade). OWC was selected for this study because it represents many coastal wetlands of the Great Lakes region. It includes 61 Ha of natural wetland area, currently consisting of four characteristic vegetation patch types. These patch types are defined through their vegetation status, i.e., (1) open water; (2) mud flats; or subclassified among types of patches with emergent vegetation according to the most common (by area) emergent vegetation species. These vegetated types are: (3) American water lily (*Nymphaea odorata*), (4) yellow lotus (*Nelumbo lutea*), (5) cattail (several species and hybrids of *Typha* spp.), and (6) reed (several species of *Phragmites* spp.). During the past 10 years, patches of reed as well as mud flats have been present, although they occupied a very small (<5%) fraction of the area. However, due to the rising water levels of Lake Erie and the consequent deepening of the wetland water, the mud flats disappeared by 2016, reed in 2018, and cattail in 2020. As our analysis centers on 2017–2020, the only patch types we considered were: open water, water lily, lotus, and cattail. Long-term meteorological, hydrological, and water quality data for OWC are available through the NERR data interface https://cdmo.baruch.sc.edu/ (31 December 2021); long-term radiation, energy, and carbon flux data are available through the Ameriflux database, site ID US-OWC [33].

2.2. Remote Sensing Data

NASA's Harmonized Landsat and Sentinel-2 (HLS) dataset [34] was used in this study to generate an NDVI time series. This dataset provides consistent surface reflectance from the Landsat 8, Sentinel-2A, and Sentinel-2B satellites. The merged product was run though a set of algorithms that include atmospheric correction, cloud and cloud-shadow masking, spatial co-registration and common gridding, illumination and view angle normalization, and spectral band-pass adjustment. The HLS dataset contains two products of surface reflectance at a 30 m resolution: HLS S30 from the Sentinel-2A and Sentinel-2B satellites, and HLS L30 from the Landsat satellite. These two products are gridded into the same tiling system, the Military Grid Reference System (MGRS), so that they can be stacked together to generate time series with more frequent observations. OWC is fully covered by the tile at location code 17TLF. We accessed the data through https://hls.gsfc.nasa.gov/data/ (31 December 2021).

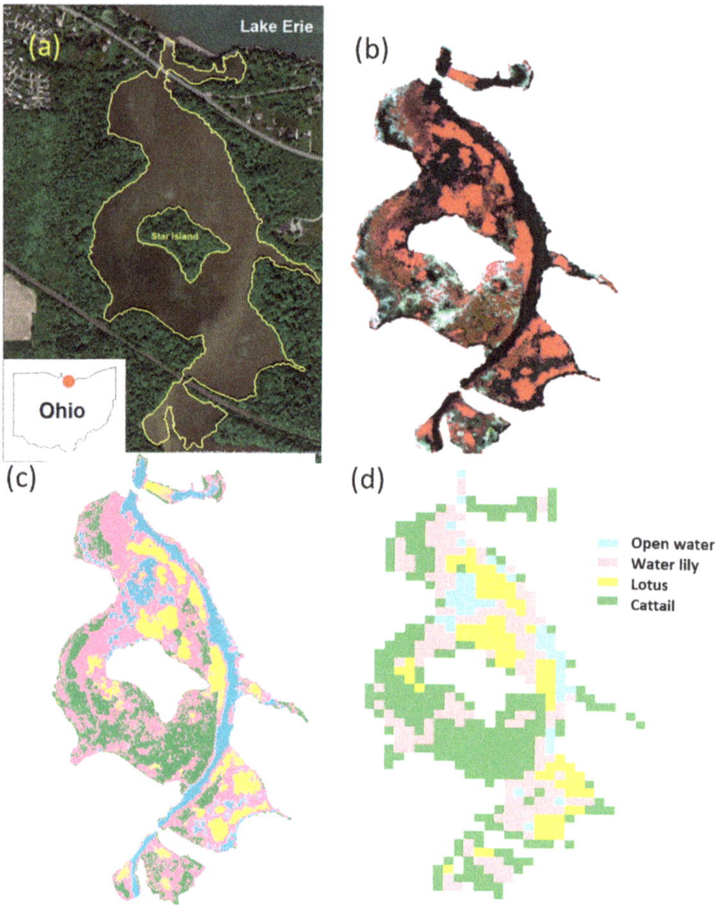

Figure 1. (**a**) The location of our wetland site—Old Woman Creek (OWC) National Estuarine Research Reserve, at the shore of Lake Erie. The boundary of OWC is highlighted in yellow. Star island is located at the center of OWC. The background image is from the imagery basemap in the Esri Arcmap software package. (**b**) The false-color composition (bands 8, 5, and 3 correspond to near-infrared, red, and green, respectively) of WorldView-3 imagery was acquired on 17 August 2017. (**c**) Supervised classification result of WorldView-3 imagery. (**d**) Classification result of the HLS dataset based on the NDVI time series of 2017.

WorldView-3 high-resolution multispectral imagery was used to provide a ground-truth reference by supervised classification. We purchased all the WorldView-3 images that included OWC, where >80% were cloud free and from during the growing season (July–September). Only one image from 17 August 2017 complied with these standards (011281167020_01_P001_MUL). The image was purchased from Apollo Mapping. It contains 8 multispectral bands with 1.2 m resolutions.

2.3. NDVI Time Series Construction

We utilized the growing season of 2017 (days 134–296 of the year), when we had a high-quality WorldView-3 image of the site at peak growing season and high confidence in the results of its expert-supervised classification based on a detailed ground-based survey we conducted that year, and obtained all the available HLS images for that period. We preprocessed the data by extraction of the red and near-infrared (NIR) bands for calculation of the NDVI values and the quality assurance (QA) band for filtering of the cloud pixels. We then clipped each image in the dataset to the extent of OWC's boundary. The NDVI for each pixel of each clipped image was calculated as follows:

$$\text{NDVI} = (\text{NIR} - \text{Red})/(\text{NIR} + \text{Red}). \quad (1)$$

The cloud cover was filtered by indexing of the cloud pixels from the QA band. This resulted in a quality-controlled time series of the NDVI values per pixel at a weekly temporal resolution. We used linear interpolation to gap-fill these NDVI time series over gaps due to clouds.

2.4. Classification Standards from WorldView-3

WorldView-3 imagery was clipped to OWC's extent and then classified into four categories by the QGIS using the "dzetsaka" classification tool plugin version 3.4.3 (https://github.com/nkarasiak/dzetsaka (31 December 2021), https://www.karasiak.net/dzetsaka-how-to-make-your-first-classification-in-qgis/ (31 December 2021), [35]) with the classifier set as a Gaussian Mixture Model (GMM), using all the multispectral bands contained in the WorldView-3 imagery with 40 training samples. We tested the other classifiers available through QGIS (Nearest Neighbor, Support Vector Machine, and Random Forest), but they resulted in lower accuracy, and thus we selected GMM. To evaluate the classification accuracy, we compared it with ground observations at 100 randomly selected points that were classified by visual interpretation and ground-based site knowledge. For the area that correspond to each HLS pixel (roughly 25 $\times$ 25 WorldView-3 pixels), we calculated the fractions of the area covered by each of the four land-cover types as classified by the WorldView-3 image. We identified HLS pixel areas that contained predominantly a single land-cover type (with a coverage fraction greater than 80%), i.e., "pure" pixel areas. Five "pure" pixel areas were selected at random for each land-cover type. The HLS pixels that corresponded to these WorldView-3 "pure"-pixel areas were used as a reference to construct a standard HLS-NDVI time series per land-cover type.

2.5. Classification Using Dynamic Time Warping

With these standard NDVI time series for each land-cover type, we used the Dynamic Time Warping (DTW) approach to calculate the similarity and distance between each HLS pixel in the dataset and the four standards. The Dynamic Time Warping algorithm was first introduced and explored in the field of speech recognition [36], but it was then applied in many other fields, including time series clustering specifically applied to land-use and land-cover mapping from remote sensing data [37]. DTW measures the similarity between time series and minimizes the effects of shifting and distortion over time. Particularly suitable to phenological time series, the approach by Maus et al. [37] can handle time series with different lengths. Considering that the length of the NDVI time series from different years can vary, DTW was selected in this study to measure the similarity between the time series to ensure that the standard NDVI time series built from one year can be applied to other years to conduct the classification. We classified the vegetation throughout the wetland during the growing season (May–September) of each year, 2016–2020. For each pixel, the land-cover that had the smallest distance was assigned to each HLS pixel as its dominant type. The complete process of classification using the NDVI time series from HLS is illustrated in Figure 2.

Figure 2. A schematic flowchart of the classification process using the NDVI time series computed from the HLS dataset. Cylinders represent datasets; rhombuses mark model outputs; and rectangles represent data analysis processes. Arrows represent the flow of data, and dashed frames identify different sub-stages of the analysis process.

2.6. HLS Classification Validation and Partial Composition of Patch Types within HLS Pixels

We used the classified WorldView-3 image to validate the HLS classification result and the classified type in HLS pixels to the most common patch type among WorldView-3 pixels within the area of the HLS pixel. The true-positive rate was calculated as the count of HLS pixels classified as the same class as the majority of the co-located pixels classified from WorldView-3 and divided by the total amount of the HLS classification pixels that correspond to that class among all the HLS pixels. We computed the fraction of each patch type within each HLS pixel by counting the number of classified WorldView-3 pixels that belong to that particular patch type within the corresponding HLS pixel area and dividing it by the total number of WorldView-3 pixels within the same HLS pixel area.

3. Results

Classification of WorldView-3 Image for Identification of "Pure" Pixels

The classification results of the WorldView-3 image (Figure 1b) are shown in Figure 1c, and a confusion matrix (Table 1) was calculated to evaluate the classification accuracy of the WorldView-3 imagery. The overall accuracy was 78%; the open water and cattail patch types had the highest precision of 100%, and the water lily patch type had the lowest precision of 64%. Most of the water lily patch type that was wrongly classified was classified as cattail. Water lily had the least contiguous distribution pattern compared to cattail and lotus (Figure 1c), which might be the reason for its misclassification. Many of the vegetation patches, especially the water lily patches, were very small, and the wetland included many transitions from one type to another even within the very high resolution of the WoldView-3 image, leading to a large number of mixed pixels that challenged the classification and led to an overall accuracy of 78%.

Table 1. Confusion matrix of the WorldView-3 classification result for 2017. The classified result was compared to 100 randomly selected points and classified by visual interpretation and ground-based site knowledge.

		Classified As				Total	User Accuracy
		Open Water	Water Lily	Lotus	Cattail		
True Patch Type	Open Water	16	0	0	0	16	100%
	Water Lily	3	36	3	14	56	64%
	Lotus	0	0	8	2	10	80%
	Cattail	0	0	0	18	18	100%
Total		19	36	11	34	100	78%
% True		84%	100%	73%	53.0%		

Using the HLS-NDVI time series of the selected pixels that were identified as "pure" pixels (>80% of the corresponding WoldView-3 pixels were of a single patch type), we defined a characteristic NDVI pattern for the four land-cover types in OWC (Figure 3). Water lily and cattail showed a very similar temporal pattern, but cattail had a consistently higher NDVI throughout the season. The lotus patch type was distinguishable from these water lily and cattail patch types as it started with a lower NDVI early in the spring and peaked at a higher NDVI early in the summer. The open water patch type showed a similar pattern to the other three vegetation types but peaked at a lower NDVI intensity and later in the summer (Figure 3). Using the temporal signature of each land-cover type NDVI time series, we classified the pixels in the weekly repeated HLS images into different types. The HLS classification map is presented in Figure 1d. Spatially, the distribution of the vegetation patches in the HLS classification agrees with that of the high-resolution WorldView-3 imagery. The cluster of lotus patches at the north part along the main channel of OWC, the open water and main channel, and the water lilies at the outflow of OWC were captured in the HLS result.

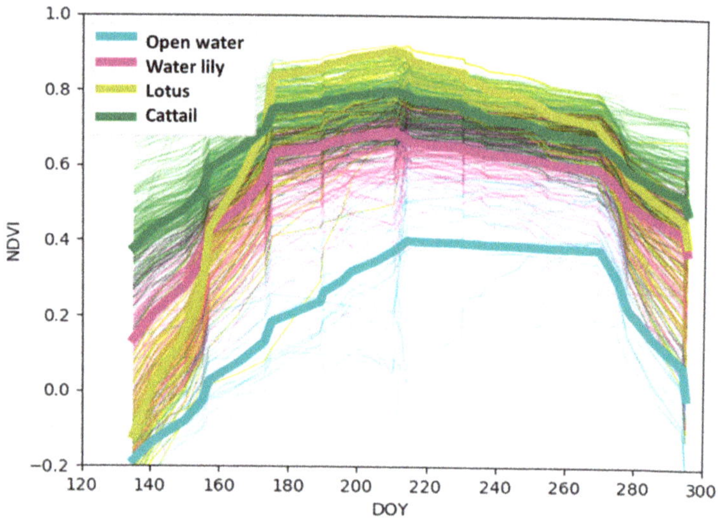

Figure 3. NDVI time series calculated from "pure" pixels in the HLS dataset of 2017. Results for individual pixels are shown in a thin line, and the mean is marked with a think line. The means were used as standards to classify the HLS pixels in other years.

Many of the HLS pixels were heterogeneous and included a mixture of different patch types, with the classified patch type only contributing to a fraction of the full area of the pixel (Figure 4). In such cases, the classified patch type was assumed to represent the type that covered the largest fraction of the pixel area. Water lilies tended to grow in small patches, which resulted in highly mixed HLS pixels for the water lily patch type (Figure 4b). Furthermore, water lily leaves float on the surface of the water and are not truly emergent. As a result, the water lily classification accuracy, at 50–64%, was lower than that of the other patch types (Figure 5). Overall, for the three other patch types, when an HLS pixel covered an area that had more than 60% of a certain type of land cover, the overall accuracy was higher than 80% (Figure 5). The classification accuracy tended to increase when the fractional cover increased, which indicates that when a pixel is less mixed, it is more likely to be classified in the correct class.

We then classified the vegetation at OWC in 2020 (Figure 6a) using the DTW approach for the HLS data and the characteristic NDVI time series we built from the 2017 HLS pure pixels. There were no cloud-free WorldView-3 images available for OWC during the growing season of 2020. The 2020 HLS-classification confusion matrix (Figure 6b) was created by comparison to the ground-based sampling of locations throughout OWC, which was conducted on 20 July 2020, and included 30 points selected at random throughout the wetland. Lotus showed a higher classification precision (80%) than water lily (66.7%). The overall accuracy for classification in 2020, when no high-resolution images were available, was 75.8% (Figure 6b), which validates the potential of using NDVI time series from HLS data to classify wetland vegetation. We used our approach to provide a consistent and continuous annual classification of the land-cover types within OWC (Figure 7a) since 2016, when HLS data became available. Our classified maps clearly show the ecological changes that the wetland has undergone due to the changes in mean water elevation (Figure 7b,c). Most notable is that the patches of emergent cattail vegetation had completely disappeared from the wetland by 2019.

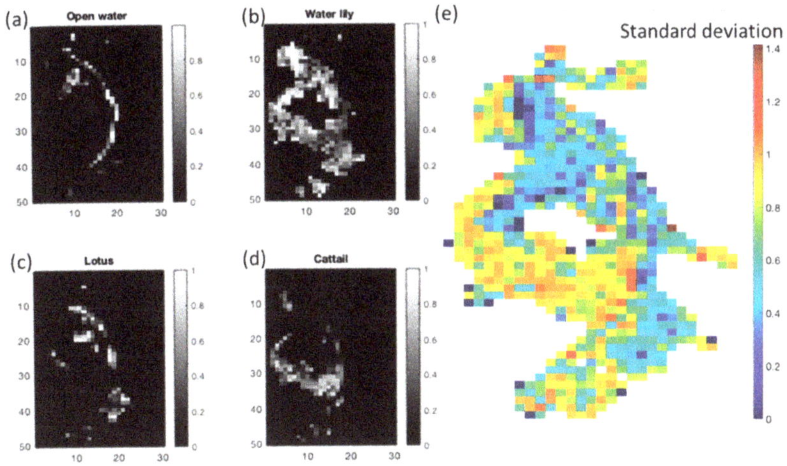

Figure 4. (**a–d**) Fractional cover. For each HLS pixel, the fractional cover indicates the fraction of pixels in the corresponding World-View3 image that are located within that HLS pixel area that are of the patch type that was classified as the HLS majority patch type (i.e., the classified HLS type of that pixel). Panels show results per HLS classified patch type (patch types are listed in panel titles). A fractional cover value of 1 indicates that all the area within the HLS pixel belongs to the classified patch type. A low fractional cover indicates that a large fraction of the area within the HLS pixel belongs to different patch types from the classified majority type. (**a**) Open water, (**b**) water lily, (**c**) lotus, (**d**) cattail. (**e**) Overall patch heterogeneity is represented as the standard deviation of patch types among the World-View3 pixels within each HLS pixel.

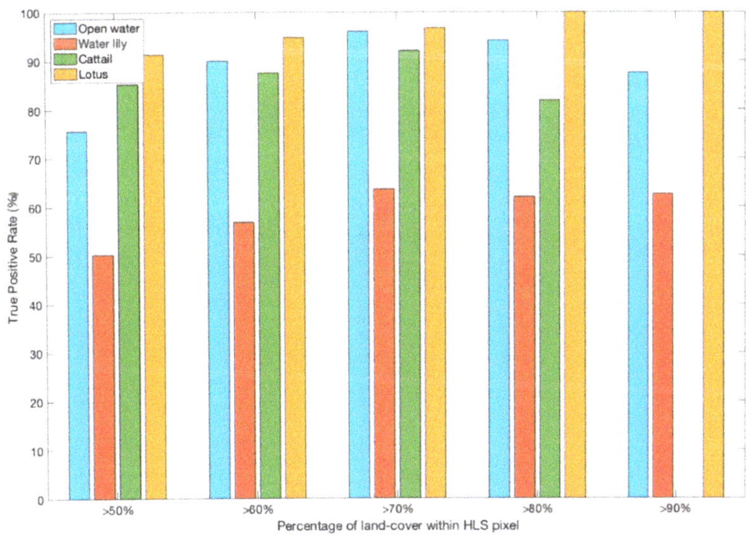

Figure 5. HLS classification accuracy for mixed pixels. The true-positive classification accuracy of the HLS classification of 2017 compared to the classified WorldView-3 image. A classification was considered correct when the majority of the WorldView-3 pixels within the area of an HLS pixel were classified as the same type as the HLS pixel. Colors correspond to the patch type of the HLS pixel. Groups of columns relate to the fractional cover of that majority patch type, from >50% (representing highly mixed pixels) to >90% (representing pure pixels), in increments of 10%.

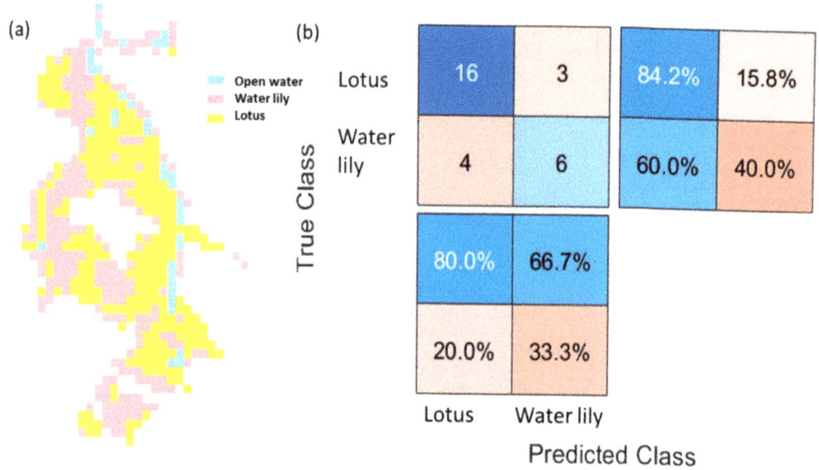

Figure 6. (**a**) HLS classification result of 2020. (**b**) A confusion matrix comparing the result with field survey points.

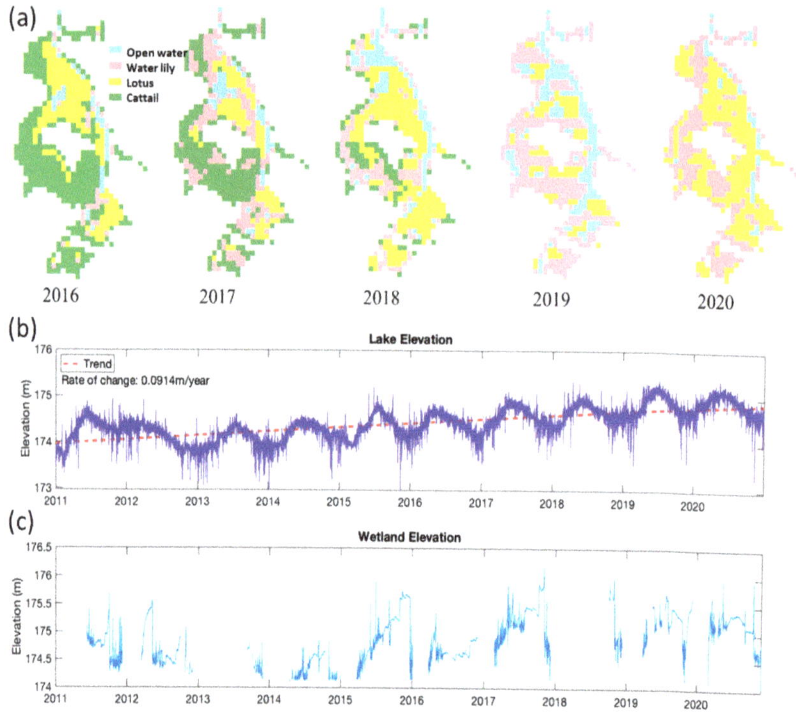

Figure 7. (**a**) Time sequence of patch location and extent in OWC, produced by the classification of HLS data from 2016 to 2020. (**b**) Rapid increases in Lake Erie's water elevation over the last decade (data from NOAA, https://tidesandcurrents.noaa.gov/ (31 December 2021)) have led to (**c**) higher water levels in the OWC wetland area (data from NERR https://cdmo.baruch.sc.edu/ (31 December 2021)) and noticeable changes in the patch type composition. Cattail, which was the most common patch type by area in 2016, had disappeared and the floating-leaf patch type (lotus and water lily) coverage area grew; in recent years, these have become the most common by area.

4. Discussion

The time series of NDVI that we observed in the pure pixels defines each patch type (Figure 3). The differences between patch types provide some ecological insight with regard to their seasonal dynamics. The cattail and water lily classes show similar NDVI trends, except that the overall mean NDVI of cattail was higher than that of water lily. Lotus started growing later than cattail and water lily, but its spring green-up was faster, such that by around day the year (DOY) 175, its NDVI value exceeded that of cattail and stayed the highest among all the land-cover types. Its fall senescence started at about DOY 210, as with all the other types, but it showed the fastest browning rate, and, by around DOY 280, its NDVI became lower than that of cattail. Open water always showed the lowest mean NDVI value and followed a similar temporal NDVI trend to that of the vegetated land-cover types. We hypothesize that the reason for this similarity is that water pixels are rarely truly pure and almost always include some vegetation within the pixel.

The reflectance of a mixed pixel in remote sensing imagery is a combination of the reflectance of the land-cover types that are present in the pixel. The linear spectral mixture model is a widely used technique to unmix remote sensing pixels [38], and it assumes that the reflectance is a linear mixture of the reflectance from all of the land cover in the pixel. Therefore, the more a certain type of land-cover is present in a pixel, the more dominant the effect it has on the mixed reflectance and thus the NDVI value. Mixed pixels could influence our confidence in the location of the identified "pure" pixels. However, there is a scaling issue to consider. The "pure" pixels correspond with pixels of the coarser HLS. They are identified as areas within the high-resolution world view images that are continuously and uniformly classified as belonging to a single patch type. By choosing "pure" pixel locations that are as far as possible from the transition between one World-View3-based classified patch type to another, we effectively minimized the effect of mixed pixels within the high-resolution image for "pure" pixel location detection.

Nonetheless, mixed pixels within the classified HLS images remain a concern and a source of error and potential bias. For water lily and cattail, the higher percentage pixels at the very north part of the wetland were classified as their corresponding classes, but for those pixels at the southern part of the wetland, the fractions of cattail and water lily were often around 50% to 60%, with some lower percentage lotus (Figure 1d). As a result, water lily showed the lowest precision of 54.5% (Figure 5). Future progress may consider a spectral unmixing approach [39], treating the NDVI time series of pure pixels, rather than a spectral reflectance series, as the signature of each end member and estimating the mixture fractions of each end member within each HLS pixel. We predict that this approach will be advantageous when the mixing is simple and includes only two end members (patch types) but will lead to larger errors in pixels of a more complex mixture. In cases when some patch types are intrinsically characterized by small or narrow patches with a typical size much smaller than the ~30 m resolution of HLS (for example, open water mostly in narrow channels), the classification of mixed pixels can lead to a bias against these patch types. Nonetheless, for many applications related to modeling of wetland processes, where patch-level processes are resolved at the sub-grid-scale without explicitly tracking the patches' locations, the purpose of the patch-type classification is to inform the model of the aerial distribution of patch types within the full wetland. In such an application, the site-level statistics over all the mixed pixels will average out classification errors, as the probability of misclassification is proportional to the fraction of the partial area of the large patch type (Figure 5). Another potential cause of misclassification that needs to be taken into consideration is the cloud detection omissions of the Sentinel-2 cloud mask [34]. An outlier-detection-and-removal method is needed in the future to improve the accuracy of the classification.

The areas most affected by mixed types within the same pixel are those where the characteristic size of the patches is small, with transitions from one type to another at a short spatial scale. For example, the south part of OWC does not show a tight alignment of the two classification results. In this part of OWC in 2017, small patches of all three types of vegetation were growing next to each other. The HLS pixels in this area of the wetland, therefore, always represent a mixture of at least two land-cover types. Most of the mixtures were identified as cattail in the HLS classification, except some of the pixels with a low cattail contribution that were classified as water lily or lotus (Figures 1d and 4). Part of the main channel was not recognized because the main channel is narrow, and therefore, in many parts of the main channel, the HLS pixel area that covers the main channel is mixed with vegetation that contributes to a high NDVI, which resulted in the pixels along the main channel being misclassified as vegetation (Figures 1d, 4 and 7). The pixels of the HLS imagery at the edge of the wetland are mixtures of water and forest (non-wetland vegetation types with a higher NDVI than the wetland vegetation), which challenged the classification of some pixels at the edge of the wetland.

OWC is a Lake Erie coastal wetland and is experiencing rapid ecological and hydrological changes due to changes in Lake Erie's water elevation. The Great Lakes system has experienced rapid water elevation increases in the past decade. Lake Erie is an extreme case of the Great Lakes, and its water elevation has increased by about 1 m since 2011 (Figure 7c). Since the wetland is hydrologically attached to the lake, its corresponding water elevation also increases, although the wetland experiences much greater short-term variation than the lake due to the river and estuary hydrodynamics. This rapid water elevation increase drives a rapid shift in the patch type composition of the wetland. The consistent availability of HLS images from 2016 up to present allow the characterization and quantification of these changes (Figure 7a).

5. Conclusions

We demonstrated an approach used to classify vegetation patch types using repeated NDVI data from HLS in a small, heterogenous wetland. We demonstrated the accuracy of our approach over multiple years and its applicability for long-term monitoring of the vegetation patch type dynamics in wetlands. Advanced models of wetland function, and particularly mechanistic models, that predict nutrient processing, sedimentation, and greenhouse gas budgets at the resolution of plant functional types require knowledge of the areal distribution of different plant functional types within the wetland, their seasonal dynamics, and their longer-term (inter-annual) dynamics (e.g., Figure 7). Remote sensing-derived input for patch type distribution could improve the accuracy of the prediction of the ecological function of wetlands, specifically where they are experiencing ecological and hydrological changes and disturbances. While high-resolution multispectral images have been previously demonstrated by others and by ourselves as a reliable source of information for vegetation patch location and identity, high-resolution images may not be consistently available for a given wetland site in every year and growing season.

We demonstrated the applicability of our approach for long-term monitoring in Old Woman Creek (OWC), a small, coastal estuarine, mineral soil marsh of Lake Erie. As Lake Erie is experiencing a rapid water level rise (roughly 1 m over the last decade), OWC, which is hydrologically connected to the lake, is becoming deeper and experiencing rapid changes in its ecological patch location and distribution. It has switched from a cattail-dominated ecosystem to one absent of cattail and dominated by floating-leaf vegetation (water lily and lotus).

The approach proposed in this study, including the characteristic time series we developed, can be directly applied to other wetlands that grow the same type of plants using our characteristic NDVI time series (Figure 3). In the case that other wetlands of interest support patch types that are characterized by different emergent plant species than the ones growing in the current study site, the method we have demonstrated in this study could be still applicable to other types of vegetation, provided that the locations and times of pure pixels are available, from which a characteristic time series could be derived for each patch type. Such pure-pixel locations and times could be identified from limited ground surveys or from expert classification of patch types at one point in time from a single available high-resolution image.

Author Contributions: The manuscript was conceived and co-written by Y.J. and G.B. Y.J. conducted the data analysis and produced the figures. All authors have read and agreed to the published version of the manuscript.

Funding: This work was supported by NOAA through OWC-NERR managed by the Ohio Department of Natural Resources (award N18B 315-11), the U.S. Department of Energy (awards DE-SC0022191 and DE-SC0021067), USGS through the Ohio Water Resources Center (award G16AP00076), and the Ohio Water Development Authority (award 7880). Funding for the data collection at US-UWC was provided in part by the U.S. Department of Energy's Office of Science through the Ameriflux Management Project.

Data Availability Statement: The data used for this study are publicly available remote sensing data. The HLS data were accessed at https://hls.gsfc.nasa.gov/data/ (31 December 2021). The World-View3 data were purchased from Apollo Mapping at https://apollomapping.com/ (31 December 2021). The water elevation data for OWC are available through the NERR data interface, https://cdmo.baruch.sc.edu/ (31 December 2021). Lake Erie water elevation observations were accessed from the NOAA Tides & Currents monitoring network, https://tidesandcurrents.noaa.gov/ (31 December 2021), location 'Marblehead, OH' (Station ID: 9063079).

Acknowledgments: We thank the OWC managers Janice Kerns and Kristine Arend for their assistance with accessing the study area, Qing Xu for her help with surveying the wetland vegetation during summer 2020, Timothy Morin and Camilo Rey Sanchez for the data of the 2017 vegetation survey, and Yvette Onyango for the assistance in data access.

Conflicts of Interest: The authors declare no conflict of interest.

References

1. Mitsch, W.J. Solving Lake Erie's harmful algal blooms by restoring the Great Black Swamp in Ohio. *Ecol. Eng.* **2017**, *108*, 406–413. [CrossRef]
2. Guntenspergen, G.R.; Stearns, F.; Kadlec, J. Wetland vegetation. In *Constructed Wetlands for Wastewater Treatment*; CRC Press: Boca Raton, FL, USA, 2020; pp. 73–88.
3. Watson, D. Hydraulic effects of aquatic weeds in UK rivers. *Regul. Rivers Res. Manag.* **1987**, *1*, 211–227. [CrossRef]
4. Boto, K.K.; Patrick, J. Role of wetlands in the removal of suspended sediments. In *Wetland Functions and Values: The State of Our Understanding*; American Water Resources Association: Minneapolis, MN, USA, 1979; pp. 479–489.
5. Heliotis, F.D. *Wetland Systems for Wastewater Treatment: Operating Mechanisms and Implications for Design*; University of Wisconsin-Madison: Madison, WI, USA, 1981; p. 206.
6. Zhao, T.; Xu, H.; He, Y.; Tai, C.; Meng, H.; Zeng, F.; Xing, M. Agricultural non-point nitrogen pollution control function of different vegetation types in riparian wetlands: A case study in the Yellow River wetland in China. *J. Environ. Sci.* **2009**, *21*, 933–939. [CrossRef]
7. Weisner, S.E.B.; Thiere, G. Effects of vegetation state on biodiversity and nitrogen retention in created wetlands: A test of the biodiversity-ecosystem functioning hypothesis. *Freshw. Biol.* **2010**, *55*, 387–396. [CrossRef]
8. Knox, S.H.; Jackson, R.B.; Poulter, B.; McNicol, G.; Fluet-Chouinard, E.; Zhang, Z.; Zona, D. FLUXNET-CH_4 Synthesis Activity: Objectives, Observations, and Future Directions. *Bull. Am. Meteorol. Soc.* **2019**, *100*, 2607–2632.
9. Villa, J.A.; Ju, Y.; Stephen, T.; Rey-Sanchez, C.; Wrighton, K.C.; Bohrer, G. Plant-mediated methane transport in emergent and floating-leaved species of a temperate freshwater mineral-soil wetland. *Limnol. Oceanogr.* **2020**, *65*, 1635–1650. [CrossRef]
10. Sha, C.; Mitsch, W.J.; Mander, Ü.; Lu, J.; Batson, J.; Zhang, L.; He, W. Methane emissions from freshwater riverine wetlands. *Ecol. Eng.* **2011**, *37*, 16–24. [CrossRef]
11. Rey-Sanchez, A.C.; Morin, T.H.; Stefanik, K.C.; Wrighton, K.; Bohrer, G. Determining total emissions and environmental drivers of methane flux in a Lake Erie estuarine marsh. *Ecol. Eng.* **2018**, *114*, 7–15. [CrossRef]
12. Belluco, E.; Camuffo, M.; Ferrari, S.; Modenese, L.; Silvestri, S.; Marani, A.; Marani, M. Mapping salt-marsh vegetation by multispectral and hyperspectral remote sensing. *Remote Sens. Environ.* **2006**, *105*, 54–67. [CrossRef]
13. Harvey, K.R.; Hill, G.J.E. Vegetation mapping of a tropical freshwater swamp in the Northern Territory, Australia: A comparison of aerial photography, Landsat TM and SPOT satellite imagery. *Int. J. Remote Sens.* **2001**, *22*, 2911–2925. [CrossRef]
14. May, A.M.B.; Pinder, J.E.; Kroh, G.C. A comparison of Landsat Thematic Mapper and SPOT multi-spectral imagery for the classification of shrub and meadow vegetation in northern California, USA. *Int. J. Remote Sens.* **1997**, *18*, 3719–3728.
15. Pengra, B.W.; Johnston, C.A.; Loveland, T.R. Mapping an invasive plant, Phragmites australis, in coastal wetlands using the EO-1 Hyperion hyperspectral sensor. *Remote Sens. Environ.* **2007**, *108*, 74–81. [CrossRef]
16. Rosso, P.H.; Ustin, S.L.; Hastings, A. Mapping marshland vegetation of San Francisco Bay, California, using hyperspectral data. *Int. J. Remote Sens.* **2005**, *26*, 5169–5191. [CrossRef]
17. Schmidt, K.S.; Skidmore, A.K. Spectral discrimination of vegetation types in a coastal wetland. *Remote Sens. Environ.* **2003**, *85*, 92–108. [CrossRef]

18. Adam, E.; Mutanga, O.; Rugege, D. Multispectral and hyperspectral remote sensing for identification and mapping of wetland vegetation: A review. *Wetl. Ecol. Manag.* **2010**, *18*, 281–296. [CrossRef]
19. Adam, E.; Mutanga, O. Spectral discrimination of papyrus vegetation (*Cyperus papyrus* L.) in swamp wetlands using field spectrometry. *ISPRS J. Photogramm. Remote Sens.* **2009**, *64*, 612–620. [CrossRef]
20. Zomer, R.J.; Trabucco, A.; Ustin, S.L. Building spectral libraries for wetlands land cover classification and hyperspectral remote sensing. *J. Environ. Manag.* **2009**, *90*, 2170–2177. [CrossRef]
21. Malthus, T.J.; George, D.G. Airborne remote sensing of macrophytes in Cefni Reservoir, Anglesey, UK. *Aquat. Bot.* **1997**, *58*, 317–332. [CrossRef]
22. Yuan, L.; Zhang, L. Identification of the spectral characteristics of submerged plant Vallisneria spiralis. *Acta Ecol. Sin.* **2006**, *26*, 1005–1010. [CrossRef]
23. Guyot, G. 2—Optical properties of vegetation canopies. In *Applications of Remote Sensing in Agriculture*; Steven, M.D., Clark, J.A., Eds.; Butterworth-Heinemann: London, UK, 1990; pp. 19–43.
24. Carle, M.V.; Wang, L.; Sasser, C.E. Mapping freshwater marsh species distributions using WorldView-2 high-resolution multispectral satellite imagery. *Int. J. Remote Sens.* **2014**, *35*, 4698–4716. [CrossRef]
25. Szantoi, Z.; Escobedo, F.; Abd-Elrahman, A.; Smith, S.; Pearlstine, L. Analyzing fine-scale wetland composition using high resolution imagery and texture features. *Int. J. Appl. Earth Obs. Geoinf.* **2013**, *23*, 204–212. [CrossRef]
26. Wang, C.; Menenti, M.; Stoll, M.P.; Belluco, E.; Marani, M. Mapping mixed vegetation communities in salt marshes using airborne spectral data. *Remote Sens. Environ.* **2007**, *107*, 559–570. [CrossRef]
27. Zhang, X.; Liu, L.; Chen, X.; Gao, Y.; Xie, S.; Mi, J. GLC_FCS30: Global land-cover product with fine classification system at 30 m using time series Landsat imagery. *Earth Syst. Sci. Data* **2021**, *13*, 2753–2776. [CrossRef]
28. Arastoo, B.; Ghazarian, S.; Avetyan, N. An approach for land cover classification system by using NDVI data in arid and semiarid region. *Elixir Remote Sens.* **2013**, *60*, 16327–16332.
29. Fernandes, M.R.; Aguiar, F.C.; Silva, J.M.; Ferreira, M.T.; Pereira, J.M. Spectral discrimination of giant reed (*Arundo donax* L.): A seasonal study in riparian areas. *ISPRS J. Photogramm. Remote Sens.* **2013**, *80*, 80–90. [CrossRef]
30. Gao, Z.G.; Zhang, L.Q. Multi-seasonal spectral characteristics analysis of coastal salt marsh vegetation in Shanghai, China. *Estuar. Coast. Shelf Sci.* **2006**, *69*, 217–224. [CrossRef]
31. Ouyang, Z.T.; Gao, Y.; Xie, X.; Guo, H.Q.; Zhang, T.T.; Zhao, B. Spectral Discrimination of the Invasive Plant Spartina alterniflora at Multiple Phenological Stages in a Saltmarsh Wetland. *PLoS ONE* **2013**, *8*, e67315.
32. Sun, C.; Liu, Y.; Zhao, S.; Zhou, M.; Yang, Y.; Li, F. Classification mapping and species identification of salt marshes based on a short-time interval NDVI time series from HJ-1 optical imagery. *Int. J. Appl. Earth Obs. Geoinf.* **2016**, *45*, 27–41. [CrossRef]
33. Bohrer, G.; Kerns, J. *AmeriFlux BASE US-OWC Old Woman Creek*; AmeriFlux AMP: Berkeley, CA, USA, 2018. [CrossRef]
34. Claverie, M.; Ju, J.; Masek, J.G.; Dungan, J.L.; Vermote, E.F.; Roger, J.C.; Justice, C. The Harmonized Landsat and Sentinel-2 surface reflectance data set. *Remote Sens. Environ.* **2018**, *219*, 145–161. [CrossRef]
35. Karasiak, N. *Dzetsaka: v3.4.3 (Version v3.4.3)*; Zenodo: Geneva, Switzerland, 2019. [CrossRef]
36. Sakoe, H.; Chiba, S. Dynamic programming algorithm optimization for spoken word recognition. *IEEE Trans. Acoust. Speech Signal Process.* **1978**, *26*, 43–49. [CrossRef]
37. Maus, V.; Câmara, G.; Cartaxo, R.; Sanchez, A.; Ramos, F.M.; De Queiroz, G.R. A time-weighted dynamic time warping method for land-use and land-cover mapping. *IEEE J. Sel. Top. Appl. Earth Obs. Remote Sens.* **2016**, *9*, 3729–3739. [CrossRef]
38. Qin, Y.; Guo, F.; Ren, Y.; Wang, X.; Gu, J.; Ma, J.; Shen, X. Decomposition of mixed pixels in MODIS data using Bernstein basis functions. *J. Appl. Remote Sens.* **2019**, *13*, 046509. [CrossRef]
39. Shi, C.; Wang, L. Incorporating spatial information in spectral unmixing: A review. *Remote Sens. Environ.* **2014**, *149*, 70–87. [CrossRef]

remote sensing

MDPI

Article

Monitoring Sand Spit Variability Using Sentinel-2 and Google Earth Engine in a Mediterranean Estuary

Mar Roca [1,*], Gabriel Navarro [1], Javier García-Sanabria [2] and Isabel Caballero [1]

[1] Instituto de Ciencias Marinas de Andalucía (ICMAN-CSIC), Campus Universitario Río San Pedro, s/n, Puerto Real, E-11510 Cadiz, Spain; gabriel.navarro@icman.csic.es (G.N.); isabel.caballero@icman.csic.es (I.C.)
[2] Facultad de Ciencias del Mar y Ambientales (CASEM), Campus Universitario Río San Pedro, s/n, Puerto Real, E-11510 Cadiz, Spain; javier.sanabria@uca.es
* Correspondence: mar.roca@csic.es

Abstract: Estuarine degradation is a major concern worldwide, and is rapidly increasing due to anthropogenic pressures. The Mediterranean Guadiaro estuary, located in San Roque (Cadiz, Spain), is an example of a highly modified estuary, showing severe negative effects of eutrophication episodes and beach erosion. The migration of its river mouth sand spit causes the closure of the estuary, resulting in serious water quality issues and flora and fauna mortality due to the lack of water renewal. With the aim of studying the Guadiaro estuary throughout a 4-year period (2017–2020), the Sentinel-2 A/B twin satellites of the Copernicus programme were used thanks to their 5-day and 10 m temporal and spatial resolution, respectively. Sea–land mapping was performed using the Normalized Difference Water Index (NDWI) in the Google Earth Engine (GEE) platform, selecting cloud-free Sentinel-2 Level 2A images and computing statistics. Results show a closure trend of the Guadiaro river mouth and no clear sand spit seasonal patterns. The study also reveals the potential of both Sentinel-2 and GEE for estuarine monitoring by means of an optimized processing workflow. This improvement will be useful for coastal management to ensure a continuous and detailed monitoring in the area, contributing to the development of early-warning tools, which can be helpful for supporting an ecosystem-based approach to coastal areas.

Keywords: sea–land mapping; cloud computing; remote sensing; coastal monitoring; Guadiaro river

Citation: Roca, M.; Navarro, G.; García-Sanabria, J.; Caballero, I. Monitoring Sand Spit Variability Using Sentinel-2 and Google Earth Engine in a Mediterranean Estuary. *Remote Sens.* **2022**, *14*, 2345. https://doi.org/10.3390/rs14102345

Academic Editor: Jacek Lubczonek

Received: 12 April 2022
Accepted: 11 May 2022
Published: 12 May 2022

1. Introduction

Estuarine eutrophication is one of the most common and harmful environmental problems in coastal areas around the globe [1]. The world has lost 50% of its natural wetlands since 1900, most of them located in highly anthropized areas of the temperate zone in the northern hemisphere [2]. Anthropogenic pressure is the main vector for estuarine ecosystem alteration, accelerating the deterioration of water quality in natural coastal wetlands. Littoral population settlements have greatly increased in the last decades, gathering more than 30% of the world population [3] and expanding faster than any other region [4]. This ongoing trend of coastal migration extends especially to flat estuarine lands, linked to urbanization processes and human activities [5]. Across the Western Mediterranean, both climate change and anthropic activities have had impacts on water quality across coastal areas. Some examples, commonly associated to river discharges, are the Mar Menor cyclic environmental collapse [6] and the permanent eutrophication in the Albufera of Valencia [7]. To mitigate anthropic impacts, adaptive policies are needed [8], and consequently, monitoring is also required, this information being crucial for optimized coastal wetland management.

Estuary systems evolve very rapidly with channel migrations, loss of sand and erosion/accretion of adjacent beaches [9]. This study is focused on the Guadiaro estuary, located in San Roque, Cádiz (Spain), which delta started to disappear during the 1970s, as seen in Figure 1, caused by multiple factors but massive extraction of sediments along the basin was decisively, consequently reducing the sedimentary input in the Guadiaro system [10]. As a complex enclave, the Guadiaro estuary has been widely studied by engineers and scientists in the last 20 years in order to characterize and effectively manage the area [9–12], albeit without using a continuous monitoring technique.

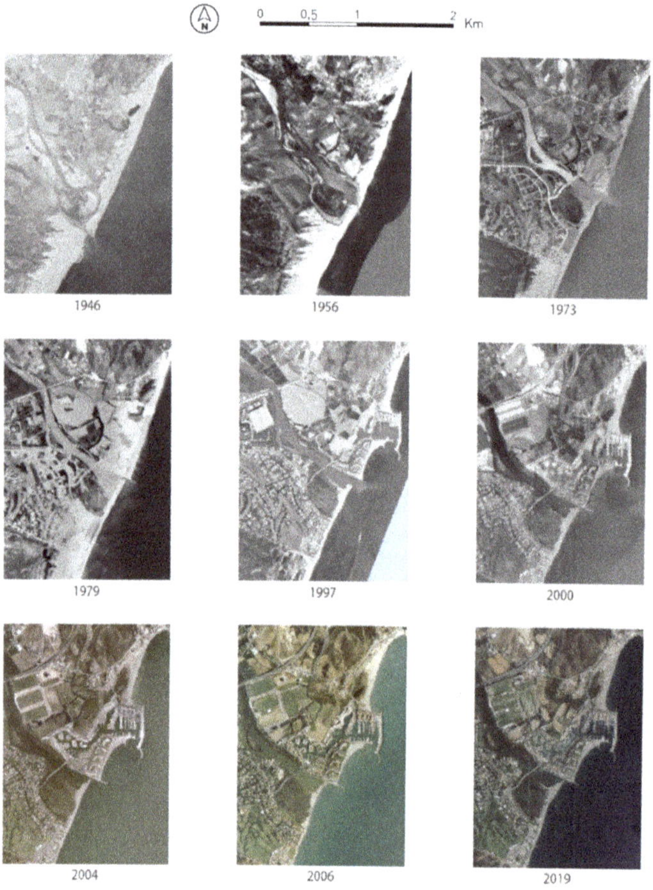

Figure 1. Satellite and aerial images from 1946 to 2019 showing the evolution of the Guadiaro estuary (source: PNOA).

In order to manage this ecosystem, it is important to understand and evaluate the sand spit patterns, where information and knowledge are essential. Therefore, improved monitoring to evaluate the different scenarios is required [13], in particular, to determine spatio-temporal variability of the sand spit closing. The context affecting the Guadiaro estuary and its surrounding requires an efficient and real-time method of monitoring the state of the sand spit. Traditional monitoring approaches are labor intensive and time consuming when long-term monitoring and management are required, and new technologies can play a powerful part. Furthermore, integrated coastal management requires information during the entire management cycle [14–17]. In practice, a monitoring tool developed with the proper technology should be able to provide long-term sustained information, required for an adaptive ecosystem-based management approach [18–20]. This means the monitoring tool can be used not only as an early-warning system, but also to evaluate responses given by the administration correlated to changes of state [21].

Earth Observation (EO) tools and remote-sensing techniques yield precise data thanks to the wide range of sensors and their spatio-temporal resolution. Satellite imagery is especially useful for shoreline and coastal monitoring [22–25], with increased public availability through open access programs. The European Copernicus programme is the EO program of the European Commission and the European Space Agency (ESA). It is composed of a set of dedicated satellites called "Sentinel" under an open and free policy. Sentinel-2A/B (S2) satellites are part of this Sentinel family, with images of 10–20–60 m and 5-day spatial and temporal resolutions, respectively, are very useful for studying coastal processes [26–29]. These satellites carry the MultiSpectral Instrument (MSI) sensor whose bands are useful for sea–land mapping. To process this huge amount of data, costly high-computational capacity and long-processing times are required [30]. Nevertheless, imagery data production has been improved in parallel to software development, which enables powerful geoprocesses resulting in more accessible and faster analysis.

The use of satellite imagery also has of the advantage of using available open-source tools for image processing, such as Google Earth Engine (GEE). GEE is a free cloud-based platform for planetary-scale geospatial analysis [31] which has a massive catalogue of satellite imagery and geospatial datasets available. This platform allows this information to be processed in the cloud in order to detect changes, map trends and quantify differences on the Earth's surface, and can develop applications for final users [32]. Some studies have been carried out using S2 and GEE platform in marine and coastal areas, including water quality studies [33,34], coastline evolution [35,36], coastal management [37], and wetland mapping [38], showing the potential of the Copernicus open data and the GEE open-source platform for coastal monitoring.

An integrated sea–land approach is needed to quantify coastal changes [39]. Some studies carried out this approach manually [40], which is more precise but time consuming and non-viable for monitoring tools, which need more automatic workflows [22]. Sea–land mapping is mostly used for coastline evolution studies, including erosion and accretion trends or sea-level rise [23,41,42], using image classification, image segmentation, and several indices. However, the use of indices based on optical properties has proven its optimized performance against image classification and segmentation methods in order to detect water and non-water pixels [43,44]. Among the different indices, the Normalized Difference Water Index (NDWI), defined by McFeeters, is widely accepted for this purpose [45], as well as the Modified NDWI (MNDWI) [45–48], although other indices have been developed, such as the AWEI index [49] or the MuWi index [50]. According to Rokni et al., the MNDWI index has been generally developed for application in urbanized areas [51], whereas the NDWI is better used in natural coasts [44].

This study evaluates the use of S2-A/B satellites for sea–land mapping in the Guadiaro estuary using the NDWI and GEE between 2017 and 2020. Using the innovative GEE platform, we explore the suitability of the S2 Level-2 A (L2A) dataset for multi-temporal studies and developed a JavaScript algorithm generating an operational monitoring tool. S2 images can serve as an input for continuous and low-cost monitoring information with GEE, which can be complemented with in situ sampling. This information would be useful for stakeholders and coastal managers in order to address the persistent problems in the Guadiaro river. The monitoring tool would help as an evaluation method for adapting management actions throughout the cycle, as an early-warning system for eutrophication episodes and as a pilot to be used in other estuaries. Therefore, this tool can be implemented not only operationally, but also as a valuable information system for strategic and adaptive coastal management. The aim is to contribute to coastal management by means of developing and evaluating a methodology in GEE for monitoring the sand spit spatio-temporal patterns as a pilot project.

2. Materials and Methods

2.1. Study Area

The Guadiaro estuary is a semi-closed Mediterranean coastal lagoon located in the Strait of Gibraltar, San Roque (Cadiz, southern Spain) (Figure 2), the only coastal wetland along 115 km of coastline [10]. This estuary is a multi-use water resource system, including fish nurseries, flood control and recreational and cultural services. Due to the micro-tidal conditions of about 40 cm, shallow bathymetry, severe wave refraction [9], and sediment transportation following the main longitudinal current right to the Strait of Gibraltar with a SW component [52]. Regarding the meteorology, the Guadiaro river is characterized by its irregular and strong dependence on rainfall regime and torrential character [11,53,54], the sand spit behavior being strongly dependent on the river, and by the sea being dependent on E winds and consequent wave behavior [11]. On the other hand, Sotogrande beach suffers from chronical erosion trends, having a regression by more than 160 m which was evident between 1960 and 2000, and is still going (−3.1 m/year) [55]. This implies an increase in vulnerability to climate change due to forecasts of increased wave energy [9–11,56].

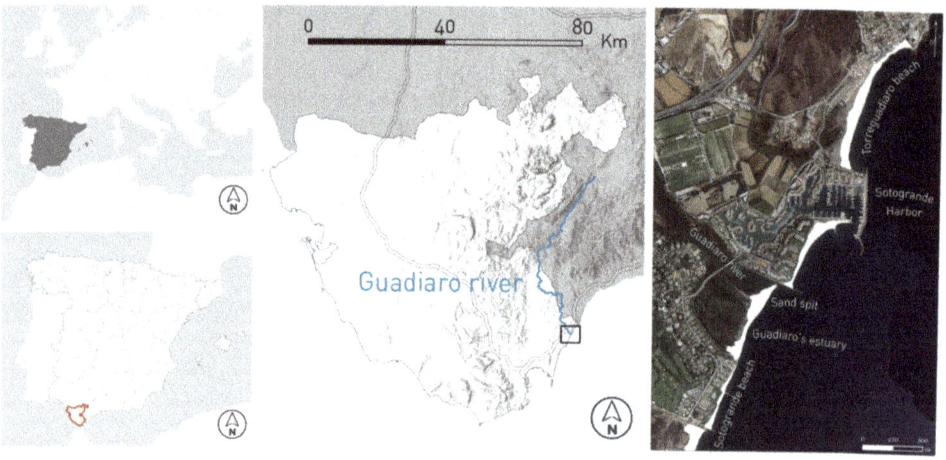

Figure 2. Study area: Guadiaro river mouth (San Roque, Cadiz), Spain.

Throughout recent decades, the Guadiaro estuary has experienced several problems, suffering from eutrophication episodes with harmful consequences for the biota and direct impacts on tourism. Its major concern is that the Guadiaro river mouth is frequently partially or totally closed by its sand spit formation. This occurs partly due to the presence of a local coastal drift current inverse to the predominant one [10,53], which transports sediment northwards and blocks water renewal in the estuary. Furthermore, during the summer, waves tend to close the Guadiaro river mouth as a result of the weak current, reducing the water renewal rate with the consequent reduction in oxygen levels and increased eutrophication [57]. Anthropogenic discharges as well as uncontrolled tourism further aggravate the problem [9]. In only 3 years (2016–2018) the Spanish Government invested more than EUR 0.5 M on dredging the Guadiaro river mouth, retrieving almost 85,000 m^3 of sand used mainly to support the Sotogrande beach, which suffers from a severe erosive trend. In extreme eutrophication episodes, the Guadiaro estuary urges from artificial sand spit openings, which are very expensive, as can be seen in Figure 3. This fact reflects the need to develop a more complete coastal management strategy for its optimization, including monitoring tools, operational early-warning systems, as well as the improvement of coordination and cooperation processes.

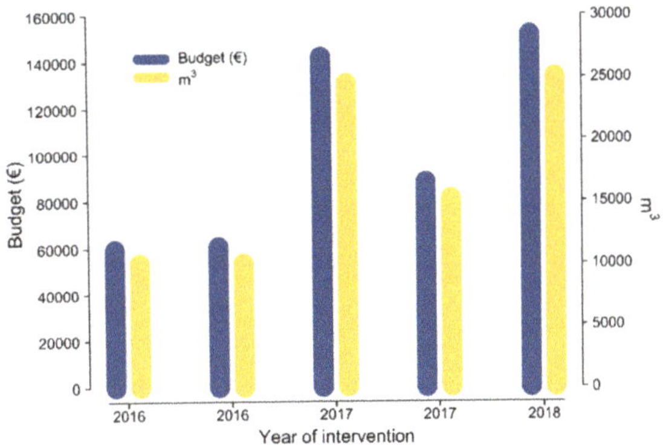

Figure 3. Guadiaro river mouth artificial openings from 2016 to 2018. Sand extraction volume (m^3) and budget involved from the Spanish Government. There was no intervention during 2019 and 2020.

Considering the different activities developed in the Guadiaro river and the littoral zone, both river and coastal processes have been highly modified since the 1970s. The Port of Sotogrande was built in 1987 to the north of to the river mouth. Two groins were built in 1973 (one removed 2 years later) and the port breakwater was constructed in 1986 and extended by 1994 [12]. Due to sand extraction activities, with 8 mining concessions by 2009, the river sediment input decreased from 100,000 m^3/year in the 1960s to 50,000 m^3/year in 2009 [14]. Furthermore, an inter-basin upstream transfer was carried out in the upper course of the Guadiaro river to the Majaceite River in 1999, with a maximum transfer capacity of 30,000 m^3/year [58], altering the sedimentary balance, decreasing river flow strength, and reducing mouth flow peaks and, consequently, favoring river mouth closure [12]. Moreover, considering other factors, such us agricultural and urban waste, extraction of water for agricultural use and the Sotogrande urbanization, the Guadiaro estuary experiences periodic biota mortality episodes with massive fish kills directly related with the sand spit closing and lack of oxygen availability [10].

2.2. Satellite Data

The Guadiaro estuary covers a local area with a high morpho-dynamic variability, requiring high spatial resolution and short-term revisit time, for which S2-A/B satellites have been used. S2 L2A refers to the level in which images are categorized in the MSI S2 database on-board S2. In GEE, two levels are currently available: Level 1 ("COPERNICUS/S2") and Level 2A ("COPERNICUS/S2_SR"). Level 1-C (L1C) contains images radiometrically and geometrically corrected at Top Of Atmosphere (TOA) reflectance, while L2A contains orthorectified atmospherically corrected surface reflectance, being a Bottom Of Atmosphere (BOA) product. Therefore, L2A datasets have been selected with the atmospheric correction applied, which determines a faster geo-processing workflow and improved enforcement of normalized indices. However, almost two years of data cannot be used for the study, with L1C available from 23 June 2015 to present and S2 L2A images from 30 March 2017 to present, which explains the lack of data in winter 2017 in the following results.

2.3. GEE & GIS

The GEE platform was used to process S2 images for its high computational and on-the-cloud capacity, having a wide variety of datasets available without depending on download. As atmospheric correction remains a challenge, we used atmospheric corrected S2 L2A images (corresponding to "COPERNICUS/S2_SR" in GEE). Using these images, we developed a workflow in JavaScript Code Editor interface in GEE in four main steps, as shown in Figure 4. (1) cloud coverage image selection in the Area Of Interest (AOI); (2) band computing according to NDWI index; (3) statistics calculation including climatology and Variation Rate (VR); (4) export results to Google Drive. As a final step, design was developed using ArcGIS Pro.

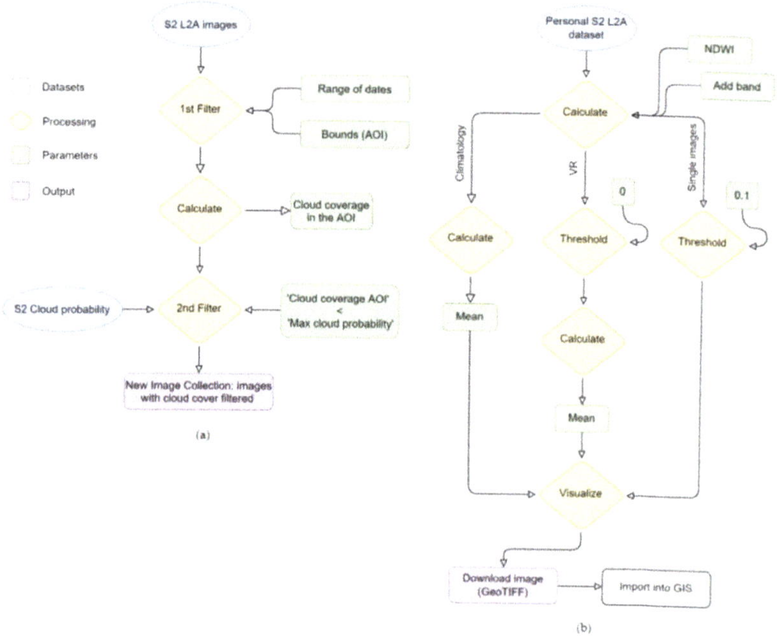

Figure 4. Geoprocessing workflow in GEE with S2 L2A images. (**a**) low cloud cover probability image selection workflow; (**b**) NDWI index application and statistical geoprocessing workflow (climatology and VR).

2.3.1. Cloud Coverage

Clouds are one of the main challenges faced by optical sensors on-board satellites in order to obtain useful data. The presence of clouds and cloud shadows in images introduces noise to data, which may result in these data being masked out. Therefore, pre-processing for cloud-free image selection has to be performed before processing any algorithm in order to obtain reliable results. In this step, we implemented a workflow developed in GEE through which cloud cover is calculated in the study area (AOI) and images can be filtered in a maximum cloud cover custom percentage from the original dataset ("COPERNICUS/S2_SR").

The geoprocessing workflow, showed in Figure 4a and available in GEE (https://code.earthengine.google.com/6363973f59918f3eacd09d890a52f5827, accessed on 1 February 2021) involved clipping the AOI from the S2 L2A image collection and filtering it by date from the first image available (30 March 2017) to December 2020. After this, cloud mask property, a general variable pre-calculated for the whole tile, was used from the S2 L2A metadata to calculate the cloud coverage in the specific AOI. With these data, the cloud cover percentage was obtained in the AOI for each image, representing valuable information for a second filtering. Finally, with the S2 Cloud probability dataset ("COPERNICUS/S2_CLOUD_PROBABILITY"), images were filtered according to the max cloud probability fixed at 15% and used in this study as a pre-processing result. However, in order to test the workflow accuracy, images with low cloud cover were selected manually from the EO Browser and compared with the automatic filtering in GEE, which is useful for the implementation of a maximum cloud probability value in order to avoid manual image selection.

2.3.2. Sea–Land Mapping

Once the images were pre-processed and stored in the new Image Collection, in order to study the morpho-dynamical variability of the Guadiaro sand spit through sea–land mapping, the NDWI was applied (Figure 4). The NDWI was proposed by McFeeters [45], designed to maximize the reflectance of the water body in the green band (ρ3) (559 nm) and minimize the reflectance of the water body in the near infrared (NIR) band (ρ8) (864 nm) (1), with a range value from −1 and +1.

$$\text{NDWI} = \frac{\rho 3 - \rho 8}{\rho 3 + \rho 8} \quad (1)$$

Two main analyses were performed using NDWI in order to characterize the sand spit of the Guadiaro river and the adjacent beaches. Firstly, mean values were calculated obtaining climatologies for the whole period (2017–2020) and for each year in the whole AOI and, seasonality mean values for the sand spit zone to analyze seasonal trends (code available in GEE: https://code.earthengine.google.com/7b779ffe4fd2526bfd7397c1a9383206, accessed on 1 February 2021). The climatology concept refers to the average for all the raster, a geoprocessed S2 L2A image with applied NDWI index in our case, available from spring 2017 to autumn 2020. Secondly, in order to identify hotspot for accretion and erosion trends, Variation Rates (VR) between sea and land pixels were performed for the AOI (code available in GEE: https://code.earthengine.google.com/414fc98a57150171490c6f1fb909ca16, accessed on 1 February 2021). In this last case, NDWI threshold was used to discretize pixels into Boolean data (0: land; 1: ocean), for which thresholds between 0.05 to 0.15 were tested, revealing the 0.1 threshold value as an optimized coastline lineation for single images. However, when computing several images, 0.1 threshold included too much data noise in the ocean and river area, not properly distinguishing the coastline, and needing to use a threshold with value 0 for multi-image mining. Hence, it is important to highlight that, when working with a single image, threshold of 0.1 obtained the most accurate segmentation of land–water.

During value mining, a NDWI value for each case was added as a new parameter in the metadata Image Collection, with continuous values from −1 to 1 for climatologies (−1: permanent land; 1: permanent water) and between 0 and 1 for VR (0: permanent land; 1: permanent water). Finally, all the results were visualized in GEE and downloaded to Google Drive as a GeoTIFF image, ready for custom visualizations to be produced in ArcGIS Pro 2.5. The computational analysis was structured in three temporal scales for both Climatologies and VR: (1) total period (2017–2020); annual results for each year (2017, 2018, 2019, 2020); and analysis for each season, each year (from spring 2017 to autumn 2020).

3. Results

3.1. *Cloud-Free Image Selection and Thresholding*

The chart (Figure 5) shows image selection from both qualitative (EO Browser) and quantitative (GEE) methods, to validate GEE script accuracy, with a 15% maximum cloud probability applied. The percentage shows the number of images selected from the original GEE "COPERNICUS/S2_SR" dataset for each year, showing an optimal performance of the method. From a total of 484 images in the original dataset, 244 images (50.4%) were selected through the GEE algorithm and 264 images (54.5%) were selected using EO Browser, as a qualitative selection. All images selected with the GEE script were also selected visually, proving a 95% reliability and effectiveness with the GEE unsupervised image selection and, therefore, validating the methodology. This selection is accurate and consistent due to the fact that the cloud cover percentage is recalculated for all images in the AOI, not using the pre-calculated cloud cover percentage of the entire tile for filtering. For more detail, cloud cover validation percentages for each season can be seen in Table 1.

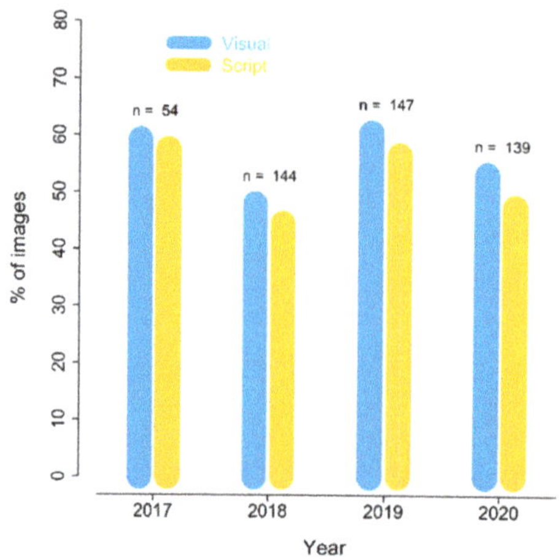

Figure 5. Workflow validation for cloud cover images. Percentage (%) of images selected in both methodologies to validate GEE script (n: number of images available in the image collection for each year. A total of 484 images were available before cloud-cover image filtering).

Table 1. Seasonal number of images used for cloud-free image filtering validation. All images available, visual selection, automatic selection in GEE and percentages.

Season	Images Available	Visual Selection	Automatic Filtering	% Images Selected Visually *	% Images Selected in GEE *
Spring 2017	15	8	7	53.3	46.7
Summer 2017	15	10	10	66.7	66.7
Autumn 2017	19	13	13	68.4	68.4
Winter 2018	34	13	13	38.2	38.2
Spring 2018	37	14	11	37.8	29.7
Summer 2018	35	25	21	71.4	60.0
Autumn 2018	36	17	17	47.2	47.2
Winter 2019	36	17	16	47.2	44.4
Spring 2019	36	25	19	69.4	52.8
Summer 2019	36	27	27	75.0	75.0
Autumn 2019	36	20	19	55.6	52.8
Winter 2020	37	14	9	37.8	24.3
Spring 2020	37	17	12	45.9	32.4
Summer 2020	37	29	28	78.4	75.7
Autumn 2020	32	15	15	46.9	46.9

* Percentage of selected images from available images in dataset ("COPERNICUS/S2_SR"). Image selection (visual or automated)/available images for corresponding season.

Concerning the evaluation of the NDWI threshold, three different values in critical cases were tested (0.05, 0.1 and 0.15) (Appendix A. Figure A1). The 0.1 value was used as an optimal NDWI, allowing us to detect the sand spit in the closed river mouth, just before closing, and the appearance of additional and smaller sand bars formed by local currents.

3.2. Climatology and Seasonality

The general climatology (2017–2020) in Figure 6 shows a mean sand spit position with a mostly closed river mouth, variating in its upper third and demonstrating its tendency to block water renewal in the estuary. Rigidity of both sides of the riverbed caused by urbanization blocked natural evolution, behaving more like a lagoon than an estuary. The NDWI index perfectly delimited the boundary between sand and water. River silting areas are observed along the riverbed extremely shallow depths during certain periods of the year, slowing water flow and increasing the accumulation of sedimentation in the river mouth.

Figure 6. Total climatology (2017–2020) using Sentinel-2 L2A images and NDWI index for the Gua-diaro river mouth. Scale 1:15,000.

Figure 7 shows annual climatologies, where high sand spit variability is clearly observed, proving how often the sand spit closes the river mouth during the year, especially for 2017, where the system is frequently working as a lagoon, a behavior that is repeated in 2020. Diversity of sand bar morphologies indicates no clear patterns. However, in both 2018 and 2019 an additional small sand bar was generated towards the ocean along the coast.

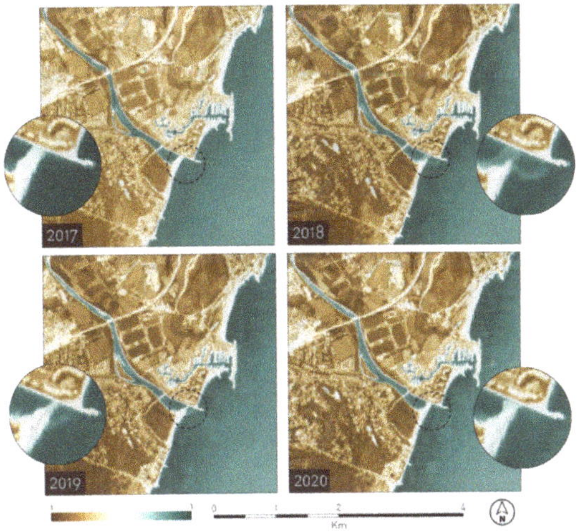

Figure 7. Annual climatologies using Sentinel-2 L2A images and NDWI index for the Guadiaro river mouth (single year means from 2017 to 2020). Scale 1:15,000.

When evaluating seasonal climatologies (Figure 8), a variety of sand spit morphologies were observed. Similar to annual climatologies, seasonal means do not show patterns for opened or closed river mouth scenarios linked to any time particular of year. During 2017, the sand spit tended to be homogeneously closed. However, in 2018, a smaller secondary sand bar was generated because of the eddy effect, both bars merging in winter 2019. From spring 2019 until the end of the year the river mouth remained partially closed until it was completely closed by summer 2019, remaining closed for the following 6 months. Finally, after being opened by torrential rains, a new stable sand spit with U-shape morphology appeared in 2020 almost completely closing the river mouth. These rapidly changing dynamics reflect the complexity of the area and the importance of addressing estuarine problems from a holistic perspective, including natural processes and human activities from the river basin, the coastal zone and the ocean, as explained later in the Discussion Section 4.2.

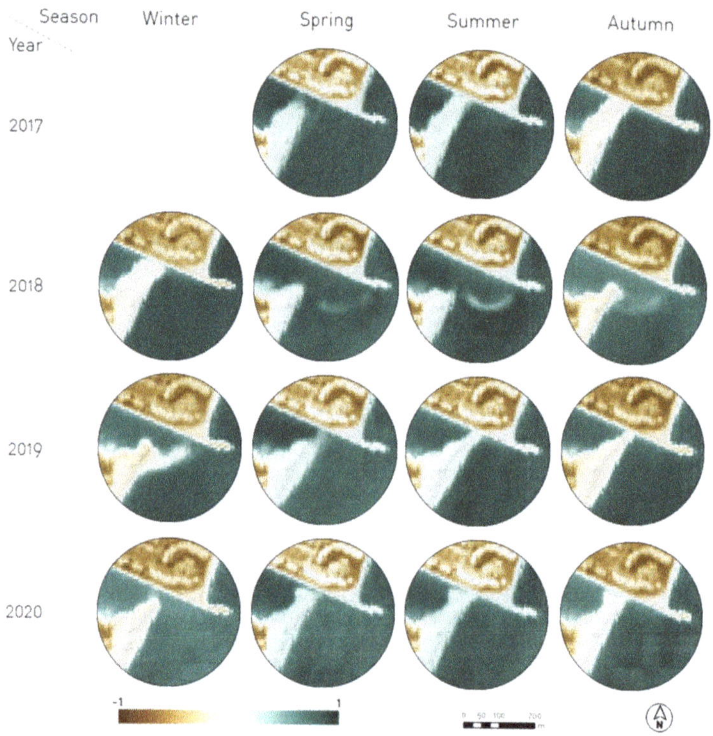

Figure 8. Seasonal climatologies using Sentinel-2 L2A images and NDWI index for the Guadiaro river mouth and its dynamic sand bar between spring 2017 and autumn 2020. Scale 1:3500.

3.3. Variation Rate

Once the spatio-temporal patterns were characterized, ranging from the riverbed to beaches and finally to the dynamic sand spit in the river mouth, in order to quantify this coastline variability, another study was carried out to evaluate its temporal and spatial variation. The VR identifies the most variable coastal areas in terms of land/water mapping pixels. This estimation has values from 0 (referred to pixels that are always land) to 1 (referred to pixels that are always water), the middle value (0.5) being areas of maximum variation. Light blue values show pixel covered by water for a longer time, while light brown values show the opposite trend. To evaluate the same temporal scales, this section follows the same structure as Section 3.2.

Variation Rate for the study period (2017–2020) (Figure 9) identifies two main hotspot areas of high variability. Firstly, the sand spit boundaries between land and water, with a water buffer in the upper third of its a pointed morphology; and secondly, Torreguadiaro beach, northwards of Sotogrande harbor. Torreguadiaro beach shows general instability due to the presence of small temporal lagoons created by maritime storms and waves, whose presence has been confirmed by RGB high-resolution images and in situ observations. Therefore, S2 images are detecting high soil moisture in the sandy areas and water accumulation in the rocky areas of Torreguadiaro beach. However, no significant variations are observed in Sotogrande beach and Guadiaro estuary between 2017 and 2020.

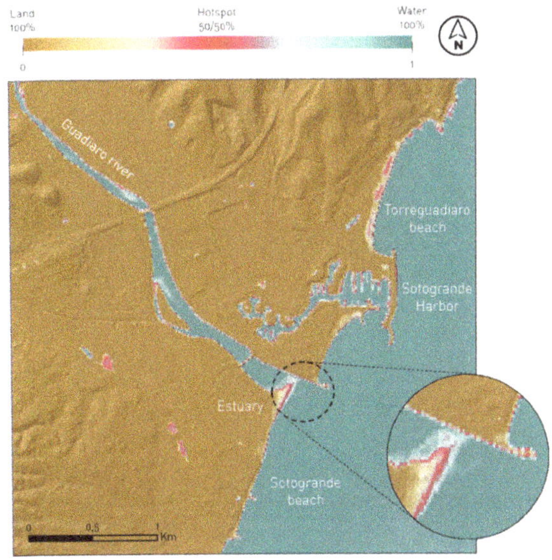

Figure 9. Variation rate of land/sea interface in the Guadiaro river mouth using Sentinel-2 L2A images and NDWI index between 2017 and 2020. Scale 1:15,000.

The correlation into temporal patterns can be very useful for coastal managers when designing adaptive policies against erosion and artificial river mouth openings. For this reason, as done in the climatology section, in order to understand whether areas of high variability followed any temporal pattern, annual and seasonal VR maps were generated between 2017 and 2020.

Generally, annual VR for the land–sea interface also shows short-term sediment stability for Sotogrande beach (Figure 10). Moreover, when looking at the adjacent beach south of Sotogrande harbor, 2017 was also the only year presenting high sediment variability, while Torreguadiaro beach showed a general instability trend for the 4 years of the study.

However, annual VR for Guadiaro coastal sand spit showed how a small secondary sand bars existing in 2018 and 2019 were unstable and of short duration. Moreover, no patterns of external or internal variability of the bar are observed, varying externally in 2019, internally in 2018 and 2020, and homogeneously in 2017.

Figure 11 shows how the sand spit varies between seasons. Summer showed a high variability between years, most likely due to artificial openings in times of increased touristic activity as an operational response to eutrophication episodes. Days with the sand spit closed, resulting in poor renewed water together with high temperatures, are associated with phytoplankton blooms, anoxia and high mortality of marine flora and fauna [59]. The days with the river mouth closed in a drought period, the typical rainfall hydrographic regime of the Mediterranean climate, the estuary works similar to a coastal lagoon [14].

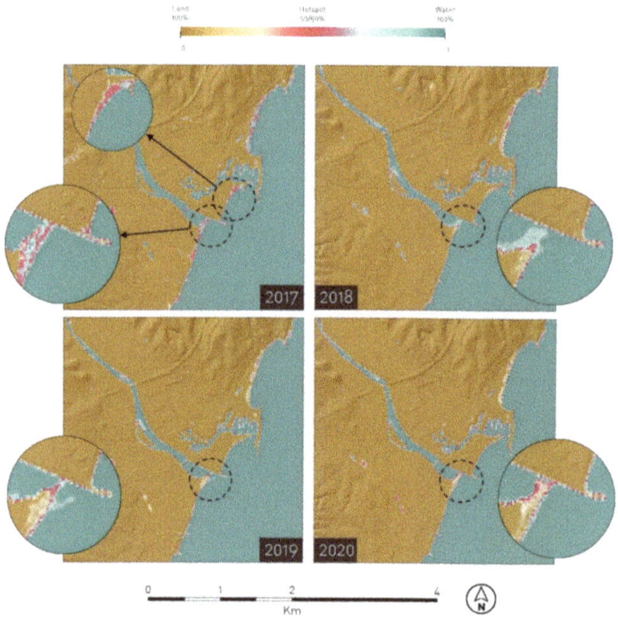

Figure 10. Annual variation rate of land/sea interface in the Guadiaro river mouth using Sentinel-2 L2A images and NDWI index (single year VR from 2017 to 2020). Scale 1:15,000.

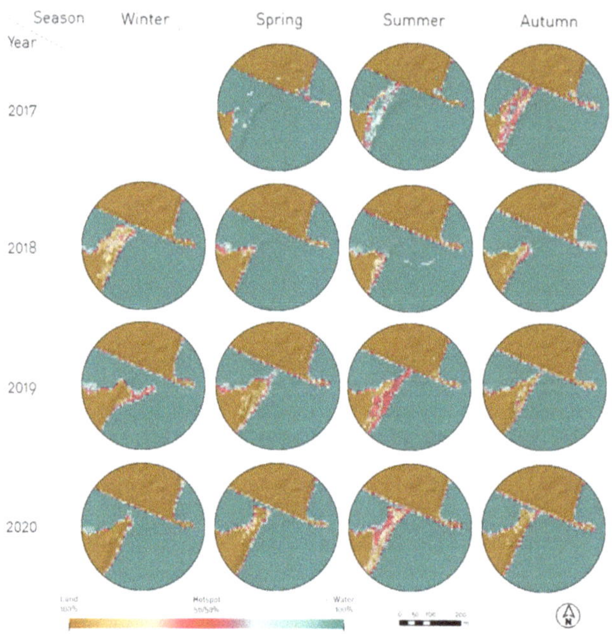

Figure 11. Seasonal variation rate of land/sea interface in the Guadiaro river mouth using Sentinel-2 L2A images and NDWI index from 2017 to 2020. Sand spit variability. Scale 1:3500.

4. Discussion and Conclusions

4.1. High Frequency Remote Sensing Data: S2, GEE and NDWI

Coastal environments are constantly changing and their continuous monitoring is key to proper evaluation and management. Estuaries and coastal lagoons are vulnerable environments affected by atrophic impacts and climate change; therefore, monitoring efforts are crucial for their conservation. By applying remote sensing techniques, consistent and robust methodologies for sea–land mapping and coastline detection were recently obtained on a global scale [22,25,29,48,60,61]. However, heterogeneity of sea–land mapping and coastline delineation methods and the difficult access to these tools by coastal managers usually restricts their use [24,35,36,40,62,63]. The potential of GEE has already been demonstrated in previous studies [37,38,60,64,65], and we used S2 images within GEE in order to ensure accuracy and robust performance, as well as a user-friendly and accessible tool for coastal managers.

Research studies of spatio-temporal variability of sand spit formations and evolution of nearby beaches are crucial for understanding the adaptation of estuarine systems under multiple anthropogenic pressures. In the Guadiaro estuary, eutrophication events have been connected to the lack of water renewal commonly promoted by sand spit closure since 1970s, where earlier studies employed low-frequency data sampling [9–12]. Thanks to the 10 m spatial resolution of S2, sea–land mapping has been successfully accomplished. Moreover, recent studies of S2 pixel accuracy have demonstrated upscaling improvements up to 2.5 m of spatial resolution [66], a substantial improvement paving the way for the use of S2 in the future of coastal monitoring. The methodology used in this study can be implemented for monitoring sand spit and beach variability for qualitative evaluation, with images available since 30 March 2017 (since 2015 with L1C). This is the first study in the Guadiaro estuary using S2-A/B imagery for coastal proposes as well as using the GEE platform for analytics. Previous studies used aerial imagery to inspect the Guadiaro river mouth modelling both northern and southern beaches [10,12,14], but not satellite imagery. Video-imaging techniques provide high-frequency data, albeit with a limited spatial coverage as fixed and site-specific ground-based technologies [67].

The method proposed in this study is completely automatic and cloud-based, avoiding the need to download each satellite image as well as intermediate steps during geoprocessing, in contrast to the CoastSat methodology [36]. Firstly, it has an automated workflow for selecting S2 L2A satellite imagery with low custom cloud cover with 95% reliability. This approach indicated consistency in performance for cloud-free image selection as a result of the recalculation of image cloud cover after cropping the AOI. Contrary to those commonly used in other studies [38], this approach does not take predefined cloud cover values from GEE for the entire tile. Furthermore, automatic sea–land mapping has many advantages over traditional methods. While manual shoreline delimitation is more accurate, it is labor-intensive and time-consuming [40], and, therefore, not the optimal option for high-frequency and long-term monitoring. However, as already demonstrated in previous studies, the use of indices for automatic sea–land mapping is biased towards false classification pixels due to clouds and breaking waves [41]. For this reason, optimization of image selection and cloud masking is key, and this study advances this field by achieved accurate results, offering a ready to use method for coastal monitoring that also generates useful information for the complete coastal management cycle on a global scale.

According to McAllister et al., the use of S2 for shoreline extraction through water and non-water pixels between 2006 and 2021 revealed a better performance of water indices and thresholding [68]. Among indices, NDWI and MNDWI are the most commonly used due to their optimized performance [43,44,49,51,69], having been subjected to different modifications [70], while thresholding is widely considered to be the simplest method of image segmentation [68]. We used the first NDWI defined by McFeeters [45], as MNDWI enhances the performance for coastal shadows (e.g., cliffs) and it can reveal more detail in open water [68,70], our study being focused on flat beaches and sand spit morphologies along the coast. Regarding NDWI threshold, this study revealed an optimized 0.1 value for single images, while the threshold value of 0 revealed less data noise for long temporal statistics. According to Xu [70], "threshold values for MNDWI to achieve best water extraction results are usually much less than those of NDWI, suggesting using zero as a default threshold value can produce better water extraction accuracy for the MNDWI than for the NDWI". However, while we demonstrated 0.1 is suitable for single-image processing, the application of a 0 threshold in NDWI for computing statistics among several images revealed better water and non-water mapping; furthermore, other studies also refer to 0 threshold as the best performance for large spatial extents [71].

4.2. Contributions to Coastal Management

Regarding the Guadiaro river mouth, this study is essential for understanding external pressures and effects, as well as the natural dynamics of the sand spit, whose spatial distribution is highly conditioned by the river and ocean systems [11]. Applying S2 L2A satellite imagery we developed a monitoring tool that allows selection of the most optimal technique available to manage and prevent the closure of the river mouth in early stages, improving its temporal and economic efficiency, as well as to re-evaluate over time the suitability of the management measures applied. Additionally, the effectiveness of artificial openings in the short-term shown in Figure 3 could not be demonstrated, since images immediately before and after the three artificial openings corresponding to 2017 and 2018 were covered by clouds, and by 2019 and 2020 dredging was not necessary, as reported by the Spanish Government. However, further monitoring over time can reveal more insights. Only relatively short-term data are available for this study (2017–2020), hindering the overview of seasonal trends, with a large portion of estuaries depending mostly on upstream anthropogenic impacts [72]. While Landsat mission imagery offers historical data, its spatial resolution (30 m) is not suitable for small sand spits that require increased spatial detail, compared to the 10 m spatial resolution provided by the S2-A/B twin satellites. Moreover, the S2 5-day high-revisit frequency at the Equator (better at higher latitude) allows more accurate detection of changes in the sand spit in the immediate term after a rainfall event, compared with the 16-day revisit of Landsat-8. These characteristics position S2-A/B satellites in the present and future for coastal monitoring.

For the Guadiaro river mouth, the sand spit, and the surrounding beaches, we achieved the following results: (1) regarding the sand spit, especially in 2020 it remained closed throughout the summer season, most likely associated with the COVID-19 pandemic, during which many international tourists were not able to visit Sotogrande, reducing anthropogenic pressure, including urban waste; (2) through climatologies we evaluate how additional small sand bars could be caused by the northward longitudinal current blocked after the construction of Sotogrande harbor; (3) [55] replenishing of beaches can also be detected and monitored by means of S2 images, as demonstrated in the variability of the beach to the south of the Port of Sotogrande in 2017; and (4) the land–sea interface of sandy beaches is more easily identified in different studies throughout the globe compared to rocky beaches or coastal cliffs [36,61,73,74], mostly due to image spatial resolution [75]. However, while S2 undoubtedly enhanced its spatial resolution up to 10 m and, consequently, enhancing the rocky beach delineation, other effects, such as the presence of natural areas of water accumulation on the rocky beach, can lead to misinterpretation, as happened on the Torreguadiaro beach. In summary, the presented optimized monitoring tool can not only be used as an early-warning system for preventing closure episodes, but also for evaluating past measures and future management strategies.

Accordingly, integrated coastal management requires information throughout the entire management cycle [14–17]. Our monitoring tool can provide the necessary information for each management stage. At the beginning, information is required to make a diagnosis of the problems and impacts to be faced by the planning initiative. This will allow the formulation of suitable tools for coastal management. Information is also needed during the implementation of the management program, so adjustments can be made in order to improve its development. As an example, it is possible to assess if a coastal or watershed plan is given the desired results in the estuarine area in the short, medium or long term. Finally, it is extremely important to have information from the monitoring of the complete planning cycle during the evaluation phase. At this stage, the results are assessed and some questions must be answered: Have we achieved the objectives? Are we in the desired scenario? What are the main difficulties we have faced? What can be improved in the next planning cycle?

Among the advantages of the monitoring tool for coastal management, the followings can be highlighted: (a) A coastal management program is a result of a public policy. In this sense, and according to Olsen [76], it is important to have well defined and unambiguous goals, so progress towards to its achievement can be measured. The monitoring tool is able to provide long-term sustained information helping to assess, adapt, evaluate and improve such policy objectives; (b) an adaptive ecosystem-based management approach requires sustained monitoring [18–20]. The presented monitoring tool can provide a cost-effective early warning system each 5 days (without cloud coverage), facilitating adaptive management in the area. In this sense, it is possible to correlate the changes of state measured in the estuarine area with the responses given by the administration [21]; (c) the Guadiaro estuarine area shows the result of several sectoral policies: ports, watershed, mining, tourism, agriculture, among others. Although the closure of the river mouth is part of a natural process due to the effect of the maritime climate and the waves in the sand spit, uncoordinated sectoral planning frequently results in a cumulative effect of closures and severe eutrophication episodes. Information about sand spit trends can be related with the effects of different policy cycles. This knowledge can be used to support the development of a joint policy in the studied area.

In conclusion, the method proposed is simple and easily transferable to other environments on a global scale, with the scripts being open access, online, and ready to use. This tool provides continuous and long-term critical information and analysis through sea–land mapping assessing coastal zone response, approaching this technology as an accessible tool for coastal managers, and supporting daily operative and strategic management. This implementation ensures continuous and detailed monitoring in order to support an ecosystem-based approach for estuarine areas, and provides an added value that can encourage optimal decision making.

Author Contributions: Conceptualization, M.R., G.N., J.G.-S. and I.C.; methodology, software, validation and formal analysis, M.R.; writing—original draft preparation, M.R.; writing—review, editing and supervision, G.N., J.G.-S. and I.C.; project administration, G.N. and I.C.; funding acquisition, M.R., G.N. and I.C. All authors have read and agreed to the published version of the manuscript.

Funding: This research was funded by RTI2018-098784-J-I00 Sen2Coast project by the MCIN/AEI/10.13039/501100011033 and by "ERDF A way of making Europe", by grant IJC2019-039382-I (Juan de la Cierva-Incorporación) by the Ministry of Science and Innovation of the Spanish Government, by grants FPU20/01294 and JAEINT_20_00462 (JAE INTRO 2020) by the Ministry of Universities of the Spanish Government and by PY20-00244 Sat4Algae project by the Andalusian Regional Government (Junta de Andalucía).

Institutional Review Board Statement: Not applicable.

Informed Consent Statement: Not applicable.

Data Availability Statement: Publicly available datasets were analyzed in this study, accessible through GEE [https://developers.google.com/earth-engine/datasets/catalog/COPERNICUS_S2_SR, accessed on 1 February 2021 and https://developers.google.com/earth-engine/datasets/catalog/COPERNICUS_S2_CLOUD_PROBABILITY, accessed on 1 February 2021]. The three GEE scripts can be found as follows: (1) for cloud-free image selection [https://code.earthengine.google.com/6363973f5918f3eacd09d890a52f5827, accessed on 1 February 2021]; (2) for climatology analysis [https://code.earthengine.google.com/7b779ffe4fd2526bfd7397c1a9383206, accessed on 1 February 2021]; and (3) for VR analysis [https://code.earthengine.google.com/414fc98a57150171490c6f1fb909ca16, accessed on 1 February 2021].

Acknowledgments: We would like to thank the European Space Agency, the European Commission and the Copernicus programme for distributing Sentinel-2 imagery; Google for developing the Google Earth Engine cloud computing platform, for its data accessibility and computational capacity freely available for researchers; Patricio Poullet Brea and Gregorio Gómez Pina, from the Coastal Office in the Atlantic Andalusia Ministry of Environment, for supporting this project as a pilot to monitor estuaries in Andalusia, as well as for the technical data provided; and Martha Bonnet Dunbar for the English language revision. This work represents a contribution to the CSIC Thematic Interdisciplinary Platform PTI TELEDETECT.

Conflicts of Interest: The authors declare no conflict of interest.

Appendix A

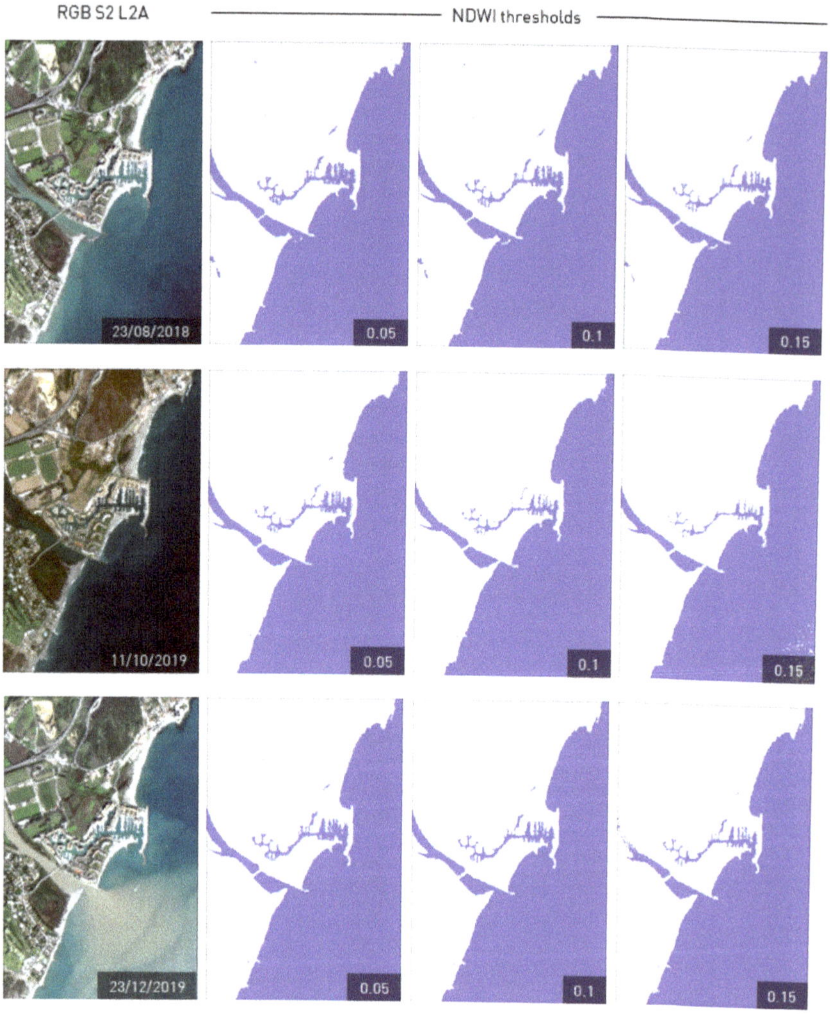

Figure A1. NDWI thresholding test showing 0.1 value as optimal. Image from 23 August 2018 shows the delineation of an additional sand bar; second image (11 October 2019) shows the detection of the sand spit closing the river mouth; and last image (23 December 2019) shows how scenarios with high turbidity do not interfere in water and non-water pixel classification.

References

1. Lotze, H.K.; Lenihan, H.S.; Bourque, B.J.; Bradbury, R.H.; Cooke, R.G.; Kay, M.C.; Kidwell, S.M.; Kirby, M.X.; Peterson, C.H.; Jackson, J.B. Depletion, degradation, and recovery potential of estuaries and coastal seas. *Science* **2006**, *312*, 1806–1809. [CrossRef]
2. Meybeck, M. Carbon, nitrogen, and phosphorus transport by world rivers. *Am. J. Sci.* **1982**, *282*, 401–450. [CrossRef]
3. Rao, N.S.; Ghermandi, A.; Portela, R.; Wang, X. Global values of coastal ecosystem services: A spatial economic analysis of shoreline protection values. *Ecosyst. Serv.* **2015**, *11*, 95–105. [CrossRef]
4. Fragkias, M.; Seto, K.C. The rise and rise of urban expansion. *Glob. Chang.* **2012**, *78*, 16–19.
5. Neumann, B.; Vafeidis, A.T.; Zimmermann, J.; Nicholls, R.J. Future coastal population growth and exposure to sea-level rise and coastal flooding-a global assessment. *PLoS ONE* **2015**, *10*, e0118571. [CrossRef]
6. Erena, M.; Domínguez, J.A.; Aguado-Giménez, F.; Soria, J.; García-Galiano, S. Monitoring Coastal Lagoon Water Quality through Remote Sensing: The Mar Menor as a Case Study. *Water* **2019**, *11*, 1468. [CrossRef]
7. Sòria-Perpinyà, X.; Miracle, M.R.; Soria, J.; Delegido, J.; Vicente, E. Remote sensing application for the study of rapid flushing to remediate eutrophication in shallow lagoons (Albufera of Valencia). *Hydrobiologia* **2019**, *829*, 125–132. [CrossRef]
8. Mills, A.K.; Bolte, J.P.; Ruggiero, P.; Serafin, K.A.; Lipiec, E.; Corcoran, P.; Stevenson, J.; Zanocco, C.; Lach, D. Exploring the impacts of climate and policy changes on coastal community resilience: Simulating alternative future scenarios. *Environ. Model. Softw.* **2018**, *109*, 80–92. [CrossRef]
9. Caballero, I.; Chapela-Bernatche, L.; Roque-Atienza, D.; Tejedor Álvarez, M.B.; Gomez-Pina, G.; Muñoz Pérez, J.J. *Influencia del Oleaje en las Condiciones de Cierre de la Desembocadura del Río Guadiaro (Cádiz)*; IX Jornadas Españolas de Ingenieria de Costas y Puertos: Cadiz, Spain, 2008; pp. 32–39.
10. Muñoz Pérez, J.J.; de la Casa, Á.; Gomez-Pina, G.; Acha Martin, A. Environmental Restoration of the Guadiaro River Estuary, Cadiz, Spain. *Period. Biol.* **2000**, *102*, 333–338.
11. Román-Sierra, J.; Navarro-Pons, M.; Muñoz Pérez, J.J.; Tejedor Álvarez, M.B.; Física, A. Variabilidad espacio-temporal de la flecha del río Guadiaro. In *Spatial and Temporal Variability in the Spif of Guadiaro River*; Ingenieriía Civil: Madrid, Spain, 2008; Volume 149, pp. 111–122. ISSN 0213-8468.
12. Diez, J.J.; Fernando, R.; Veiga, E.M. (Eds.) Coastal Impacts Around Guadiaro River Mouth (Spain). In *Engineering Geology for Society and Territory*; Springer International Publishing: Cham, Switzerland, 2014; Volume 4.
13. Aleksandrov, S.V. Biological production and eutrophication of Baltic Sea estuarine ecosystems: The Curonian and Vistula Lagoons. *Mar. Pollut. Bull.* **2010**, *61*, 205–210. [CrossRef]
14. Chica Ruiz, J.A.; Barragán Muñoz, J.M. *Estado y Tendencia de los Servicios de los Ecosistemas Litorales de Andalucía*; Universidad de Cádiz Rectorado: Cádiz, Spain, 2014.
15. GESAMP. *The Contributions of Science to Integrated Coastal Management*; FAO: Rome, Italy, 1996.
16. Barragán Muñoz, J.M. Coastal management and public policy in Spain. *Ocean. Coast. Manag.* **2010**, *53*, 209–217. [CrossRef]
17. Barragán, J.M. *Política, Gestión y Litoral: Una Nueva Visión de la Gestión Integrada de Áreas Litorales;* Flores: Cadiz, Spain, 2014; 685p.
18. Barragán, J.M.; Lazo, Ó. Policy progress on ICZM in Peru. *Ocean Coast. Manag.* **2018**, *157*, 203–216. [CrossRef]
19. Barragán Muñoz, J.M. Progress of coastal management in Latin America and the Caribbean. *Ocean Coast. Manag.* **2020**, *184*, 105009. [CrossRef]
20. Nava Fuentes, J.C.; Arenas Granados, P.; Martins, F.C. Coastal management in Mexico: Improvements after the marine and coastal policy publication. *Ocean Coast. Manag.* **2017**, *137*, 131–143. [CrossRef]
21. Elliott, M. The role of the DPSIR approach and conceptual models in marine environmental management: An example for offshore wind power. *Mar. Pollut. Bull.* **2002**, *6*, 3–7. [CrossRef]
22. Pardo-Pascual, J.E.; Sánchez-García, E.; Almonacid-Caballer, J.; Palomar-Vázquez, J.M.; Priego de los Santos, E.; Fernández-Sarría, A.; Balaguer-Beser, Á. Assessing the Accuracy of Automatically Extracted Shorelines on Microtidal Beaches from Landsat 7, Landsat 8 and Sentinel-2 Imagery. *Remote Sens.* **2018**, *10*, 326. [CrossRef]
23. Taveneau, A.; Almar, R.; Bergsma, E.W.J.; Sy, B.A.; Ndour, A.; Sadio, M.; Garlan, T. Observing and Predicting Coastal Erosion at the Langue de Barbarie Sand Spit around Saint Louis (Senegal, West Africa) through Satellite-Derived Digital Elevation Model and Shoreline. *Remote Sens.* **2021**, *13*, 2454. [CrossRef]
24. Quang Tuan, N.; Cong Tin, H.; Quang Doc, L.; Anh Tuan, T. Historical Monitoring of Shoreline Changes in the Cua Dai Estuary, Central Vietnam Using Multi-Temporal Remote Sensing Data. *Geosciences* **2017**, *7*, 72. [CrossRef]
25. Splinter, K.D.; Harley, M.D.; Turner, I.L. Remote Sensing is Changing Our View of the Coast: Insights from 40 Years of Monitoring at Narrabeen-Collaroy, Australia. *Remote Sens.* **2018**, *10*, 1744. [CrossRef]
26. Caballero, I.; Fernández, R.; Escalante, O.M.; Mamán, L.; Navarro, G. New capabilities of Sentinel-2A/B satellites combined with in situ data for monitoring small harmful algal blooms in complex coastal waters. *Sci. Rep.* **2020**, *10*, 8743. [CrossRef]
27. Caballero, I.; Stumpf, R.P. On the use of Sentinel-2 satellites and lidar surveys for the change detection of shallow bathymetry: The case study of North Carolina inlets. *Coast. Eng.* **2021**, *169*, 103936. [CrossRef]
28. Traganos, D.; Reinartz, P. Mapping Mediterranean seagrasses with Sentinel-2 imagery. *Mar. Pollut. Bull.* **2018**, *134*, 197–209. [CrossRef]
29. Cabezas-Rabadán, C.; Pardo-Pascual, J.E.; Palomar-Vázquez, J.; Fernández-Sarría, A. Characterizing beach changes using high-frequency Sentinel-2 derived shorelines on the Valencian coast (Spanish Mediterranean). *Sci. Total Environ.* **2019**, *691*, 216–231. [CrossRef]

30. Whitehead, B.; Andrews, D.; Shah, A.; Maidment, G. Assessing the environmental impact of data centres part 1: Background, energy use and metrics. *Build. Environ.* **2014**, *82*, 151–159. [CrossRef]
31. Gorelick, N.; Hancher, M.; Dixon, M.; Ilyushchenko, S.; Thau, D.; Moore, R. Google Earth Engine: Planetary-scale geospatial analysis for everyone. *Remote Sens. Environ.* **2017**, *202*, 18–27. [CrossRef]
32. Amani, M.; Ghorbanian, A.; Ahmadi, S.A.; Kakooei, M.; Moghimi, A.; Mirmazloumi, S.M.; Moghaddam, S.H.A.; Mahdavi, S.; Ghahremanloo, M.; Parsian, S.; et al. Google Earth Engine Cloud Computing Platform for Remote Sensing Big Data Applications: A Comprehensive Review. *IEEE J. Sel. Top. Appl. Earth Obs. Remote Sens.* **2020**, *13*, 5326–5350. [CrossRef]
33. Caballero, I.; Navarro, G. Monitoring cyanoHABs and water quality in Laguna Lake (Philippines) with Sentinel-2 satellites during the 2020 Pacific typhoon season. *Sci. Total Environ.* **2021**, *788*, 147700. [CrossRef]
34. Bioresita, F.; Ummah, M.H.; Wulansari, M.; Putri, N.A. (Eds.) Monitoring Seawater Quality in the Kali Porong Estuary as an Area for Lapindo Mud Disposal leveraging Google Earth Engine. In *IOP Conference Series: Earth and Environmental Science*; IOP Publishing: Bristol, UK, 2021.
35. Vos, K.; Harley, M.D.; Splinter, K.D.; Simmons, J.A.; Turner, I.L. Sub-annual to multi-decadal shoreline variability from publicly available satellite imagery. *Coast. Eng.* **2019**, *150*, 160–174. [CrossRef]
36. Vos, K.; Splinter, K.D.; Harley, M.D.; Simmons, J.A.; Turner, I.L. CoastSat: A Google Earth Engine-enabled Python toolkit to extract shorelines from publicly available satellite imagery. *Environ. Model. Softw.* **2019**, *122*, 104528. [CrossRef]
37. Terres de Lima, L.; Fernández-Fernández, S.; Gonçalves, J.F.; Magalhães Filho, L.; Bernardes, C. Development of Tools for Coastal Management in Google Earth Engine: Uncertainty Bathtub Model and Bruun Rule. *Remote Sens.* **2021**, *13*, 1424. [CrossRef]
38. Hird, J.N.; DeLancey, E.R.; McDermid, G.J.; Kariyeva, J. Google Earth Engine, Open-Access Satellite Data, and Machine Learning in Support of Large-Area Probabilistic Wetland Mapping. *Remote Sens.* **2017**, *9*, 1315. [CrossRef]
39. De Muro, S.; Ibba, A.; Simeone, S.; Buosi, C.; Brambilla, W. An integrated sea-land approach for mapping geomorphological and sedimentological features in an urban microtidal wave-dominated beach: A case study from S Sardinia, western Mediterranean. *J. Maps* **2017**, *13*, 822–835. [CrossRef]
40. Rodríguez-Santalla, I.; Roca, M.; Martínez-Clavel, B.; Pablo, M.; Moreno-Blasco, L.; Blázquez, A.M. Coastal changes between the harbours of Castellón and Sagunto (Spain) from the mid-twentieth century to present. *Reg. Stud. Mar. Sci.* **2021**, *46*, 101905. [CrossRef]
41. Adebisi, N.; Balogun, A.-L.; Mahdianpari, M.; Min, T.H. Assessing the Impacts of Rising Sea Level on Coastal Morpho-Dynamics with Automated High-Frequency Shoreline Mapping Using Multi-Sensor Optical Satellites. *Remote Sens.* **2021**, *13*, 3587. [CrossRef]
42. Nazeer, M.; Waqas, M.; Shahzad, M.I.; Zia, I.; Wu, W. Coastline Vulnerability Assessment through Landsat and Cubesats in a Coastal Mega City. *Remote Sens.* **2020**, *12*, 749. [CrossRef]
43. Fisher, A.; Flood, N.; Danaher, T. Comparing Landsat water index methods for automated water classification in eastern Australia. *Remote Sens. Environ.* **2016**, *175*, 167–182. [CrossRef]
44. Latella, M.; Luijendijk, A.; Moreno-Rodenas, A.M.; Camporeale, C. Satellite Image Processing for the Coarse-Scale Investigation of Sandy Coastal Areas. *Remote Sens.* **2021**, *13*, 4613. [CrossRef]
45. McFeeters, S.K. The use of the Normalized Difference Water Index (NDWI) in the delineation of open water features. *Int. J. Remote Sens.* **1996**, *17*, 1425–1432. [CrossRef]
46. Kaplan, G.; Avdan, U. Object-based water body extraction model using Sentinel-2 satellite imagery. *Eur. J. Remote Sens.* **2017**, *50*, 137–143. [CrossRef]
47. Murray, N.J.; Phinn, S.R.; Clemens, R.S.; Roelfsema, C.M.; Fuller, R.A. Continental scale mapping of tidal flats across East Asia using the Landsat archive. *Remote Sens.* **2012**, *4*, 3417–3426. [CrossRef]
48. Wu, Q.; Miao, S.; Huang, H.; Guo, M.; Zhang, L.; Yang, L.; Zhou, C. Quantitative Analysis on Coastline Changes of Yangtze River Delta Based on High Spatial Resolution Remote Sensing Images. *Remote Sens.* **2022**, *14*, 310. [CrossRef]
49. Feyisa, G.L.; Meilby, H.; Fensholt, R.; Proud, S.R. Automated Water Extraction Index: A new technique for surface water mapping using Landsat imagery. *Remote Sens. Environ.* **2014**, *140*, 23–35. [CrossRef]
50. Wang, Z.; Liu, J.; Li, J.; Zhang, D.D. Multi-spectral water index (MuWI): A native 10-m multi-spectral water index for accurate water mapping on Sentinel-2. *Remote Sens.* **2018**, *10*, 1643. [CrossRef]
51. Rokni, K.; Ahmad, A.; Selamat, A.; Hazini, S. Water Feature Extraction and Change Detection Using Multitemporal Landsat Imagery. *Remote Sens.* **2014**, *6*, 4173–4189. [CrossRef]
52. Hayes, M.O. Barrier island morphology as a function of tidal and wave regime. In *Barrier Islands*; Leatherman, S.P., Ed.; Academic Press: New York, NY, USA, 1979.
53. Muñoz-Pérez, J.J.; Caballero, I.; Tejedor, B.; Gomez-Pina, G. Reversal in longshore sediment transport without variations in wave power direction. *J. Coast. Res.* **2010**, *26*, 780–786. [CrossRef]
54. Martín-Rodríguez, J.F.; Mudarra, M.; Andreo, B.; de la Torre, B.; Gil-Márquez, J.M.; Martín-Arias, J.; Nieto-López, J.M.; Prieto-Mera, J.; Rodríguez-Ruize, M.D. (Eds.) *Monitoring and Preliminary Analysis of the Natural Responses Recorded in a Poorly Accessible Streambed Spring Located at a Fluviokarstic Gorge in Southern Spain*; Springer International Publishing: Cham, Switzerland, 2020.
55. Del Río, L.; Benavente, J.; Gracia, F.J.; Anfuso, G.; Aranda, M.; Montes, J.B.; Puig, M.; Talavera, L.; Plomaritis, T.A. Beaches of Cadiz. In *The Spanish Coastal Systems: Dynamic Processes, Sediments and Management*; Morales, J.A., Ed.; Springer International Publishing: Cham, Switzerland, 2019; pp. 311–334.

56. Reguero, B.G.; Losada, I.J.; Méndez, F.J. A recent increase in global wave power as a consequence of oceanic warming. *Nat. Commun.* **2019**, *10*, 205. [CrossRef]
57. Universidad de Cádiz (UCA). *Gestión Integrada de Zonas Costeras y Cuencas Hidrográficas: Introducción a un Caso de Estudio. El Río Guadiaro. Grupo de Investigación en Gestión Integrada de Áreas Litorales. Convenio UCA-DGCOSTAS*; Universidad de Cádiz (UCA): Cádiz, Spain, 2009.
58. Rodríguez-Alarcón, R.; Lozano, S. A complex network analysis of Spanish river basins. *J. Hydrol.* **2019**, *578*, 124065. [CrossRef]
59. Glibert, P.M.; Burkholder, J.M. The Complex Relationships Between Increases in Fertilization of the Earth, Coastal Eutrophication and Proliferation of Harmful Algal Blooms. In *Ecology of Harmful Algae*; Granéli, E., Turner, J.T., Eds.; Springer: Berlin/Heidelberg, Germany, 2006; pp. 341–354.
60. Ding, Y.; Yang, X.; Jin, H.; Wang, Z.; Liu, Y.; Liu, B.; Zhang, J.; Liu, X.; Gao, K.; Meng, D. Monitoring Coastline Changes of the Malay Islands Based on Google Earth Engine and Dense Time-Series Remote Sensing Images. *Remote Sens.* **2021**, *13*, 3842. [CrossRef]
61. Hagenaars, G.; de Vries, S.; Luijendijk, A.P.; de Boer, W.P.; Reniers, A.J.H.M. On the accuracy of automated shoreline detection derived from satellite imagery: A case study of the sand motor mega-scale nourishment. *Coast. Eng.* **2018**, *133*, 113–125. [CrossRef]
62. Hu, X.; Wang, Y. Coastline Fractal Dimension of Mainland, Island, and Estuaries Using Multi-temporal Landsat Remote Sensing Data from 1978 to 2018: A Case Study of the Pearl River Estuary Area. *Remote Sens.* **2020**, *12*, 2482. [CrossRef]
63. Smith, K.E.L.; Terrano, J.F.; Pitchford, J.L.; Archer, M.J. Coastal Wetland Shoreline Change Monitoring: A Comparison of Shorelines from High-Resolution WorldView Satellite Imagery, Aerial Imagery, and Field Surveys. *Remote Sens.* **2021**, *13*, 3030. [CrossRef]
64. Dong, D.; Wang, C.; Yan, J.; He, Q.; Zeng, J.; Wei, Z. Combing Sentinel-1 and Sentinel-2 image time series for invasive Spartina alterniflora mapping on Google Earth Engine: A case study in Zhangjiang Estuary. *J. Appl. Remote Sens.* **2020**, *14*, 044504.
65. Zhang, K.; Dong, X.; Liu, Z.; Gao, W.; Hu, Z.; Wu, G. Mapping Tidal Flats with Landsat 8 Images and Google Earth Engine: A Case Study of the China's Eastern Coastal Zone circa 2015. *Remote Sens.* **2019**, *11*, 324. [CrossRef]
66. Galar, M.; Sesma, R.; Ayala, C.; Albizua, L.; Aranda, C. Super-Resolution of Sentinel-2 Images Using Convolutional Neural Networks and Real Ground Truth Data. *Remote Sens.* **2020**, *12*, 2941. [CrossRef]
67. Ribas, F.; Simarro, G.; Arriaga, J.; Luque, P. Automatic Shoreline Detection from Video Images by Combining Information from Different Methods. *Remote Sens.* **2020**, *12*, 3717. [CrossRef]
68. McAllister, E.; Payo, A.; Novellino, A.; Dolphin, T.; Medina-Lopez, E. Multispectral satellite imagery and machine learning for the extraction of shoreline indicators. *Coast. Eng.* **2022**, *174*, 104102. [CrossRef]
69. Du, Y.; Zhang, Y.; Ling, F.; Wang, Q.; Li, W.; Li, X. Water Bodies' Mapping from Sentinel-2 Imagery with Modified Normalized Difference Water Index at 10-m Spatial Resolution Produced by Sharpening the SWIR Band. *Remote Sens.* **2016**, *8*, 354. [CrossRef]
70. Xu, H. Modification of normalised difference water index (NDWI) to enhance open water features in remotely sensed imagery. *Int. J. Remote Sens.* **2006**, *27*, 3025–3033. [CrossRef]
71. Bishop-Taylor, R.; Sagar, S.; Lymburner, L.; Alam, I.; Sixsmith, J. Sub-Pixel Waterline Extraction: Characterising Accuracy and Sensitivity to Indices and Spectra. *Remote Sens.* **2019**, *11*, 2984. [CrossRef]
72. Sharp, J.H.; Yoshiyama, K.; Parker, A.E.; Schwartz, M.C.; Curless, S.E.; Beauregard, A.Y.; Ossolinski, J.E.; Davis, A.R. A Biogeochemical View of Estuarine Eutrophication: Seasonal and Spatial Trends and Correlations in the Delaware Estuary. *Estuaries Coasts* **2009**, *32*, 1023–1043. [CrossRef]
73. Hagenaars, G.; Luijendijk, A.; de Vries, S.; de Boer, W. Long term coastline monitoring derived from satellite imagery. *Proc. Coast. Dyn.* **2017**, *12–16*, 1551–1562.
74. Sekovski, I.; Stecchi, F.; Mancini, F.; Del Rio, L. Image classification methods applied to shoreline extraction on very high-resolution multispectral imagery. *Int. J. Remote Sens.* **2014**, *35*, 3556–3578. [CrossRef]
75. Chen, D.; Stow, D.A.; Gong, P. Examining the effect of spatial resolution and texture window size on classification accuracy: An urban environment case. *Int. J. Remote Sens.* **2004**, *25*, 2177–2192. [CrossRef]
76. Olsen, S.B. A Practitioner's perspective on coastal ecosystem governance. In *Integrated Coastal Zone Management*; Wiley-Blackwell: Oxford, UK, 2009.

MDPI
St. Alban-Anlage 66
4052 Basel
Switzerland
Tel. +41 61 683 77 34
Fax +41 61 302 89 18
www.mdpi.com

Remote Sensing Editorial Office
E-mail: remotesensing@mdpi.com
www.mdpi.com/journal/remotesensing

www.ingramcontent.com/pod-product-compliance
Lightning Source LLC
Chambersburg PA
CBHW061829210926
51850CB00006B/839
* 9 7 8 3 0 3 6 5 6 6 7 4 0 *